Lecture Notes in Computer Science

Lecture Notes in Bioinformatics

15276

The series Lecture Notes in Bioinformatics (LNBI) was established in 2003 as a topical subseries of LNCS devoted to bioinformatics and computational biology.

The series publishes state-of-the-art research results at a high level. As with the LNCS mother series, the mission of the series is to serve the international R & D community by providing an invaluable service, mainly focused on the publication of conference and workshop proceedings and postproceedings.

Luigi Cerulo · Francesco Napolitano ·
Francesco Bardozzo · Lu Cheng ·
Annalisa Occhipinti · Stefano M. Pagnotta
Editors

Computational Intelligence Methods for Bioinformatics and Biostatistics

19th International Meeting, CIBB 2024
Benevento, Italy, September 4–6, 2024
Revised Selected Papers

 Springer

Editors
Luigi Cerulo
University of Sannio
Benevento, Italy

Francesco Bardozzo
University of Salerno
Fisciano, Italy

Annalisa Occhipinti
Teesside University
Middlesbrough, UK

Francesco Napolitano
University of Sannio
Benevento, Italy

Lu Cheng
Aalto University
Espoo, Finland

Stefano M. Pagnotta
University of Sannio
Benevento, Italy

ISSN 0302-9743   ISSN 1611-3349 (electronic)
Lecture Notes in Bioinformatics
ISBN 978-3-031-89703-0   ISBN 978-3-031-89704-7 (eBook)
https://doi.org/10.1007/978-3-031-89704-7

LNCS Sublibrary: SL8 – Bioinformatics

Preface

This volume presents selected scientific papers from the 19th Conference on Computational Intelligence Methods for Bioinformatics and Biostatistics (CIBB 2024), held for the first time at the University of Sannio in Benevento, Italy, from September 4 to 6, 2024. The conference gathered researchers, scholars, and students from around the world to share advancements in computational intelligence, bioinformatics, biostatistics, and medical informatics.

The CIBB 2024 conference program featured a rich mix of activities, including oral presentations, thematic sessions, and keynote lectures delivered by leading experts in the field. Keynote speakers included Michele Ceccarelli (University of Miami, USA), who leads research in computational oncology and cancer bioinformatics; Ruth Heller (Tel Aviv University, Israel), recently awarded the prestigious Rousseeuw Prize for Statistics, expert in multiple comparisons, selective inference, and replicability analysis; and Marieke Kuijjer (University of Oslo, Norway), who focuses on network-based modeling approaches in molecular biology. One Special Session (see the list below) involved the invited speaker Claudio Angione, Teesside University, UK, who presented his research on interpretable deep learning for biomedical applications. Their presentations offered cutting-edge insights into machine learning applications, statistical methods for biomedical data, and computational models for healthcare solutions.

This edition saw an especially rich presence of Special Sessions, fostering discussions on emerging challenges and opportunities within the scientific community. In particular, besides three traditional main tracks (Bioinformatics, Biostatistics, and Medical Informatics), the following Special Session were held:

- Computational intelligence in personalized medicine, chaired by Caroline König and Alfredo Vellido.
- Natural language processing (NLP) and large language models (LLM) for unstructured data in health informatics, chaired by Lerina Aversano, Antonella Madau, Debora Montano, and Chiara Verdone.
- Informatics research in bioinformatics: contributions from the CINI-Info-Life network, chaired by Vincenzo Bonnici, Simone Pernice, Giacomo Baruzzo, Bruno Galuzzi, Mikele Milia, Leonardo Pellegrina, and Roberto Pagliarini.
- Machine learning for structured data in clinical informatics and medical biology, chaired by Davide Chicco, Giuseppe Jurman, Giuseppe Agapito, Wei Liu, and Matthijs Warrens.
- Advancements in biological network analysis, uncertainty in deep learning applications, chaired by Annamaria Carissimo, Dario Righelli, and Valeria Policastro.
- Statistical modelling and machine learning to support cancer research, chaired by Antonella Iuliano and Monica Franzese, including invited speaker Claudio Angione.
- Modeling and simulation methods for computational biology and systems medicine, chaired by Simone Avesani, Marco Beccuti, Vincenzo Bonnici, Chiara Damiani,

Bruno Giovanni Galuzzi, Rosalba Giugno, Gospel Ozioma Nnadi, Simone Pernice, Michele Tebaldi, Manuel Tognon, and Eva Viesi.
- Computational structural bioinformatics, chaired by Michela Quadrini, Luca Tesei, and Alexander Miguel Monzon.

Overall, the conference received 63 short paper submissions. After a rigorous blind review process conducted by the Program Committee and external reviewers, 54 papers were accepted for presentation. Each paper was assigned at least to 3 reviewers. In this CIBB edition, we also experimented with a limited poster session, which collected 7 extended abstract submissions, of which 5 were presented at the conference and 3 are included in this volume. The contributions submitted to the conference represent a wide array of high-quality research, encompassing both theoretical advancements and practical applications.

For this edition, we also organized a "Best Paper Award" with the collaboration of the Program Committee, which was asked to vote for the best papers during peer review. Based on these votes, a restricted committee including the general chairs, the main track chairs and the chair of the special sessions finally awarded the prize ex aequo to Maria Claudia Costa, Luciano Garofano, and Michele Ceccarelli for "Uncovering the Activity of Master Kinases in Cancers by System Biology Approaches", and Dali Wang for "CellMigrationGym: An Open Deep Reinforcement Learning System for Cell Migration Study".

The success of CIBB 2024 would not have been possible without the dedication and support of numerous individuals and organizations. The editors extend their deepest gratitude to all authors, keynote speakers, reviewers, session chairs, and participants for their invaluable contributions. Special thanks are extended to the organizing committee, the advisory board, and the technical staff for their tireless efforts in creating a stimulating and productive conference environment. Generous support was granted by: Department of Science and Technology of the University of Sannio, Benevento, Italy; BioGeM Institute of Ariano Irpino (AV), Italy; Centro Regionale for Information Communication Technology (CeRICT), Campania, Italy.

Finally, the achievements of CIBB 2024 build on the foundations laid by previous editions of the conference. The editors would like to recognize and thank the general chairs and organizers of past conferences, whose dedication has shaped the success of the CIBB series. The spirit of collaboration and innovation fostered at CIBB 2024 will undoubtedly contribute to further advancements in computational biology, biostatistics, and health informatics.

Luigi Cerulo
Francesco Napolitano
Francesco Bardozzo
Lu Cheng
Annalisa Occhipinti
Stefano M. Pagnotta

Organization

General Chairs

Cerulo, Luigi	University of Sannio, Italy
Napolitano, Francesco	University of Sannio, Italy

Program Committee Chairs

Bardozzo, Francesco	University of Salerno, Italy
Cheng, Lu	University of Eastern Finland, Finland
Occhipinti, Annalisa	Teesside University, UK
Pagnotta, Stefano M.	University of Sannio, Italy

Special Session Chairs

Monzon, Alexander Miguel	University of Padua, Italy
Agapito, Giuseppe	Magna Græcia University of Catanzaro, Italy
Aversano, Lerina	University of Foggia, Italy
Avesani, Simone	University of Verona, Italy
Baruzzo, Giacomo	University of Padua, Italy
Beccuti, Marco	University of Turin, Italy
Bonnici, Vincenzo	University of Parma, Italy
Carissimo, Annamaria	IAC/CNR, Italy
Chicco, Davide	University of Toronto, Canada
Damiani, Chiara	University of Milano-Bicocca, Italy
Franzese, Monica	IRCCS SYNLAB SDN, Italy
Galuzzi, Bruno Giovanni	Institute of Bioimaging and Complex Biological Systems, Italy
Giugno, Rosalba	University of Verona, Italy
Iuliano, Antonella	University of Basilicata, Italy
Jurman, Giuseppe	Bruno Kessler Foundation, Italy
König, Caroline	Polytechnic University of Catalunya, Spain
Liu, Wei	Fujian Agriculture and Forestry University, China
Madau, Antonella	University of Sannio, Italy
Milia, Mikele	University of Padua, Italy

Montano, Debora   Regional Center Information Communication
                                Technology (CeRICT), Italy
Nnadi, Gospel Ozioma   University of Verona, Italy
Pagliarini, Roberto   University of Udine, Italy
Pellegrina, Leonardo   University of Padua, Italy
Pernice, Simone   University of Turin, Italy
Policastro, Valeria   University of Naples "Federico II", Italy
Quadrini, Michela   University of Camerino, Italy
Righelli, Dario   University of Naples Federico II, Italy
Tebaldi, Michele   University of Verona, Italy
Tesei, Luca   University of Camerino, Italy
Tognon, Manuel   University of Verona, Italy
Vellido, Alfredo   University of Catalunya, Spain
Verdone, Chiara   University of Sannio, Italy
Viesi, Eva   University of Verona, Italy
Warrens, Matthijs   Rijksuniversiteit Groningen, Netherlands

Program Committee

Agapito, Giuseppe   Magna Græcia University of Catanzaro, Italy
Aidos, Helena   University of Lisbon, Portugal
Alameer, Abbas   Kuwait University, Kuwait
Ammendola, Antonio   University of Sannio, Italy
Angelini, Claudia   IAC National Research Council (CNR), Italy
Angione, Claudio   Teesside University, UK
Auephanwiriyakul, Sansanee   Chiang Mai University, Thailand
Aversano, Lerina   University of Foggia, Italy
Avesani, Simone   University of Verona, Italy
Baldan, Matteo   University of Padua, Italy
Bartoszek, Krzysztof   Linköping University, Sweden
Baruzzo, Giacomo   University of Padua, Italy
Baumann, Petra   Medical University of Graz, Austria
Beccuti, Marco   University of Turin, Italy
Bernasconi, Anna   Polytechnic University of Milan, Italy
Besozzi, Daniela   University of Milano-Bicocca, Italy
Birolo, Giovanni   University of Turin, Italy
Blum, Christian   Spanish National Research Council (CSIC), Spain
Bonnici, Vincenzo   University of Parma, Italy
Bressan, Davide   University of Trento, Italy
Calderaro, Salvatore   University of Palermo, Italy
Campagner, Andrea   Università degli Studi di Milano-Bicocca, Italy

Caruso, Francesca Pia	University of Naples "Federico II", Italy
Cazzaniga, Paolo	University of Bergamo, Italy
Cesaro, Giulia	University of Padua, Italy
Ciaramella, Angelo	University of Naples Parthenope, Italy
Cicceri, Giovanni	University of Palermo, Italy
Cicolini, Giancarlo	"G.d'Annunzio" University of Chieti, Italy
Colaprico, Antonio	University of Miami, USA
Costa, Maria Claudia	University of Naples "Federico II", Italy
Cumbo, Fabio	Lerner Research Institute - Cleveland Clinic, USA
D'Angelo, Fulvio	University of Miami, USA
Damiani, Chiara	University of Milano-Bicocca, Italy
Dannhauser, David	Italian Institute of Technology - iit@CRIB, Italy
De Falco, Antonio	Biogem, Italy
Di Rocco, Lorenzo	Sapienza University of Rome, Italy
Etminani, Kobra	Halmstad University, Sweden
Faretra, Luca	University of Sannio, Italy
Ferraro, Luigi	University of Naples "Federico II", Italy
Ferraro Petrillo, Umberto	Sapienza University of Rome, Italy
Formenti, Enrico	Université Côte d'Azur, France
Friedrich, Christoph	Fachhochschule Dortmund, Germany
Galuzzi, Bruno Giovanni	Institute of Bioimaging and Complex Biological Systems, Italy
Garofano, Luciano	University of Miami, USA
Gatta, Roberto	University of Brescia, Italy
Geraci, Filippo	IIT National Research Council (CNR), Italy
Giugno, Rosalba	University of Verona, Italy
Goldberg, Yair	Israel Institute of Technology, Israel
Graudenzi, Alex	National Research Council of Italy (IBFM-CNR), Italy
Halgamuge, Saman	University of Melbourne, Australia
Holden, Sean	University of Cambridge, UK
Imberg, Henrik	University of Gothenburg, Sweden
Iuliano, Antonella	University of Basilicata, Italy
Jurman, Giuseppe	Bruno Kessler Foundation, Italy
Liquet, Benoit	Macquarie University, Australia
Lo Bosco, Giosuè	University of Palermo, Italy
Longato, Enrico	University of Padua, Italy
Longhin, Francesca	University of Padua, Italy
López Fernández, Hugo	Universidade de Vigo, Spain
Maccarone, Francesca	University of Milano-Bicocca, Italy
Mahmoudi, Zeinab	Novo Nordisk, Denmark
Marturano, Francesca	University of Padua, Italy

Mauri, Giancarlo	University of Milano-Bicocca, Italy
Merelli, Emanuela	University of Camerino, Italy
Migliozzi, Simona	University of Miami, USA
Milia, Mikele	University of Padua, Italy
Militello, Carmelo	National Research Council of Italy, Italy
Monaco, Gianni	Sonata Therapeutics, USA
Nigro, Marco	University of Sannio, Italy
Noviello, Teresa Maria Rosaria	University of Miami, USA
Orini, Stefania	University of Brescia, Italy
Paglialonga, Alessia	IEIIT CNR, Italy
Pennisi, Marzio	University of Eastern Piedmont, Italy
Pernice, Simone	University of Turin, Italy
Podda, Marco	University of Pisa, Italy
Popescu, Mihail	University of Missouri School of Medicine, USA
Prinzi, Francesco	University of Palermo, Italy
Raposo, Maria	New University of Lisbon, Portugal
Ribeiro, Paulo	New University of Lisbon, Portugal
Righelli, Dario	University of Naples Federico II, Italy
Rizzo, Riccardo	Institute for High-Performance Computing and Networking, National Research Council of Italy
Rovetta, Stefano	University of Genoa, Italy
Rueda, Luis	University of Windsor, Canada
Saibene, Aurora	University of Milano-Bicocca, Italy
Sanavia, Tiziana	University of Turin, Italy
Sanges, Remo	International School for Advanced Studies (SISSA), Italy
Sato-Ilic, Mika	University of Tsukuba, Japan
Scala, Giovanni	University of Naples "Federico II", Italy
Sibilio, Pasquale	Sapienza University of Roma, Italy
Simeone, Ines	Humanitas Research Hospital, Italy
Staiano, Antonino	University of Naples Parthenope, Italy
Tavazzi, Erica	University of Padua, Italy
Tebaldi, Michele	University of Verona, Italy
Treccani, Mirko	University of Verona, Italy
Tufano, Rossella	University of Sannio, Italy
Utro, Filippo	IBM Research, USA
Vella, Filippo	National Research Council of Italy, Italy
Vellido, Alfredo	University of Catalunya, Spain
Veschetti, Laura	IRCCS San Raffaele Scientific Institute, Italy
Vettoretti, Martina	University of Padua, Italy
Viesi, Eva	University of Verona, Italy

Vinciotti, Veronica	University of Trento, Italy
Wong, Ka-Chun	City University of Hong Kong, China
Zoppoli, Pietro	University of Naples "Federico II", Italy
Zotti, Tiziana	University of Sannio, Italy

Steering Committee

Di Serio, Clelia	Vita-Salute San Raffaele University, Italy
Baldi, Pierre	University of California Irvine, USA
Biganzoli, Elia	University Statale di Milano, Italy
Chicco, Davide	University of Toronto, Canada
Floares, Alexandru	Solutions of Artificial Intelligence Applications Institute, Romania
Garibaldi, Jon	University of Nottingham, UK
Kasabov, Nikola	Auckland University of Technology, New Zealand
Masulli, Francesco	University of Genoa, Italy
Peterson, Leif	Rice University, USA
Tagliaferri, Roberto	University of Salerno, Italy

Local Organizing Committee

Ammendola, Antonio	University of Sannio, Italy
Di Meo, Maria Chiara	University of Sannio, Italy
Faretra, Luca	University of Sannio, Italy
Stilo, Romania	University of Sannio, Italy
Zotti, Tiziana	University of Sannio, Italy

Reviewers

Ammendola, Antonio	University of Sannio, Italy
Angelini, Claudia	IAC National Research Council (CNR), Italy
Avesani, Simone	University of Verona, Italy
Baldan, Matteo	University of Padua, Italy
Bardozzo, Francesco	University of Salerno, Italy
Bartoszek, Krzysztof	Linköping University, Sweden
Baruzzo, Giacomo	University of Padua, Italy
Bonnici, Vincenzo	University of Parma, Italy
Burrone, Giulia	University of Turin, Italy
Caruso, Francesca Pia	University of Naples "Federico II", Italy

Cerulo, Luigi	University of Sannio, Italy
Cheng, Lu	University of Eastern Finland, Finland
Chicco, Davide	University of Toronto, Canada
Costa, Maria Claudia	University of Naples "Federico II", Italy
D'auria, Daniela	Free University of Bozen-Bolzano, Italy
Damiani, Chiara	University of Milano-Bicocca, Italy
Di Rocco, Lorenzo	Sapienza University of Rome, Italy
Faretra, Luca	University of Sannio, Italy
Fiore, Pierpaolo	University of Salerno, Italy
Galuzzi, Bruno Giovanni	Institute of Bioimaging and Complex Biological Systems, Italy
Holden, Sean	University of Cambridge, UK
Iuliano, Antonella	University of Basilicata, Italy
Jurman, Giuseppe	Bruno Kessler Foundation, Italy
Liu, Wei	Fujian Agriculture and Forestry University, China
Lo Bosco, Giosue'	University of Palermo, Italy
Longhin, Francesca	University of Padua, Italy
López Fernández, Hugo	Universidade de Vigo, Spain
Maccarone, Francesca	University of Milano-Bicocca, Italy
Marabotti, Anna	University of Salerno, Italy
Mauri, Giancarlo	University of Milano-Bicocca, Italy
Merelli, Emanuela	University of Camerino, Italy
Milia, Mikele	University of Padua, Italy
Monaco, Gianni	Sonata Therapeutics, USA
Napolitano, Francesco	University of Sannio, Italy
Nazzaro, Angelo	University of Salerno, Italy
Nigro, Marco	University of Sannio, Italy
Nnadi, Gospel Ozioma	University of Verona, Italy
Noviello, Teresa Maria Rosaria	University of Miami, USA
Pernice, Simone	University of Turin, Italy
Podda, Marco	University of Pisa, Italy
Quadrini, Michela	University of Camerino, Italy
Rodriguez-Martinez, Iosu	University of Navarra, Spain
Scala, Giovanni	University of Naples "Federico II", Italy
Staiano, Antonino	University of Naples Parthenope, Italy
Terlizzi, Andrea	University of Salerno, Italy
Treccani, Mirko	University of Verona, Italy
Tufano, Rossella	University of Sannio, Italy
Vellido, Alfredo	University of Catalunya, Spain
Viesi, Eva	University of Verona, Italy
Vinciotti, Veronica	University of Trento, Italy
Warrens, Matthijs	Rijksuniversiteit Groningen, Netherlands

Zhang, Yingbo	Chinese Academy of Tropical Agricultural Sciences, China
Zoppoli, Pietro	University of Naples "Federico II", Italy

Respiratory Disease Classification Using Novel Lung Sound Data

Holly Burrows, Mahdi Maktabdar Oghaz, and Lakshmi Babu Saheer

Anglia Ruskin University, Cambridge, UK
hb643@pgr.aru.ac.uk

Abstract. High death rates around the world each year are reportedly caused by respiratory diseases such as asthma and Chronic Obstructive Pulmonary Disease (COPD). Auscultation remains a key diagnostic tool for clinicians in the assessment and monitoring of respiratory diseases and symptoms. Advanced Deep Learning techniques such as transformer-based models can be leveraged to detect respiratory diseases from lung auscultation sounds. Currently, there lacks publicly available lung sound data from asthmatics, and research in this field is typically reliant on a heavily imbalanced dataset containing only a single sample of asthma. As part of this work, we are in the process of collecting longitudinal lung sound and spirometry data from healthy and asthmatic people using digital stethoscopes and spirometers. Participants are asked to collect their lung sounds morning and evening for at least two weeks, from twelve different auscultation points, one minute at a time. These samples will be combined with existing, publicly available lung sounds to create a novel, combined dataset. We are investigating various methods to represent these lung sounds as images, to use Computer Vision based models for disease classification. These include melspectrograms, scalograms, and mel frequency cepstral coefficients, experimenting with different parameter values to create these images from sounds. We employ this method as raw waveforms are uninformative of signal frequency content, which can be a valuable feature for classification. This work employs various deep learning architectures for classification of COPD, healthy, and asthma. Firstly, as baseline, we use a simple Convolutional Neural Network with four convolutional layers, containing a comparatively small number of parameters. We then evaluate this performance against previously State-Of-The-Art models within this domain, namely ConvNeXt architectures in several configurations. Finally, we demonstrate the performance of transformer networks for this task, namely Vision and Swin transformers. To evaluate each model performance, we use metrics including loss and accuracy, sensitivity/recall and precision, specificity, F1 score, and confusion matrices obtained using unseen data. Taking the best performing model and data representation technique we perform hyperparameter optimisation using techniques such as random search. The hyperparameters used in this process include learning rate, weight decay, optimiser, and batch size. Presently, the results we have obtained thus far demonstrate the vision transformer architecture with melspectrograms is superior in performance, achieving 0.87 F1 score with a small but growing dataset. Deep learning techniques for lung disease classification are well explored in the literature, however, most make use of a dataset that contains just one sample of asthma, meaning that the methods are constructed and evaluated with synthetic samples and augmentation techniques. The development of a novel dataset containing many asthma samples could allow for improvements in model prediction and robustness.

Keywords: Respiratory disease classification · Lung sounds · Vision transformer

Contents

Machine Learning for Structured Data in Clinical Informatics and Medical Biology

Computational Intelligence in Personalized Medicine

Computational Structural Bioinformatics

Short Papers

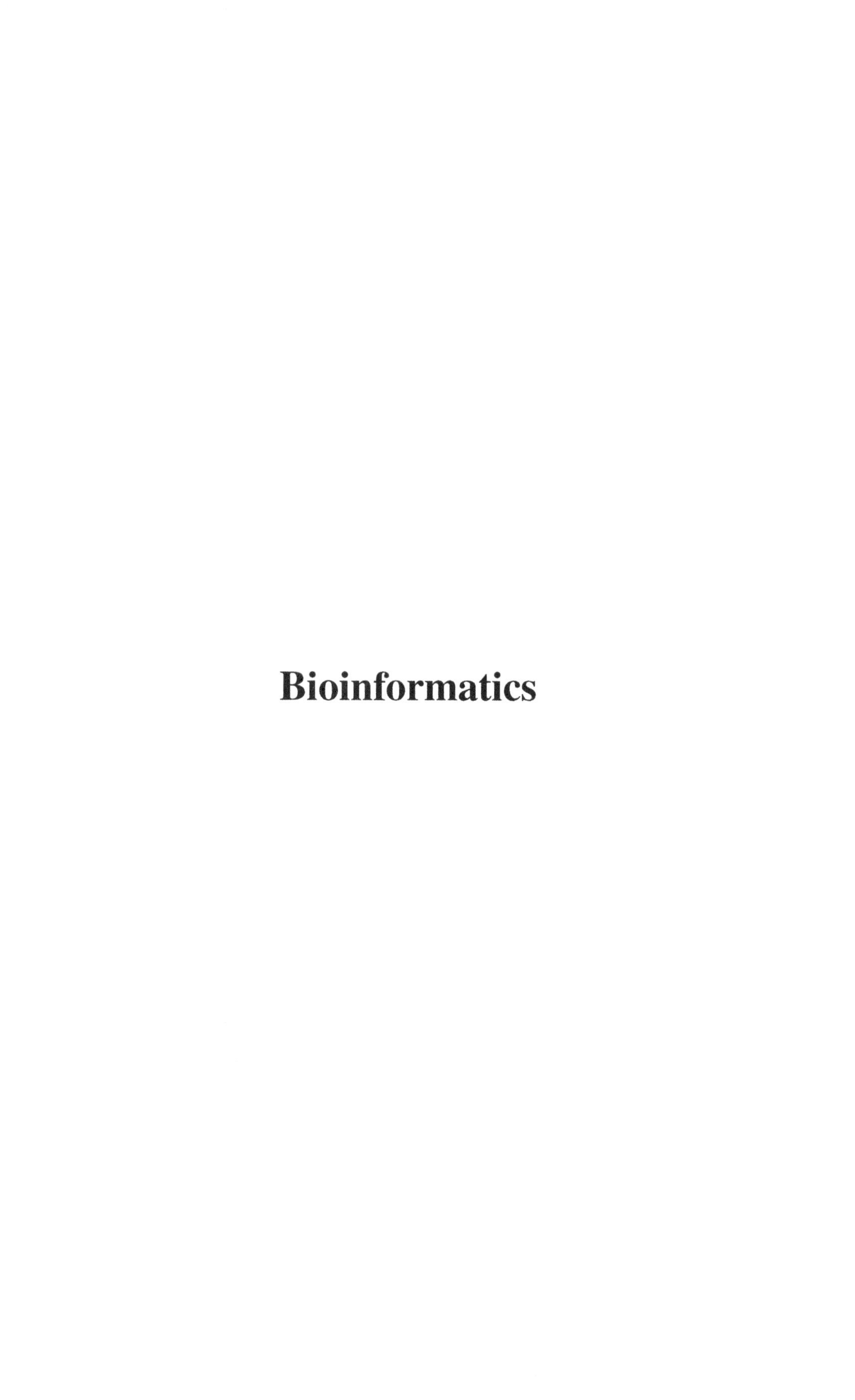

Bioinformatics

Clustering-Based Negative Sampling Approaches for Protein-Protein Interaction Prediction

Zehra Kesemen[1]([envelope]) [iD], İlknur Karadeniz[1,2] [iD], and Reyhan Aydoğan[1,2,3] [iD]

[1] Computer Science, Özyeğin University, Istanbul, Turkey
zehra.kesemen@ozu.edu.tr,
{ilknur.karadeniz,reyhan.aydogan}@ozyegin.edu.tr
[2] Artificial Intelligence and Data Engineering, Özyeğin University, Istanbul, Turkey
[3] Interactive Intelligence Group, Delft University of Technology,
Delft, The Netherlands

Abstract. The lack of confirmed negative interactions poses a major challenge to the prediction of protein-protein interactions. The reliable selection of these negative samples within a dataset is crucial for a better understanding of the underlying patterns and dynamics. The random sampling method is the most widely used negative sampling method, where negative pairs are randomly selected from unlabelled samples (i.e., samples not experimentally confirmed as positive interactions). However, they tend to introduce inaccurately labelled negative samples, resulting in less reliable predictions, which may affect the efficiency of the learning process. Our study aims to assess the reliability of clustering-based negative sampling methods and highlight their fundamental differences from the widely used random sampling method. To achieve this goal, we propose a hierarchical clustering-based algorithm that uses different mechanisms to select negative instances from unlabelled instances. We investigated the effectiveness of our proposed approach compared to existing clustering-based negative sampling methods and random sampling on four different datasets. The results indicate that clustering-based methods surpass the commonly used random sampling method.

Keywords: Host-pathogen interactions · Negative sampling strategy · Machine learning methods · Binary classification

1 Introduction

Viral infections result from the interaction between pathogen proteins and host proteins. The extraction of these protein interactions is crucial for understanding the mechanisms of infection. However, the experimental extraction of protein interactions faces several challenges including the complicated nature of the domain that requires expert knowledge, high experimental costs, and time constraints. Therefore, automated extraction of pathogen-host protein interactions (PHI) using computational methods has become an increasingly important

L. Cerulo et al. (Eds.): CIBB 2024, LNBI 15276, pp. 3–14, 2025.
https://doi.org/10.1007/978-3-031-89704-7_1

research topic in recent years [9]. One of the major challenges in the automated extraction of PHI interactions is the lack of experimentally verified negative samples for non-interacting protein pairs, although positive samples with experimentally confirmed interacting protein pairs are available. Therefore, the selection of reliable negative samples is crucial for building satisfactorily generalisable prediction models to gain a better understanding of pathogen-host protein interactions. In the literature, most of the currently available studies use a random sampling method [7, 8, 10, 13, 14, 18], where negative pairs are randomly selected from the unlabelled samples. However, this approach tends to introduce inaccurately labelled negative samples, resulting in less reliable predictions, which may affect the efficiency of the learning process and reduce the performance of predictive models [6]. In addition, the reproduction of the results might been affected negatively due to the randomization in the selection process.

In this paper, we present a novel cluster-based approach for negative sampling. Then, we explore existing cluster-based negative sampling methods in the literature to compare them with the widely used random sampling method. The intuition of our method is to avoid selecting an unknown positive sample as a negative sample as much as possible. We aim to select unknown samples with different characteristics by forming clusters of positive samples and selecting a few examples from each cluster. To see how the similarity of the negative samples to the positive samples is affected, we investigate four different alternative selection mechanisms in which the samples are selected from the clusters depending on their distance from the centroid of the positive sample clusters. These mechanisms include selecting the closest samples, the farthest samples, samples selected uniformly from cluster centroids, and selecting both the closest and the farthest samples. We empirically compare our approach with the existing cluster-based selection approaches and the random sampling method. The results show that clustering-based methods perform better than the random sampling method.

The rest of the paper is organized as follows. Section 2 introduces the problem and highlights a recent study that serves as the basis for this paper. Section 3 describes the data sets used in this study. The proposed new clustering-based method (CNS) is explained in detail in Sect. 4. Section 5 presents an overview of the benchmark methods, experimental results, and performance comparisons. Finally, Sect. 6 concludes the paper and points to future research directions.

2   Problem Statement

As widely used random sampling method generates negative samples for host-pathogen protein-protein interactions, which randomly selects from the unlabelled pairs (i.e. the pairs we cannot ensure about their interactions as they were not experimentally verified). However, this method may lead to a higher number of false negatives, potentially affecting the learning process and reducing the sensitivity in predicting protein-protein interactions. Regarding the negative sampling strategy, the biggest risk is to select unlabelled samples as negative, although there is an interaction between these protein pairs that has not yet been discovered by scientists. This can mislead the prediction model and lead to potential interactions being overlooked. To address this issue, the main task of this study is to identify unlabelled protein pairs through a novel cluster-based negative sampling approach and evaluate their performance against existing methods.

This study is based on a recently published article Koca *et al.* [7], which obtained state-of-the-art results on the prediction of protein-protein interactions between human and virus proteins. Their study incorporates the topological properties into the amino acid embeddings as part of the graph convolution process using the GraphSAGE model [4]. For encoding of protein sequences, they used Doc2Vec [11] along with the Byte Pair Encoding (BPE) method [2] to convert variable-length text segments into vector representations as the amino acid sequences are considered as documents.

The overall workflow, as illustrated in Fig. 1, shows how these topological properties and sequence embeddings are integrated into the prediction model. Our study adopted this workflow for the protein-protein interaction prediction problem in order to examine the effect of negative sampling strategies.

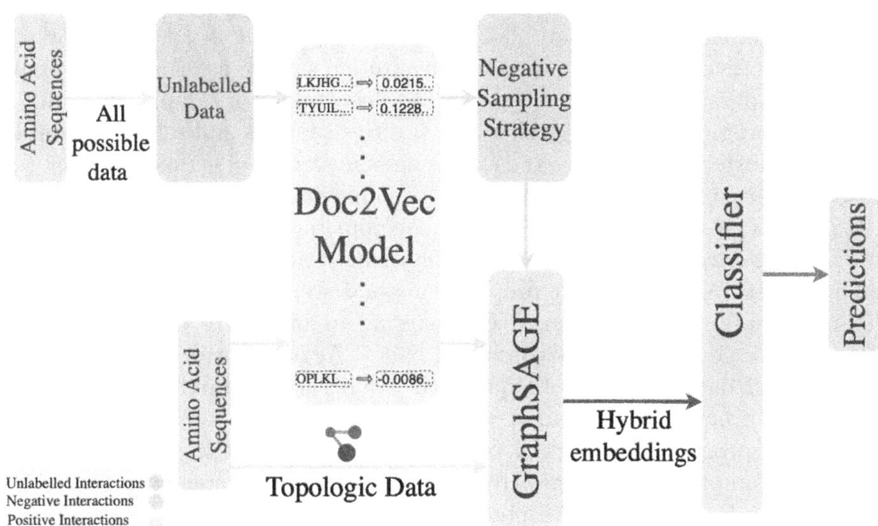

Fig. 1. Prediction framework for protein-protein interaction between human and virus proteins, as proposed by Koca *et al.* [7].

3   Datasets

The main dataset of this study, as used by Koca et al. [7], is the PHISTO dataset [1], which contains $39,544$ interactions, including $6,571$ viral proteins and

1,715 human proteins. Although this dataset provides a comprehensive basis for the analysis, additional datasets were utilized to ensure the reliability and generalizability of the methods, as detailed below:

- **Gordon's SARS-CoV-2 dataset** is based on the SARS-CoV-2 and human protein-protein interaction, as detailed in the study by [3]. This study identified 332 high-confidence protein-protein interactions, involving 332 SARS-CoV-2 proteins and 27 human proteins.
- **Human-Virus dataset** which includes 8,929 interactions from the DeepTrio study [5] is used as a benchmark dataset. It originally compiled DeepViral study by Liu-Wei et al. [10].
- **Tsukiyama et al. dataset** from the study [15] includes 22,383 interactions, comprising 5,882 viral proteins and 996 human proteins.

4   Proposed Approach: Clustering-Based Negative Sampling Strategy (CNS)

In this section, we propose a novel clustering-based sampling method to select reliable negative samples from the unlabelled samples. Our intuition for this method is to avoid selecting an unknown positive sample as a negative sample as much as possible. Therefore, we aim to select unknown samples with different characteristics by forming clusters of positive samples and selecting samples from each cluster. To minimize the randomness in sample selection, we propose a clustering-based negative sampling strategy employing Agglomerative Clustering [12]. In contrast to the K-means method, Agglomerative Clustering is deterministic, consistently producing the same clustering structure when applied to the same dataset. We explain our proposed method step by step, as illustrated in Fig. 2. We apply the Agglomerative Clustering Algorithm to the positive sample set P (i.e., experimentally confirmed pairs). Agglomerative Clustering is an unsupervised data mining technique used to create a hierarchy of clusters. It operates in a bottom-up manner, where each sample starts as its cluster, and clusters are progressively merged based on similarity, resulting in a hierarchy. Our algorithm organizes the positive sample set P into different k clusters. The positive sample set P is represented in Eq. 1, where each cluster is denoted as C_n. The centroid of each cluster, $C_{n\text{-center}}$, serves as a representative of the corresponding cluster, which is calculated by averaging the positive samples within the cluster.

$$P = \bigcup_{n=1}^{k} C_n, \quad n \in \{1, 2, \ldots, k\}. \tag{1}$$

The next step is to assign the unlabelled protein sequences from the set U to the closest cluster. Since Agglomerative Clustering does not support assigning unseen data to existing clusters and is only capable of clustering the training data, we assigned those sequences in a similar way to K-means. That

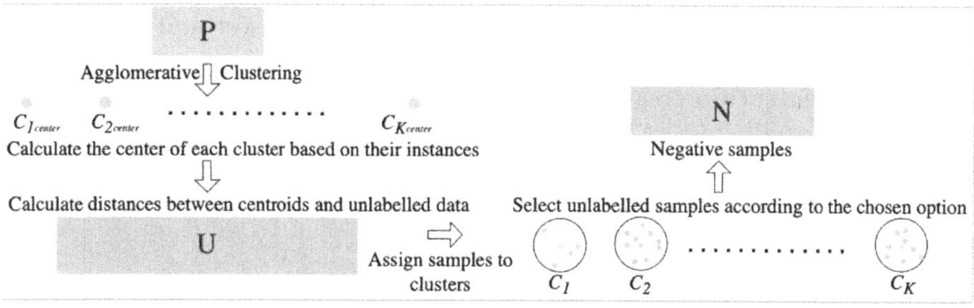

Fig. 2. The process of our clustering-based sampling approach.

is, we assigned the protein sequences to the clusters based on their similarity to the cluster centroids. To assess the similarity between positive protein sequences x and unlabelled protein sequences y, we use three distance metrics such as $d_{\mathrm{Canberra}}(x, y)$, $d_{\mathrm{Euclidean}}(x, y)$ and $d_{\mathrm{Cosine}}(x, y)$, which are calculated according to the corresponding equations (See Eqs. 2–4). In all equations, $x = (x_1, x_2, \ldots, x_m)$ stands for the vector of attributes of the cluster centroid and $y = (y_1, y_2, \ldots, y_m)$ for the vector of attributes of the unlabelled sample.

- **Canberra distance** normalizes the differences between elements by their sum, making it effective for measuring relative differences.
- **Euclidean distance** measures the straight-line distance between two points in a multidimensional space and calculates the absolute geometric distance between the cluster centroid and the unlabelled sample.
- **Cosine distance** evaluates how the vectors are in terms of direction. It assesses the cosine of the angle between two vectors x and y, where $x \cdot y$ is the dot product of the vectors, and $|x|$ and $|y|$ represent Euclidean norms.

$$d_{\mathrm{Canberra}}(x, y) = \sum_{i=1}^{m} \frac{|x_i - y_i|}{|x_i| + |y_i|} \tag{2}$$

$$d_{\mathrm{Euclidean}}(x, y) = \sqrt{\sum_{i=1}^{m} (x_i - y_i)^2} \tag{3}$$

$$d_{\mathrm{Cosine}}(x, y) = 1 - \frac{x \cdot y}{\|x\| \|y\|} \tag{4}$$

We pose the question of which unlabelled samples should be selected as negative samples from each cluster. To address this question, we propose four methods for utilizing the cluster structure in the selection process:

- **Closest selection** selects the elements closest to the centroids to ensure that the most representative elements of each cluster are selected.
- **Farthest selection** selects the elements farthest from the centroids, therefore, the more outlier instances within each cluster are captured.

- **Closest and Farthest selection** selects both the closest and farthest elements to create a balance between representative and diverse elements.
- **Uniform selection** ensures an equitable distribution by selecting elements evenly across all clusters (See Fig. 3 for detailed information).

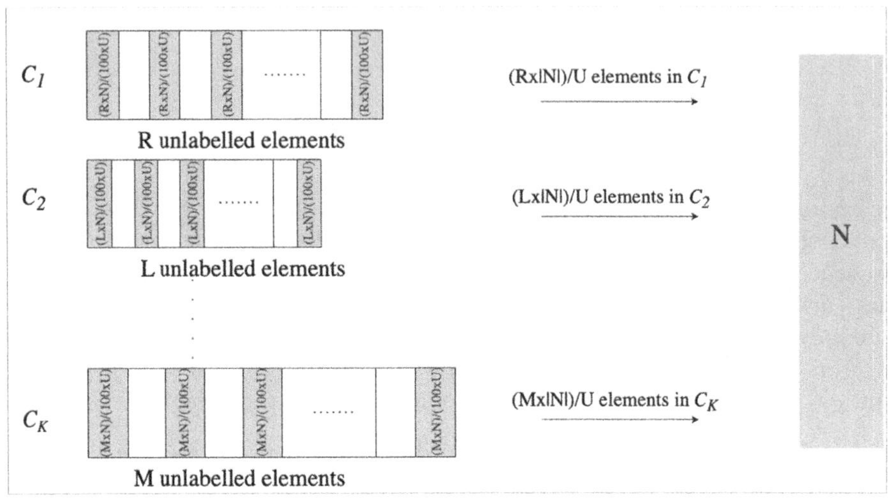

Fig. 3. The sets $R, L, \ldots, M$ denote unlabelled samples, with a total sum of U. The equation $\frac{R \times |N|}{U} + \frac{L \times |N|}{U} + \ldots + \frac{M \times |N|}{U} = |N|$ ensures proportional selection from each set, where $|N|$ is the total number of negative samples required.

We investigate the degree of similarity between the unlabelled samples and their positive centroids according to the selection mechanism to understand how these unlabelled samples are positioned within their assigned clusters. In each selection method, the number of elements chosen from each cluster is adjusted to be proportionate to its size. Here, $|N|$ denotes the total number of negative samples we seek to select, while $|C_n|$ signifies the number of elements within a given cluster where $n \in (1, k)$. The number of elements chosen from this cluster is determined by the $|N|$ to $|C_n|$ ratio, guaranteeing proportional representation relative to cluster size.

5   Evaluation

There are two notable studies in the literature [16,17] that use clustering-based methods for negative sample selection, which we compare with our proposed model. We compare our approach with these clustering-based strategies for negative sampling. *Wang et al.* [16] propose a clustering-based negative sampling strategy, in which all unlabelled data U are clustered using the K-means clustering algorithm, while our approach clusters the positive samples P using hierarchical clustering. In Wang's study, the number of unlabelled samples to be selected from each cluster is determined based on the proportion of samples within the cluster relative to the total unlabelled data. This ensures that the negative samples are proportionally distributed across clusters. Finally, the samples closest to the center of each cluster are selected based on the estimated number of instances. On the other hand, *Wei et al.* [17] suggest applying the MiniBatchKMeans clustering algorithm to the entire dataset consisting of both positive interactions P and unlabelled interaction sets U. The ratio of unlabelled examples within each cluster for the entire cluster is calculated and the clusters are sorted in descending order according to this ratio. The unlabelled instances are extracted from the clusters whose unlabelled instance density is the highest regarding the aforementioned sort operation.

To assess the effectiveness of each clustering-based approach for negative sample selection, a binary classifier is trained to predict protein interaction (i.e., positive and negative samples are labelled 1 and 0 respectively). To increase the reliability of the performance evaluation, we applied 5-fold cross-validation technique in which the dataset is divided into five subsets and each subset in turn serves as a test set, to obtain a comprehensive evaluation of the overall performance of the model. We chose a ratio of 1 : 10 for positive and negative samples, as recommended in previous research in this field reflecting the typical imbalance in real-world data. For the Agglomerative Clustering Algorithm, we used the average linkage method, which tends to produce more balanced clusters.

5.1 Effect of Distance Metrics on Method's Performance

The performance of the proposed sampling methods may vary depending on parameters such as the number of clusters, distance metrics, and selection mechanisms. We conducted further analyses to understand the impact of these parameters. First, the impact of distance metrics on the performance of our method was evaluated using the PHISTO dataset. In this phase, the GA^2M classifier was selected aligning with the work of Koca *et al.* [7]. Figure 4(a) shows the recall metric (i.e., the percentage of true positive predictions compared to the confirmed positive samples) where the y-axis represents the selection criterion, the x-axis shows the number of clusters (k) and the type of distance metric used. As can be seen in Fig. 4, the choice of distance metric has significant effects on recall and F1 scores. Regardless of the number of clusters, it can be seen that the Canberra distance performs better than other methods with the farthest selection. Based on these results, the Canberra distance was used as the distance metric for our method in the further experimental studies below.

5.2 Effect of Cluster Numbers

We analyzed the effects of different cluster numbers, including 2, 4, and 8, on the PHISTO dataset. As can be seen in Table 1, the optimal cluster number varied depending on the method. For example, the best performance for the Closest, Closest and Farthest, and Uniform methods was achieved with 2 clusters, while

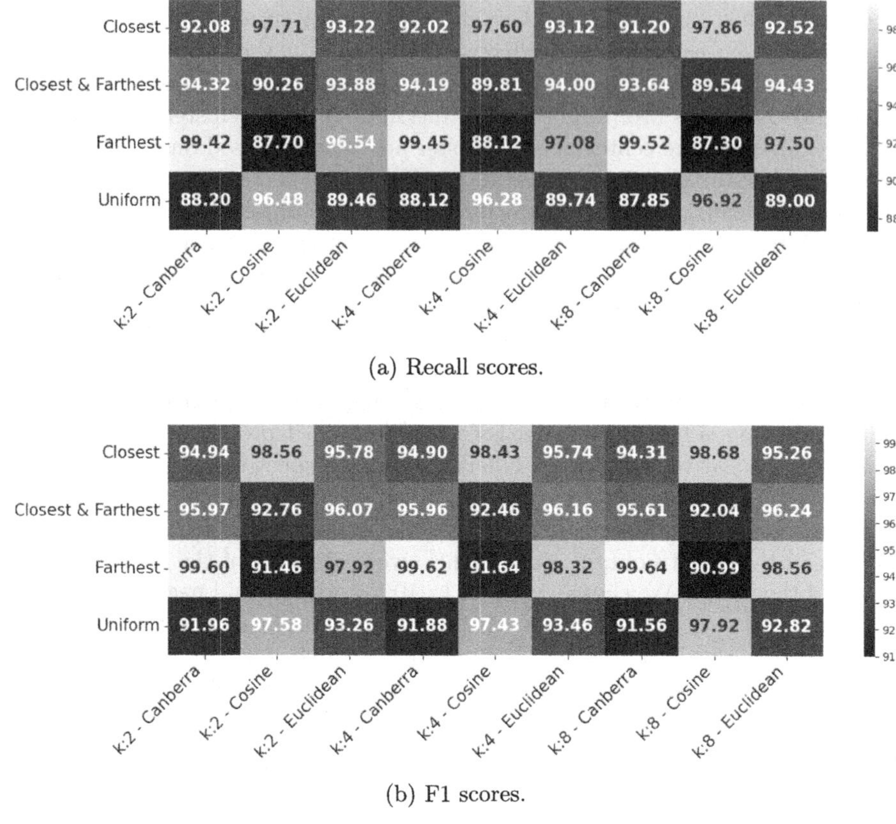

(a) Recall scores.

(b) F1 scores.

Fig. 4. Performance metrics of GA^2M on the PHISTO dataset.

the Farthest method performed best with 8 clusters. Similarly, Wang's method achieved the highest F1 and recall values with 4 clusters, while Wei's method showed the best results with 8 clusters. Consequently, the optimal number of clusters for each method was fixed based on their performance in the following analyses.

5.3   Analysis of Classifier Performance

In this section, we assess the performance of different classifiers with the sampling strategies. Table 2 shows the F1 and recall values of each classifier. We used the classifiers aligned with Koca's study [7], including Random Forest (RF), Logistic Regression (LR), Support Vector Machine (SVM) and Generalized Additive 2 Model (GA^2M). The results show that the proposed approach with the farthest selection slightly outperforms the other strategies. Of the four classifiers, the Random Forest classifier outperformed the others, showing slightly better results than GA^2M. Therefore, we advocated using the Random Forest.

Table 1. The F1 and recall values in percent for the PHISTO dataset with different cluster numbers.

Cluster Numbers	$k = 2$		$k = 4$		$k = 8$	
	F1	Recall	F1	Recall	F1	Recall
Closest	**94.94**	**92.08**	94.90	92.02	94.31	91.20
Closest & Farthest	**95.97**	**94.32**	95.96	94.19	95.61	93.64
Farthest	99.60	99.42	99.62	99.45	**99.64**	**99.52**
Uniform	**91.96**	**88.20**	91.88	88.12	91.56	87.85
Wang's method	91.56	87.85	**96.18**	**93.95**	94.28	91.58
Wei's method	97.73	95.81	97.84	96.08	**99.22**	**98.90**

Table 2. The F1 and recall values in percent for the PHISTO dataset with different classifiers and the optimal cluster number for each method.

Classifiers	GA^2M		LR		SVM		RF	
	F1	Recall	F1	Recall	F1	Recall	F1	Recall
Closest	**94.94**	**92.08**	92.76	89.30	94.33	91.28	94.90	91.46
Closest & Farthest	95.97	**94.32**	93.42	91.18	95.97	94.20	**96.34**	94.05
Farthest	**99.64**	**99.52**	99.49	99.24	99.58	99.40	**99.64**	99.44
Uniform	91.96	**88.20**	88.60	83.67	91.30	87.05	**92.36**	87.72
Wang's method	96.18	**93.95**	94.00	90.76	95.45	92.96	**96.44**	93.91
Wei's method	99.22	98.90	98.88	98.37	98.99	98.36	**99.36**	**99.10**
Random sampling	73.32	67.94	62.34	54.44	70.97	63.06	**79.39**	**73.52**

5.4 Performance Across Other Datasets

To evaluate the effectiveness of the sampling strategies across different pathogen-host protein-protein interaction datasets, we further analyze the performance of the Random Forest classifier on three additional datasets, as described in Sect. 3. Table 3 presents the results for cluster numbers 2, 4, and 8 across all datasets. Based on the results in Table 3, Wei's method consistently achieved the highest performance across most datasets and cluster numbers for Random Forest. In the PHISTO dataset, our method with the farthest selection outperformed other methods and reached an F1 score of 99.64% with Random Forest at $k = 8$.

It is worth noting that there is no superior sampling strategy resulting in terms of prediction accuracy in all datasets. On the other hand, the results show that all cluster-based sampling approaches outperformed random sampling. Therefore, we suggest that the clustering-based method is preferable to the random sampling method in this domain.

Table 3. Performance of the Random Forest classifier across all datasets with different cluster numbers.

Datasets	Cluster Numbers	Farthest		Wang's Method		Wei's Method		Random Sampling	
		F1	Recall	F1	Recall	F1	Recall	F1	Recall
PHISTO	$k = 2$	**99.62**	**99.30**	95.96	93.01	97.15	94.71		
	$k = 4$	**99.62**	**99.40**	96.44	93.91	97.39	95.36	79.39	73.52
	$k = 8$	**96.64**	**99.44**	94.53	90.94	99.36	99.10		
Tsukiyama et al.	$k = 2$	96.02	92.68	96.17	93.06	**96.62**	**93.78**		
	$k = 4$	95.85	92.62	96.42	93.44	**96.50**	**93.39**	68.84	59.80
	$k = 8$	96.06	92.62	93.78	89.24	**97.60**	**95.54**		
Human-Virus	$k = 2$	92.34	87.04	92.12	86.52	**95.10**	**91.36**		
	$k = 4$	92.06	86.88	89.76	83.53	**96.54**	**93.68**	72.32	64.16
	$k = 8$	92.31	87.12	89.00	82.48	**97.52**	**95.74**		
SARS-CoV-2	$k = 2$	69.26	55.92	70.96	55.70	**87.32**	**79.84**		
	$k = 4$	67.58	55.48	68.80	53.96	**80.89**	**69.30**	67.40	55.51
	$k = 8$	84.08	75.79	76.74	64.20	**91.99**	**85.64**		

6   Conclusion

In this study, we address the challenge of selecting negative samples from unlabelled samples for pathogen-host protein-protein interactions. We propose a novel clustering-based approach to enhance the reliability of negative sample selection. We highlight the fundamental differences between our method and other clustering-based approaches, as well as the widely used random sampling method. Our experiments, conducted on four different datasets of virus-human protein interactions, employ four selection methods and four classifiers. The results demonstrate that clustering-based approaches outperform the widely used random sampling method. We believe that providing a more precise definition of protein feature representation will result in more distinguishable outcomes in the selection of negative samples. As future work, it would be interesting to analyze the characteristics of the datasets and their correlation with the obtained results on that dataset regarding the effectiveness of the negative sampling methods.

Acknowledgments. The authors would like to thank Mehmet Burak Koca for his support in this work. Special thanks to Eymen Küçükçakır for his contribution and to Amin Deldari Alamdari for his technical support. Zehra Kesemen was supported by the Scientific and Technological Research Council of Türkiye (TÜBİTAK) through the 2210-C National M.Sc. Scholarship Program in the Priority Fields and Science, and the project grant 120N680.

References

1. Durmuş Tekir, S., et al.: PHISTO: pathogen-host interaction search tool. Bioinformatics **29**(10), 1357–1358 (2013). https://doi.org/10.1093/bioinformatics/btt137
2. Gage, P.: A new algorithm for data compression. C Users J. Arch. **12**, 23–38 (1994). https://api.semanticscholar.org/CorpusID:59804030
3. Gordon, D.E., et al.: A SARS-CoV-2 protein interaction map reveals targets for drug repurposing. Nature **583**(7816), 459–468 (2020). https://doi.org/10.1038/s41586-020-2286-9
4. Hamilton, W.L., Ying, R., Leskovec, J.: Inductive representation learning on large graphs. arXiv:1706.02216 (2018)
5. Hu, X., Feng, C., Zhou, Y., Harrison, A., Chen, M.: Deeptrio: a ternary prediction system for protein-protein interaction using mask multiple parallel convolutional neural networks (2021). https://doi.org/10.1093/bioinformatics/btab737
6. Iuchi, H., et al.: Bioinformatics approaches for unveiling virus-host interactions. Comput. Struct. Biotechnol. J. **21**, 1774–1784 (2023). https://doi.org/10.1016/j.csbj.2023.02.044
7. Koca, M.B., Nourani, E., Abbasoğlu, F., Karadeniz, İ., Sevilgen, F.E.: Graph convolutional network based virus-human protein-protein interaction prediction for novel viruses. Comput. Biol. Chem. **101**, 107755 (2022). https://doi.org/10.1016/j.compbiolchem.2022.107755, https://www.sciencedirect.com/science/article/pii/S1476927122001359
8. Kshirsagar, M., Carbonell, J., Klein-Seetharaman, J.: Multitask learning for host-pathogen protein interactions. Bioinformatics **29**(13), i217–i226 (2013). https://doi.org/10.1093/bioinformatics/btt245
9. Lian, X., Yang, X., Yang, S., Zhang, Z.: Current status and future perspectives of computational studies on human–virus protein-protein interactions. Briefings Bioinform. **22**(5), bbab029 (2021). https://doi.org/10.1093/bib/bbab029
10. Liu-Wei, W., Kafkas, Ş., Chen, J., Dimonaco, N.J., Tegnér, J., Hoehndorf, R.: DeepViral: prediction of novel virus-host interactions from protein sequences and infectious disease phenotypes. Bioinformatics **37**(17), 2722–2729 (2021). https://doi.org/10.1093/bioinformatics/btab147
11. Mikolov, T., Chen, K., Corrado, G., Dean, J.: Efficient estimation of word representations in vector space. arXiv:1301.3781 (2013)
12. Murtagh, F., Legendre, P.: Ward's hierarchical agglomerative clustering method: which algorithms implement ward's criterion? J. Classif. **274–295**(3) (2014). https://doi.org/10.1007/s00357-014-9161-z
13. Sledzieski, S., Singh, R., Cowen, L., Berger, B.: D-script translates genome to phenome with sequence-based, structure-aware, genome-scale predictions of protein-protein interactions. Cell Syst. **12**(10), 969-982.e6 (2021). https://doi.org/10.1016/j.cels.2021.08.010
14. Song, B., Luo, X., Luo, X., Liu, Y., Niu, Z., Zeng, X.: Learning spatial structures of proteins improves protein–protein interaction prediction. Briefings Bioinform. **23**(2), bbab558 (2022). https://doi.org/10.1093/bib/bbab558
15. Tsukiyama, S., Hasan, M.M., Fujii, S., Kurata, H.: LSTM-PHV: prediction of human-virus protein–protein interactions by LSTM with word2vec. Briefings in Bioinformatics **22**(6) (2021). https://doi.org/10.1093/bib/bbab228
16. Wang, B., et al.: Imbalance data processing strategy for protein interaction sites prediction. IEEE/ACM Trans. Comput. Biol. Bioinform. **18**(3), 985–994 (2021). https://doi.org/10.1109/tcbb.2019.2953908

17. Wei, Z., Yao, D., Zhan, X., Zhang, S.: A clustering-based sampling method for mirna-disease association prediction. Front. Genet. **13** (2022). https://doi.org/10.3389/fgene.2022.995535
18. Yang, X., Yang, S., Li, Q., Wuchty, S., Zhang, Z.: Prediction of human-virus protein-protein interactions through a sequence embedding-based machine learning method. Comput. Struct. Biotechnol. J. **18**, 153–161 (2020). https://doi.org/10.1016/j.csbj.2019.12.005

Proteins Transcription Factor Prediction Using Graph Neural Networks

Domenico Amato[1], Salvatore Calderaro[1(✉)], Giosué Lo Bosco[1],
Filippo Vella[2], and Riccardo Rizzo[2]

[1] Department of Mathematics and Computer Science, University of Palermo,
90123 Palermo, Italy
{domenico.amato01,salvatore.calderaro01,giosue.lobosco}@unipa.it
[2] Institute for High-Performance Computing and Networking, National Research
Council of Italy, 90146 Palermo, Italy
{filippo.vella,riccardo.rizzo}@icar.cnr.it

Abstract. A transcription factor is a regulatory protein that modulates
the transcriptional velocity of genetic information from DNA to mRNA.
This protein orchestrates the activation and deactivation of specific
genes, thereby regulating their expression to coincide precisely with requisite phases and quantities throughout the cellular lifecycle. This paper
introduces an innovative deep-learning methodology designed to ascertain the presence of transcription factors within protein sequences. This
method, through a graph-based representation of the sequence, employs a
Graph Neural Network for protein classification. We evaluated this novel
approach across four datasets derived from the Swiss-Prot database; we
achieved promising results that surpass those obtained through traditional fixed-length encodings and conventional machine-learning techniques and are comparable with state-of-the-art deep models.

Keywords: Transcription Factor · Graph Neural Networks · Proteins

1 Introduction

Transcription factors (TFs) are specific proteins that bind to DNA and are typically recognized by RNA polymerases to initiate transcription. Transcription
factors exert precise control over the expression levels of downstream genes by
selectively binding to certain DNA segments in the promoter regions under specific physiological conditions [14,18]. In general, biochemical experiments are
used to identify transcription factors as specific sorts of DNA-binding proteins.
Nevertheless, conducting these investigations can be laborious, costly, and challenging to execute on a large scale. To overcome these limitations, computational
methods using conventional machine learning models have been proposed [11,31].
These models necessitate a feature selection procedure that, most of the time,
involves computationally costly homology measurements in biological sequences
performed via alignment algorithms. In addition, TFs not showing homologies

with the known ones are very difficult to predict. To overcome this limitation, deep learning can help due to its automatic feature discovery property, as demonstrated by the approach called DeepTFactor [15]. In this paper, the prediction of TF is performed using a network that processes the input protein represented in one-hot encoding. The input has a fixed length. Three convolutional networks in parallel process the coded sequence, and the concatenation of their output representation is classified as TF or not TF. The method has been tested on Eukaryote and Prokaryote protein sequences. The problem of TF prediction can be considered as a specific case of sequence classification, and for this general topic several contributions based on deep models have been proposed so far [3–5,9,10,13,19–21,24,25]. Recent advances in machine learning show that graph representations of data can help in supervised and unsupervised classification by using so-called Graph Neural Networks (GNN) [1,7]. For instance, in the latter they are used for breast cancer grade identification, representing the tumor image with a graph and using a GNN to predict tumor grade. These networks provide a compact representation of complex data, commonly known as embeddings. Embeddings have been effectively applied to model real-world data, particularly through metric learning techniques, for the classification of histological and X-ray images [2]. It is noteworthy that a DNA or protein sequence can be represented as a graph using the De Bruijn representation, with the advantage of being independent of sequence length. Following this consideration, we want to investigate the possibility of using Graph Convolutional Neural Networks (GCNN) to identify Transcription Factor proteins in the presented study. The GCNN allows the processing of protein sequences of any length using De Bruijn graph representation. A graph representation does not need padding to obtain a fixed-length representation, such as the case of DeepTFactor [15].

We have performed experiments on four datasets obtained from the Swiss-Prot database, using a GCNN and a baseline classifier using a Support Vector Machine (SVM) and a fixed-length encoding and using a state-of-the-art deep neural network named DeepTFactor [15]. The experimental results are promising since they show the superiority of our GCNN concerning the baseline approach and comparable ones concerning the DeepTFactor model.

2   Data and Methods

In this section, we describe the procedure for building the datasets generated for our experimental activity, the overall mechanism of Graph Neural Networks computation, and, finally, the approach we used to represent a protein sequence using a graph.

2.1   Datasets Generation

We generated four different datasets during our experimental activity to test the proposed approach's effectiveness. These datasets were generated from the Swiss-Prot dataset downloaded in April 2024[1]. Swiss-Prot is a database containing

---

[1] https://www.uniprot.org/help/downloads.

manually annotated protein sequences, providing information about functional notations, the three-dimensional structure of proteins, and protein-protein interactions. The considered version of the dataset includes 571,282 protein sequences. We used the same approach reported in [15] to establish if a protein sequence was annotated as TF. We annotated a protein as a TF sequence if it satisfied the following criteria: 1) has a Gene Ontology (GO) annotations for the TF activity; 2) has both a DNA binding-related GO annotation and a transcription regulation-related GO annotation (see right part of block diagram in Fig. 1). For a more in-depth analysis, we categorised proteins based on their organism type: eucaryotic, procaryotic, or virus. To perform this operation we defined the following lists containing the taxonomic terms:

1. eucaryotic terms: Eukaryota;
2. prokaryotic terms: Bacteria and Archaea;
3. viral terms: Virus.

and then, starting from the protein ID, we retrieved a set of taxonomic terms. If this set contained one of the terms in the lists defined previously, then the protein was annotated accordingly. In Fig. 1, we depicted the block diagram relative to the procedure for TF annotation (right side) and organism identification (left side). The protein sequences in the Swiss-Prot dataset have a minimum length of 2, a maximum length of 35213 amino acids, and a mean length of about 362 amino acids. As can be seen from the histogram reported in Fig. 2, most sequences (about 96%) have a length in the range $[2, 1000]$. For this reason, we decided to filter out the sequences with more than 1000 amino acids, not considering further those that contain no standard amino acids (B, O, U, Z). After this processing, we found 19886 TF protein sequences and 531853 non-TF ones. The dataset is highly unbalanced, so we discarded many non-TF proteins. After establishing if a protein contains a transcription factor and finding the organism type, we generate four datasets:

- *all dataset*: contains 3000 TF and 9000 no-TF sequences sampled without considering the organism type;
- *eukaryotic dataset*: contains 3000 TF and 9000 no-TF sequences sampled from the sequences belonging to the eukaryotic organisms;
- *prokaryotic dataset*: contains 3000 TF and 9000 no-TF sequences sampled from the sequences belonging to the prokaryotic organisms.
- *virus dataset*: contains 538 TF and 1614 no-TF sequences sampled from the sequences belonging to the virus organisms.

The number of sequences in these datasets was decided using the same approach reported in [15].

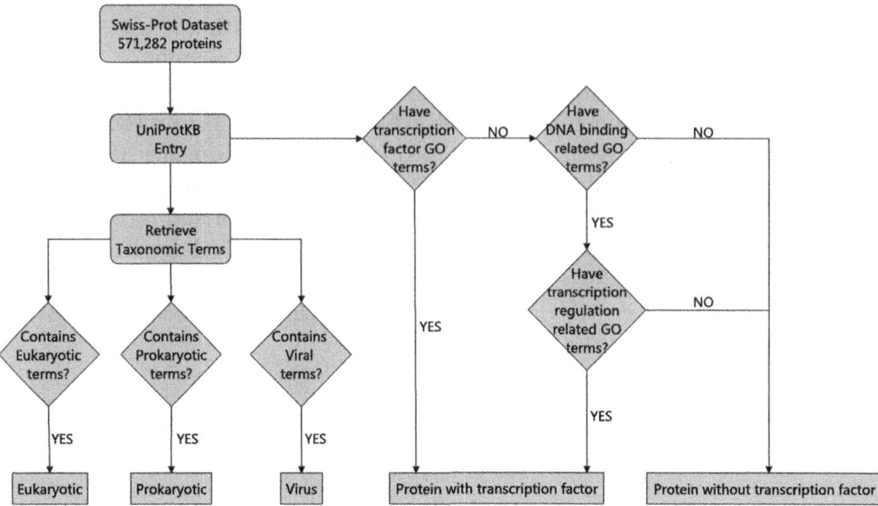

Fig. 1. A block diagram that reports the procedure used for the protein sequence TF annotation and organism identification.

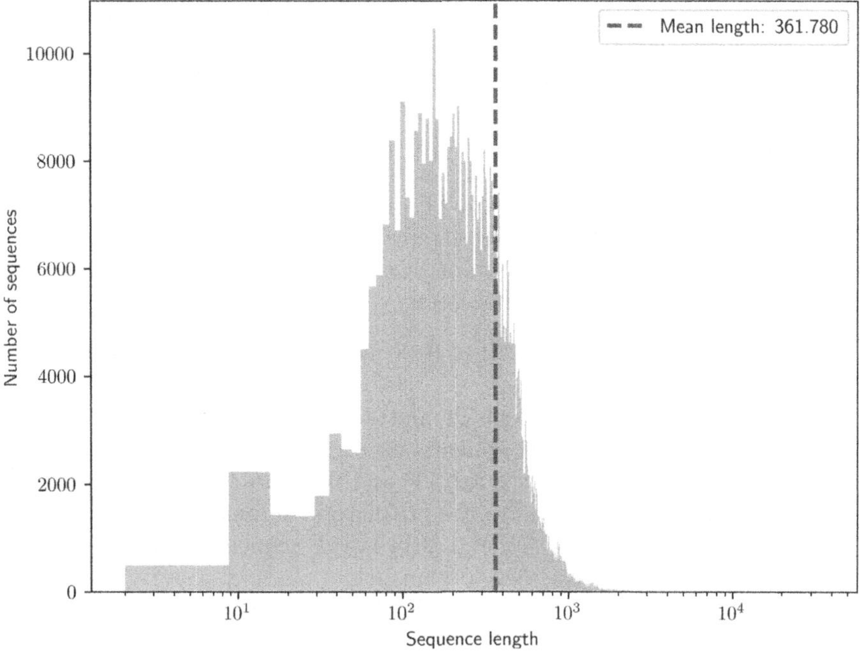

Fig. 2. Histogram that reports the distribution of the sequence lengths. The red dotted line represents the mean sequence length. (Color figure online)

2.2   Graph Neural Networks

Graph Neural Networks (GNNs) are deep neural networks that process data with graph structure. These networks can perform various tasks, such as node classification, link prediction, and graph classification. In this paper, we face the problem of determining whether a sequence contains a transcription factor as a graph classification problem. A GNN structure is a set of stacked layers. The fundamental idea behind its computation is to enhance the features of each node by integrating them with the attributes of its neighbouring nodes. Mathematically, each layer of a GNN performs two operations: *AGGREGATE* tries to aggregate the information from the neighbours of each node in the graph, and *COMBINE* updates the node representation by combining the aggregated information from neighbours with the current node representations [28]. To summarise, in the *AGGREGATE* step, each node collects information from its neighbours by applying a defined function during the aggregate step. This step leverages node features and graph topology and allows the node to capture insights from its local neighbourhood. In the *COMBINE* step, the node features are updated, integrating the aggregate information with the node's features and refining the node representation by blending its individual properties with contextual neighbours' information. In graph classification, obtaining a vector representing the whole graph is necessary. This operation is called *READOUT*, and common *READOUT* operations are summing, averaging, and finding the maximum or minimum across all nodes. For the implementation of our network, we use Graph Convolution Networks (GCN) [17], one of the most used GNN due to its effectiveness in many applications. A GCN layer transforms the node embeddings, employing a non-linear transformation on the new embeddings. Each new embedding is computed as the mean of the neighbours of all nodes. The proposed GNN consists of three stacked GCN layers. After every GCN layer, we apply dropout to prevent overfitting. The final part of the net consists of an average *READOUT* layer and a linear layer with one unit with the Sigmoid activation function, which performs the final classification. The architecture of this net is in Fig. 3.

2.3   Graph Construction

The De Bruijn graph is an alternative representation of a string of symbols, in this case, a sequence containing a protein's primary structure. These graphs are widely used in bioinformatics to develop genome assembly algorithms, as they allow the reconstruction of a DNA sequence starting from short reads by representing overlaps between them. In this work, we used a De Bruijn graph to represent a protein sequence composed of amino acids. This encoding allows us to represent protein sequences of all dimensions and analyze and process very long sequences. This representation has already been successfully used as input for a deep model to categorize bacterial nucleotide sequences [1]. A graph represents a given sequence as a composition of their subparts, the so-called k-mers. In a De Bruijn graph, the nodes are the $(k-1)$-mers of the sequence, and an edge is added if the sequence contains some k-mers whose prefix is the former and whose suffix

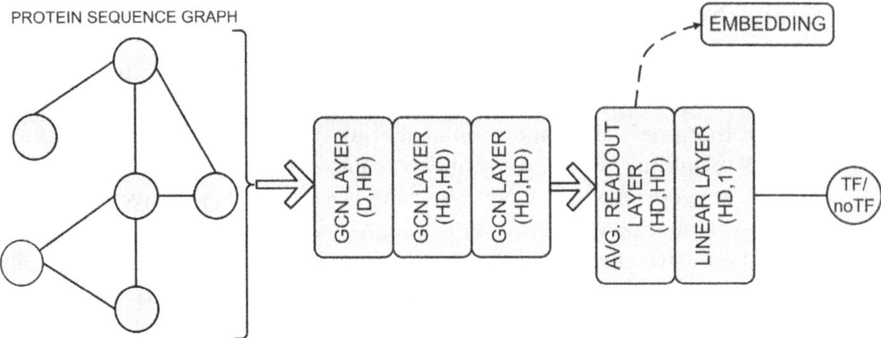

Fig. 3. The neural network used for our experiments. The values within the parenthesis represent the number of input and output features of each layer. D represents the initial node features dimension, while HD stands for Hidden Dimension and represents the number of units of every GCN layer.

is the latter [8]. It is essential to underline that in our implementation, to avoid the presence of isolated nodes in the graph, we do not consider all possible $(k-1)$-mers but only those that occur in the sequence. Given a protein sequence S on the alphabet $\Sigma = \{A, C, D, E, F, G, H, I, K, L, M, N, P, Q, R, S, T, V, W, Y, X\}$ and an integer $k \geq 2$, the procedure for the De Bruijn graph building is the following:

- identication of the k-mers of the sequence S (the number of possible k-mers of S is $L - k + 1$ where $L = |S|$)
- assignment to $(k - 1)$-mers to the nodes
- connect one node to another if the $(k - 1)$-mer overlaps another.

In the GNN computation, assigning initial node features is crucial. To achieve this, we create a feature vector of dimension D by combining the one-hot encoding (OHE) of the $(k - 1)$-mer with its positional information. The first part of this vector, with dimensionality $m = |\Sigma| \times k - 1$ represents the flattened one-hot encoding of the $(k - 1)$-mer. The second part, with dimensionality $d = D - m$ captures the positional encoding of the $(k - 1)$-mer within the sequence, computed as described in [27]. Given the position i of the $(k - 1)$-mer within the sequence and the dimension d the j-th component of the vector is computed as follow:

$$PE(i, 2j) = \sin \left(\frac{i}{10000^{\frac{2j}{d}}} \right) \tag{1}$$

$$PE(i, 2j + 1) = \cos \left(\frac{i}{10000^{\frac{2j}{d}}} \right) \tag{2}$$

It is essential to underline that if a $(k - 1)$-mer occurs more time within the sequence, its positional encoding vector is computed as the mean of vectors relative to its positions. In Fig. 4, we reported the De Bruijn graph built for a short sequence S with $k = 3$. The obtained graph has 11 nodes and 12 edges;

every node has a feature vector with D components. In the figure, we report the k-mers on the edges to make the overlap between $(k-1)$-mers clearer.

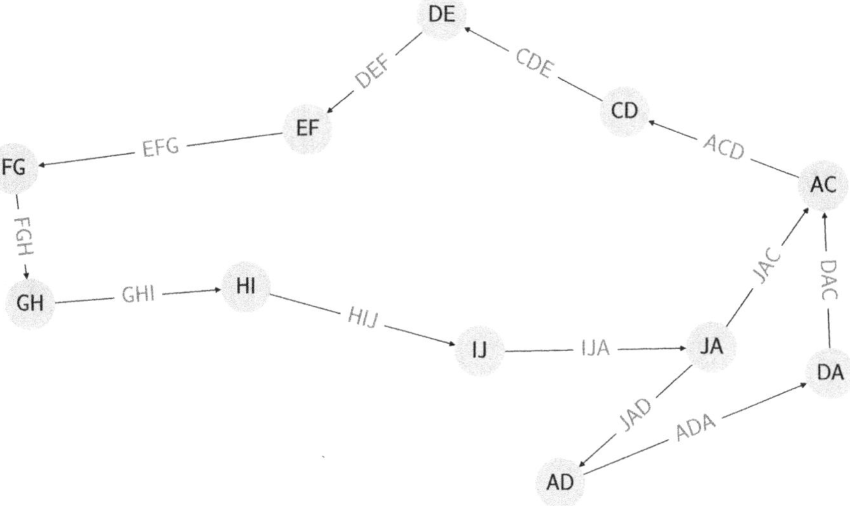

Fig. 4. Example of a De Bruijn graph built for a protein sequence.

3 Results

In our experimental activity, we evaluated the effectiveness of the proposed approach through three distinct experiments. The first experiment was aimed to assess whether a simple one-hot encoding combined with a baseline classifier, such as a Support Vector Machine (SVM), could predict the presence of a transcription factor in protein sequences. The second experiment employed DeepT-Factor, a model based on Convolutional Neural Networks (CNNs), as described in [15]. Finally, the third experiment focused on our proposed method, involving representing the protein sequences as De Bruijn graphs and using a Graph Convolutional Neural Network (GCNN) for classification. This section outlines the experiments performed on the datasets previously mentioned. All experiments were executed on a workstation equipped with an NVIDIA GeForce RTX 3090 Ti GPU and a 12th Gen Intel® CoreTM i9-12900KS CPU. The system encompasses 32 GB of DDR5 RAM for its main memory. The GPU leverages the CUDA parallel computing framework and provides 24 GB of DDR6 memory. The operating system installed is Ubuntu 24.04. The implementation of the model employed the PyTorch and PyTorch Geometric libraries in Python

[12,23]. For reliable validation, we used 10-fold cross-validation across all experiments and adopted the widely used AUROC metric to assess the performance of each approach. In the first experiment, since the SVM requires a fixed-length input, protein sequences shorter than 1000 amino acids were padded with zeros, resulting in a matrix of size 21 × 1000. Given the high dimensionality of this representation, we applied Principal Component Analysis (PCA) to reduce the number of features to 512. To find the best SVM configuration, we performed a grid search over key hyperparameters (kernel type, the regularization parameter C, the kernel coefficient γ, and the degree of the polynomial). Table 1 shows the best hyperparameter configurations after the grid search application. In most cases, the kernel that allows us better results is the Gaussian one, except for the Prokaryotic dataset, in which the kernel assumes the form of a three-degree polynomial.

Table 1. Hyperparameters used for the SVM models. The table lists the best kernel type, polynomial degree (if applicable), regularization parameter (C), and kernel coefficient (γ) found through grid search.

dataset/hyperparameters	kernel	degree	C	γ
All	rbf	–	10	$\frac{1}{\#features}$
Eukaryotic	rbf	–	10	$\frac{1}{\#features \times Var(X)}$
Prokaryotic	poly	3	100	$\frac{1}{\#features}$
Virus	rbf	–	10	$\frac{1}{\#features}$

In the second experiment, DeepTFactor was fed with input matrices of size 1000×21, where each row corresponds to an amino acid. Zero padding was again applied to sequences shorter than 1000 amino acids. The model was trained using the same hyperparameters as in [15]. The third experiment evaluated the effectiveness of our proposed graph-based method. We constructed De Bruijn graphs with $k = 5$ and $D = 128$ to represent protein sequences and trained a GNN (described in Sect. 2 and illustrated in Fig. 3) to obtain concise sequence embeddings for final classification. Given the proposed architecture for all the considered datasets, we conducted Bayesian optimization [26] using 50 trials to explore the hyperparameter space to find optimal net hyperparameters. The optimization process aimed to fine-tune key parameters critical to GNN performance. The hidden dimension HD was chosen from a set of values $[64, 512, 1024]$ to control the network's capacity, enabling it to learn complex representations effectively. To mitigate overfitting, dropout rates were explored in the range $[0.2, 0.3, 0.4, 0.5, 0.6]$. The learning rate lr, an essential parameter for controlling the optimization process, was selected from a log-scaled range $[1 \times 10^{-4}, 1 \times 10^{-2}]$, ensuring stability and fast convergence. This procedure was carried out considering graphs built with $D = 128$ and $k = 5$. Additionally, the number of epochs chosen from $[10, 20, 30, 40, 50, 60, 70, 80, 90, 100]$ allowed us to evaluate the trade-off between sufficient training and potential overfitting. The optimal configurations found are in Table 2 and are the ones used to perform the experiments. The network was trained using the Adam optimizer [16] with a mini-batch size of 128.

Table 2. Hyperparameters used for the GNN models. The table lists the hidden dimension HD, the learning rate lr, the dropout rate, and the number of epochs.

dataset/hyperparameters	HD	lr	dropout rate	num. of epochs
All	1024	0.00097	0.3	60
Eukaryotic	1024	0.00123	0.6	100
Prokaryotic	512	0.00288	0.6	90
Virus	512	0.00316	0.5	50

As a further experiment, we considered different k-mers size values for the De Bruijn graph construction. We choose k from the set of values $[3, 5, 7, 9, 11, 13]$ since fixing $D = 256$ these were the only allowed values to compute the positional encoding of the (k-1)-mers. We choose $D = 256$ since this node feature dimension enabled us to explore a wide range of k values. The obtained results in terms of AUROC are in Fig. 5. Analyzing the results reported in the figure, we can observe that AUROC improves as the k-mer size increases; this suggests that a larger k-mer size provides a more powerful representation, likely for the capability to capture richer sequence patterns. For all the considered datasets, we obtained the best results for $k = 13$.

Table 3 compares the results obtained using the three approaches to the considered datasets, considering the AUROC evaluation criteria. **OHE+PCA+SVM** refers to the first experiment, while **DBG+GNN** refers to our proposal for the configuration with $D = 256$ and $k = 13$. The **OHE+PCA+SVM** method yields the lowest AUROC scores, suggesting that, while it offers a straightforward and interpretable framework, its performance may be hindered by a simplistic feature representation. This limitation could lead to the omission of critical sequence dependencies and interactions. Moreover, the reliance on fixed-length sequences (achieved through zero-padding) and the necessity for dimensionality reduction further indicate that this method may not be well-suited for analyzing complex or highly variable protein sequences. **DeepTFactor** achieves the highest AUROC values across all datasets comparable with one obtained using the graph-based approach. These results suggest that CNN is capable of capturing sequence patterns. However, introducing massive padding is a limitation, particularly for sequences with less than 1000 amino acids, where the padded regions may play a crucial role in the final net response. Finally, our proposal demonstrates comparable and sometimes higher performance concerning **DeepTFactor**. Representing protein sequences via De Bruijn graphs allows the model to operate on sequences of any length $(\geq k)$ without needing zero-padding or arbitrary truncation. Our approach performs similarly to DeepTFactor, probably because the graph representation can capture

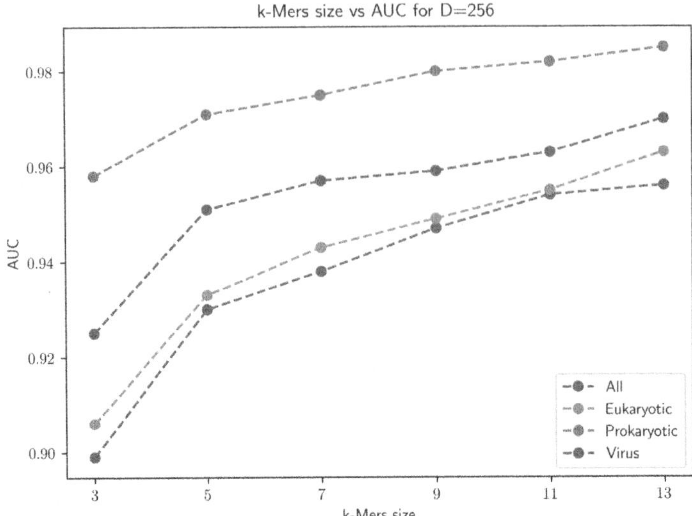

Fig. 5. AUC performance for different k-mer sizes ($k = 3, 5, 7, 9, 11, 13$) at initial node feature dimension $D = 256$ evaluated across all the considered datasets.

complex sequence interactions and dependencies without relying on fixed-length encodings. This capability allows the GNN model to avoid potential noise introduced by zero padding in shorter sequences or sequences with less than 1000 amino acids, which can affect DeepTFactor performances. Additionally, while DeepTFactor employs convolutional layers to extract localized sequence features, the GNN models global relationships and contextual sequences through its graph representation. The proposed method performance highlights the efficacy of graph-based methods in capturing biologically relevant patterns, suggesting that GNNs can serve as a scalable and interpretable alternative to traditional deep learning models like CNNs, especially for datasets with variable-length sequences.

Table 3. Mean and standard deviation of the AUROC using the discussed approaches. OHE+PCA+SVM refers to the first experiment, DeepTFactor [15] has a slightly better performance with a heavier representation, DBG+GNN refers to our proposal.

dataset	OHE+PCA+SVM	DeepTFactor	DBG+GNN
all	0.917 ± 0.009	0.965 ± 0.007	0.956 ± 0.005
eukaryotic	0.911 ± 0.010	0.963 ± 0.007	0.963 ± 0.004
prokaryotic	0.966 ± 0.004	0.990 ± 0.003	0.985 ± 0.003
virus	0.932 ± 0.018	0.967 ± 0.016	0.970 ± 0.012

In conclusion, the obtained results demonstrate the effectiveness of our proposal for classifying protein sequences, particularly in its ability to deal with sequences of any length without the need for fixed-length representation or extensive pre-processing like zero-padding. This flexibility can offer an advantage over classical machine learning algorithms and DeepTFactor, which rely on fixed-length encodings. Moreover, our approach is scalable and adaptable for various sequence lengths and types, making it well-suited for real-world applications. These advantages highlight the potential of graph-based neural networks in capturing complex interactions in biological sequences.

4   Conclusions

The paper introduces a novel deep-learning approach for detecting transcription factors in protein sequences by utilizing a De Bruijn graph representation combined with a Graph Neural Network (GNN). This method demonstrated strong performance and generalization across multiple datasets. Future work will focus on employing explainability tools to identify critical sequence segments that influence the detection of transcription factors, enhancing the interpretability of the model and potentially providing insights into the biological significance of these sequences. Tools such as GNNExplainer [29], PGExplainer [22], SubgraphX [30] and GECo [6] will be explored for their ability to identify key subgraphs and provide interpretable explanations of model predictions. By integrating these methods, the approach could reveal biologically meaningful substructures, such as conserved sequence motifs or critical interaction patterns, offering valuable insights into the mechanisms underlying transcription factor binding and function.

Data Availability. The generated datasets in FASTA format, the code to convert a protein sequence into a graph, and the Graph Neural Network model implementation can be found in the following GitHub repository.

Disclosure of Interests. The authors have no competing interests to declare that are relevant to the content of this article.

References

1. Amato, D., Calderaro, S., Lo Bosco, G., Rizzo, R., Vella, F.: Bacteria taxonomic classification using graph neural networks. In: 2024 IEEE International Conference on Evolving and Adaptive Intelligent Systems (EAIS), pp. 1–6. IEEE (2024)
2. Amato, D., Calderaro, S., Lo Bosco, G., Rizzo, R., Vella, F.: Leveraging Deep Embeddings for Explainable Medical Image Analysis, pp. 225–261. Springer Nature Switzerland, Cham (2024)
3. Amato, D., et al.: Classification of Sequences with Deep Artificial Neural Networks: Representation and Architectural Issues, pp. 27–59. Springer International Publishing, Cham (2021)

4. Amato, D., Di Gangi, M.A., Lo Bosco, G., Rizzo, R.: Recurrent deep neural networks for nucleosome classification. In: Raposo, M., Ribeiro, P., Sério, S., Staiano, A., Ciaramella, A. (eds.) Computational Intelligence Methods for Bioinformatics and Biostatistics, pp. 118–127. Springer International Publishing, Cham (2020)
5. Amato, D., Lo Bosco, G., Rizzo, R.: Corenup: a combination of convolutional and recurrent deep neural networks for nucleosome positioning identification. BMC Bioinform. **21**(8), 326 (2020)
6. Calderaro, S., Amato, D., Lo Bosco, G., Rizzo, R., Vella, F.: The geco algorithm for graph neural networks explanation. arXiv e-prints pp. arXiv–2411 (2024)
7. Calderaro, S., Lo Bosco, G., Vella, F., Rizzo, R.: Breast cancer histologic grade identification by graph neural network embeddings. In: International Work-Conference on Bioinformatics and Biomedical Engineering, pp. 283–296. Springer (2023)
8. Compeau, P.E., Pevzner, P.A., Tesler, G.: How to apply de bruijn graphs to genome assembly. Nat. Biotechnol. **29**(11), 987–991 (2011)
9. Di Gangi, M.A., Gaglio, S., La Bua, C., Lo Bosco, G., Rizzo, R.: A deep learning network for exploiting positional information in nucleosome related sequences. In: Rojas, I., Ortuño, F. (eds.) IWBBIO 2017. LNCS, vol. 10209, pp. 524–533. Springer, Cham (2017). https://doi.org/10.1007/978-3-319-56154-7_47
10. Di Gangi, M., Lo Bosco, G., Rizzo, R.: Deep learning architectures for prediction of nucleosome positioning from sequences data. BMC Bioinform. **19-S**(14), 127–135 (2018)
11. Eichner, J., Topf, F., Dräger, A., Wrzodek, C., Wanke, D., Zell, A.: Tfpredict and sabine: sequence-based prediction of structural and functional characteristics of transcription factors. PLoS ONE **8**(12), e82238 (2013)
12. Fey, M., Lenssen, J.E.: Fast graph representation learning with pytorch geometric (2019)
13. Fiannaca, A., et al.: Deep learning models for bacteria taxonomic classification of metagenomic data. BMC Bioinform. **19-S**(7), 61–76 (2018)
14. Karin, M.: Too many transcription factors: positive and negative interactions. New Biol. **2**(2), 126–131 (1990)
15. Kim, G.B., Gao, Y., Palsson, B.O., Lee, S.Y.: Deeptfactor: a deep learning-based tool for the prediction of transcription factors. Proc. Natl. Acad. Sci. **118**(2), e2021171118 (2021)
16. Kingma, D.P., Ba, J.: Adam: a method for stochastic optimization. In: Bengio, Y., LeCun, Y. (eds.) 3rd International Conference on Learning Representations, ICLR 2015, San Diego, CA, USA, May 7–9, 2015, Conference Track Proceedings (2015)
17. Kipf, T.N., Welling, M.: Semi-supervised classification with graph convolutional networks. In: International Conference on Learning Representations (ICLR) (2017)
18. Latchman, D.S.: Transcription factors: an overview. Int. J. Biochem. Cell Biol. **29**(12), 1305–1312 (1997)
19. Lo Bosco, G., Di Gangi, M.A.: Deep learning architectures for DNA sequence classification. In: Petrosino, A., Loia, V., Pedrycz, W. (eds.) Fuzzy Logic and Soft Computing Applications. Springer International Publishing, Cham (2017)
20. Lo Bosco, G., Rizzo, R., Fiannaca, A., La Rosa, M., Urso, A.: A deep learning model for epigenomic studies. In: Yétongnon, K., Dipanda, A., Chbeir, R., Pietro, G.D., Gallo, L. (eds.) 12th International Conference on Signal-Image Technology & Internet-Based Systems, SITIS 2016, Naples, Italy, November 28 - December 1, 2016, pp. 688–692. IEEE Computer Society (2016)

21. Lo Bosco, G., Rizzo, R., Fiannaca, A., La Rosa, M., Urso, A.: Variable ranking feature selection for the identification of nucleosome related sequences. In: Benczúr, A., et al. (eds.) New Trends in Databases and Information Systems, pp. 314–324. Springer International Publishing, Cham (2018)
22. Luo, D., et al.: Parameterized explainer for graph neural network. In: Larochelle, H., Ranzato, M., Hadsell, R., Balcan, M., Lin, H. (eds.) Advances in Neural Information Processing Systems, vol. 33, pp. 19620–19631. Curran Associates, Inc. (2020)
23. Paszke, A., et al.: Pytorch: an imperative style, high-performance deep learning library. In: Advances in Neural Information Processing Systems, vol. 32 (2019)
24. Rizzo, R., Fiannaca, A., La Rosa, M., Urso, A.: Classification experiments of DNA sequences by using a deep neural network and chaos game representation. In: CompSysTech '16, Proceedings of the 17th International Conference on Computer Systems and Technologies 2016, pp. 222–228. Association for Computing Machinery, New York, NY, USA (2016)
25. Rizzo, R., Fiannaca, A., La Rosa, M., Urso, A.: A deep learning approach to DNA sequence classification. In: Computational Intelligence Methods for Bioinformatics and Biostatistics: 12th International Meeting, CIBB 2015, Naples, Italy, September 10–12, 2015, Revised Selected Papers 12, pp. 129–140. Springer (2016)
26. Snoek, J., Larochelle, H., Adams, R.P.: Practical bayesian optimization of machine learning algorithms. In: Pereira, F., Burges, C., Bottou, L., Weinberger, K. (eds.) Advances in Neural Information Processing Systems, vol. 25. Curran Associates, Inc. (2012)
27. Vaswani, A., et al.: Attention is all you need. In: Guyon, I., et al. (eds.) Advances in Neural Information Processing Systems, vol. 30. Curran Associates, Inc. (2017)
28. Xu, K., Hu, W., Leskovec, J., Jegelka, S.: How powerful are graph neural networks? (2019)
29. Ying, Z., Bourgeois, D., You, J., Zitnik, M., Leskovec, J.: Gnnexplainer: generating explanations for graph neural networks. In: Wallach, H., Larochelle, H., Beygelzimer, A., d' Alché-Buc, F., Fox, E., Garnett, R. (eds.) Advances in Neural Information Processing Systems, vol. 32. Curran Associates, Inc. (2019)
30. Yuan, H., Yu, H., Wang, J., Li, K., Ji, S.: On explainability of graph neural networks via subgraph explorations. In: Meila, M., Zhang, T. (eds.) Proceedings of the 38th International Conference on Machine Learning. Proceedings of Machine Learning Research, vol. 139, pp. 12241–12252. PMLR (2021)
31. Zheng, G., et al.: The combination approach of SVM and ECOC for powerful identification and classification of transcription factor. BMC Bioinform. **9** (2008)

Identification of Differential Alternative Splicing Events: Assessing Tools Performance with Different Sequencing Parameters

Luca Faretra[1], Antonio Ammendola[1], Massimo Pancione[1], Francesco Napolitano[1,2], and Luigi Cerulo[1,2,3]

[1] Department of Science and Technology, University of Sannio, Benevento, Italy
lcerulo@unisannio.it
[2] Biogem, Institute of Genetic Research, Avellino, Italy
[3] GRGS, Genomic Research Center for Health, University of Salerno, Salerno, Italy

Abstract. Alternative splicing (AS) is a crucial process in transcript maturation, enabling the generation of multiple transcript variants and the control of transcript abundance. In gene expression analysis, accounting for AS is crucial as different transcripts, arising from the same gene, encode distinct proteins. Moreover, the identification of differences in the proportions of AS events enrich the classical differential expression analysis with additional insight regarding cell behaviour.

Most emerging AS tools focus on individual splicing events and are able to quantify the proportion of AS events from RNA-seq data. The systematic evaluation of such tools is lacking as: i) proposed benchmark studies do not consider the distinction between various types of AS events, neglecting the different performances that a tool may have on different events; ii) the proposed benchmark studies adopt single sample RNA-seq data and do not consider the potential impact of sequencing parameters on tools' performance; and iii) confounding factors such as gene expression level may be taken into consideration as they may affect tools performance.

We propose an extensive benchmark study to evaluate the performance of three widely used event-based AS tools: SUPPA, rMATS, and MAJIQ, selected based on their citations. Moreover, we propose a novel reusable benchmark dataset generator to create gold-standards from real RNA-seq data. We show that SUPPA consistently outperforms others, followed by rMATS and MAJIQ. Performance were evaluated by varying sequencing parameters, such as library size, fragment length, read length and sequencing type (paired-end or single-end). Results show that SUPPA's performance remains generally stable regardless of and is not affected by sequencing parameters. In contrast, the performance of rMATS and MAJIQ is moderately affected by sequencing parameters.

Keywords: Alternative splicing events · Synthetic data · Tool performance evaluation

L. Cerulo et al. (Eds.): CIBB 2024, LNBI 15276, pp. 28–42, 2025.
https://doi.org/10.1007/978-3-031-89704-7_3

1   Introduction

In eukaryotes, when coding genes are transcribed, initially immature transcripts are produced [1]. These transcripts consist of both introns, which are non-coding regions, and exons, which contain protein coding information. The splicing process removes introns and bound exons to form the mature transcript. Throughout the splicing process, a range of AS events can occur, reshaping the structure of the mature transcript [2]. These include: Skipped Exons (SE), where an exon is omitted from the mature transcript; Retained Introns (RI), where an intron persists in the mature transcript; Alternative 5' Splice Site usage (A5SS), resulting in a change of the cutting site at the 5' splice site; Alternative 3' Splice Site usage (A3SS), leading to a change of the cutting site at the 3' splice site; and Mutually eXclusive Exons (MXE), where only one of two contiguous exons can be retained in the mature transcript.

AS tools can be classified into three major groups [3]: Isoform-based, Exon-based, and Event-based tools. The key distinction between these categories lies in their approach to AS quantification and identification. Isoform-based tools reconstruct isoforms prior to quantification, with the resulting entities referred to AS transcripts. Exon-based tools exclusively quantify exons or both exons and junctions, while event-based tools quantify individual AS events.

For example, to quantify an SE event within a gene, event-based tools calculate the percentage of transcripts generated by the gene that retain the exon. This percentage is estimated by the following parameter, called Percent Spliced In (PSI) [4]:

$$PSI = \frac{I_C/(ExonLength + ReadLength - 1)}{I_C/(ExonLength + ReadLength - 1) + E_C/(ReadLength - 1)} \quad (1)$$

where, I_C represents the counts associated to the inclusion event and E_C are the counts associated to the skipping event.

To set up our benchmark study we rely on two literature studies. The first, from Muller *et al.* [5], conducts a comprehensive analysis of 3 event-based tools, evaluating the job time, performance changes with the number of replicates, and the concordance between tool outputs. The second, from Jiang *et al.* [6], assesses the performance of 21 distinct event-based tools including an initial analysis of known splicing events followed by the identification and assessment of novel splicing events. Both studies evaluate the tools without considering the specific parameters of the sequencing files, they do not report enough emphasis for the performance for single AS events and they do not consider genes with low expression levels. Incorporating sequencing parameters is crucial as they might influence tools performance differently, while differentiating between distinct types of splicing events provides more insightful examination of performance; finally, low-expression genes may yield lower performance compared to genes with moderate or higher expression levels.

In this work, we extend the two studies by considering 3 more recent tools, that are SUPPA, rMATS, and MAJIQ, selected based on their citations, and we propose a novel reusable benchmark dataset generator to create gold-standards from real RNA-seq data. Performance were evaluated by varying sequencing parameters, such as library size, fragment length, read length and sequencing type (paired-end or single-end). We show that SUPPA consistently outperforms others, followed by rMATS and MAJIQ. SUPPA also demonstrates remarkable stability in performance, both across different splicing events and when sequencing parameters are varied, as well as when considering events in low-expression genes. In contrast, rMATS and MAJIQ exhibit much more variable performance with changes in these factors. Results obtained from this study offer valuable guidance for selecting tools tailored to experimental needs in RNA-seq analyses.

2   Data and Methods

2.1   Tools Selection

We have selected three widely adopted AS tools: rMATS [7], SUPPA [8] and MAJIQ [9]. These tools were chosen because they are highly popular in literature and they use different approaches to quantify AS events.

rMATS implements a hierarchical approach that accounts for two levels of variability in sequencing data. To manage the uncertainty associated with individual replicates, it first models the read counts of isoforms for exon inclusion and exclusion using a binomial distribution, where the probability of success corresponds to the PSI. Secondly, to address variability among replicates, rMATS employs a logit-normal distribution.

MAJIQ focuses on quantifying local splicing variations (LSVs), which represent different combinations of splice junctions within a genomic region. It employs a probabilistic model to estimate the PSI for each junction involved in an LSV, using read rate modeling and a Bayesian approach. To address variability in read counts, MAJIQ applies bootstrapping techniques to generate an empirical distribution of the PSI estimates. This methodology enhances the robustness of the splicing analysis and provides insights into the dynamics of alternative splicing events.

SUPPA calculates the PSI directly from the transcript abundances associated with each splicing event. Specifically, SUPPA assumes that the PSI of an event corresponds to the relative abundance of the isoforms that include a specific form of the event. This simplified approach allows for rapid estimation of PSI without the need for complex models.

2.2   Gold-Standard Generation

Our aim was to generate three replicates each for control and treatment groups, with the only difference between them being the TPM levels of the two most highly expressed transcripts (T1 and T2) of certain selected genes. Specifically, these transcripts had their TPM values switched between control and treatment. This means that for the selected genes, transcript T1 was more highly expressed than T2 in the control, while T2 was more highly expressed than T1 in the treatment. Obviously, T1 and T2 have structural differences resulting from alternative splicing events. For example, T1 might retain an exon that is skipped in T2. This expression switch results in the exon being predominantly retained in the control group and more frequently spliced out in the treatment group. Such a splicing event is therefore classified as positive. In summary, all events distinguishing the switched transcripts were categorized as positive, while other splicing events unaffected by the switch were considered negative.

To generate this gold-standard, we developed the workflow shown in Fig. 1 that extends the pipeline developed by Jiang *et al.*. We started with three files provided by the study of Jiang *et al.* [6]: the transcript expression matrix obtained from the RSEM [10] analysis of experiment SRR493366, which reports counts, IsoPct, TPM, and FPKM values of all transcripts detected in the sequencing file (transcripts with zero counts were removed); the mean-dispersion matrix, which reports the mean and dispersion of various genes, derived from sequencing data analyzed by Pickrell *et al.* [11] and Cheung *et al.* [12] across a heterogeneous set of samples; and the events matrix, which reports all the events that differ among the various transcripts of the same gene, obtained through the analysis of version 32 of the human genome annotation analyzed using the Astalavista tool [13].

From the transcript expression matrix, we derived the gene expression matrix by summing the contributions of transcripts belonging to the same gene to calculate gene counts. We also estimated the effective length of the genes by calculating the weighted average of the isoform lengths, using IsoPct as the weight.

Next, we assigned a dispersion parameter to the count of each gene using the mean-dispersion matrix. Specifically, the count of each gene was compared to the means in the matrix, and to each gene was assigned a dispersion corresponding to the nearest mean value. Using the rnbinom function in R, we generated six simulated replicates for each gene by sampling from a negative binomial distribution based on the obtained mean and dispersion parameters. For each replicate, we calculated the TPM using the estimated lengths.

Starting from the transcript expression matrix, we also identified the two highly expressed isoforms per gene, for genes with a TPM of at least equal to 1 and an IsoPct difference of at least 5 between the two isoforms. These transcript pairs were categorized based on the TPM distribution of the genes into four groups: low expression ($TPM < 7.35$), low to moderate expression ($7.35 \leq TPM < 19.48$), moderate to high expression ($19.48 \leq TPM < 49.20$), and high expression ($TPM \geq 49.20$).

From each of the obtained groups, we identified AS events that differed between the transcript pairs using the events matrix, and selected 375 AS events distributed as the percentages commonly observed in higher eukaryotes [3]. Counting all groups, we selected 1495 events distributed as 800 SE, 103 RI, 400 A3SS, 160 A5SS, and 32 MXE.

We also obtained the IsoPct of transcripts for six replicates by applying the Dirichlet distribution to the original IsoPct values of the transcripts expression matrix.

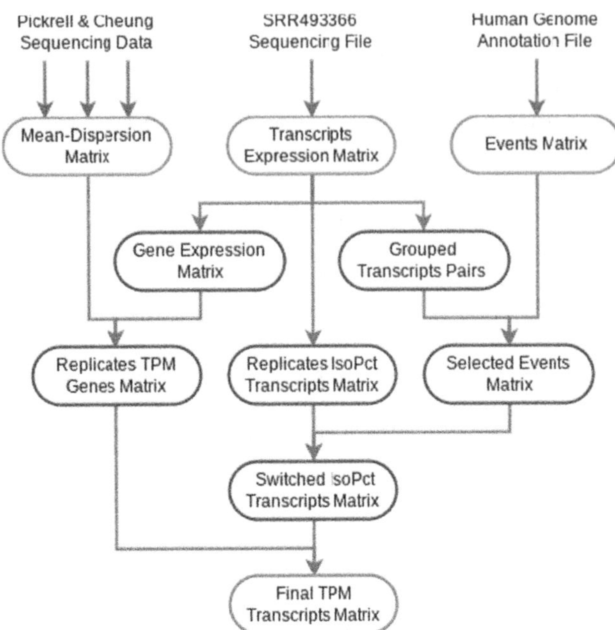

Fig. 1. Workflow of the gold-standard generation. We generated the TPMs for transcripts from three treated and three control samples, which differ by 1,495 splicing events that are switched between the two conditions.

To simulate the differences between treated and control samples, from the replicates IsoPct transcripts matrix, we switched the IsoPct of the transcripts couples associated with the selected events in replicates 4-6, obtaining three control replicates and three treatment replicates.

Finally, we obtained the TPM of the isoforms by multiplying the genes TPM of the replicates TPM gene matrix by the IsoPct of the switched IsoPct transcripts matrix.

The modifications made to the pipeline developed by Jiang *et al.* were aimed at ensuring that the TPM distribution of the selected genes was not random, but reflected that observed in real samples, and that the relative frequencies of the events affecting these genes were representative of the biological context, more accurately reflecting the splicing variability observed under natural conditions.

The obtained TPM values for each transcript were used to simulate the sequencing files for each replicate. To do that, we proceeded by adopting Rsubread [14] instead of RSEM. This allowed us to control important parameters that are not supported by RSEM. Table 1 illustrates the sequencing parameters that can be controlled using Rsubread, along with the values chosen in our experiments.

Table 1. Parameters considered by Rsubread. Each parameter is reported with the values selected for our analysis.

Parameter	Values
Library Size	20×10^6 / 50×10^6 / 80×10^6 / 100×10^6
Fragment Length	$200\,bp$ / $300\,bp$ / $400\,bp$ / $500\,bp$
Read Length	$75\,bp$ / $100\,bp$ / $125\,bp$ / $150\,bp$
Sequenecing Type	*Single-End* / *Paired-End*

2.3   Experimental Parameters and Settings

We initially establish a baseline by setting each parameter to the mean value of its intended range. Subsequently, we iterate over the baseline setting, systematically changing one parameter at a time to assess its impact on the ability of AS tools to detect both general and specific types of AS events. The array of experimental parameters is reported in Table 2.

Table 2. Parameters considered by Rsubread. Parameters used for each simulation. Beginning from the baseline, we systematically altered one parameter at a time to assess its impact on tool performance.

Experiment	Library Size	Fragments (bp)	Reads (bp)	Sequencing Type
00	50×10^6	500	100	*Paired-End*
01	10×10^6	500	100	*Paired-End*
02	30×10^6	500	100	*Paired-End*
03	70×10^6	500	100	*Paired-End*
04	100×10^6	500	100	*Paired-End*
05	50×10^6	300	100	*Paired-End*
06	50×10^6	400	100	*Paired-End*
07	50×10^6	700	100	*Paired-End*
08	50×10^6	900	100	*Paired-End*
09	50×10^6	500	50	*Paired-End*
10	50×10^6	500	75	*Paired-End*
11	50×10^6	500	125	*Paired-End*
12	50×10^6	500	150	*Paired-End*
13	50×10^6	500	100	*Single-End*

For the sequencing file analysis, we generally followed the pipelines recommended by the respective tool developers with only a few necessary modifications.

The human genome annotation was filtered to include only protein-coding genes, ensuring that the analysis focused specifically on the relevant transcripts.

For rMATS analysis, reads were first aligned using STAR [15], followed by alternative splicing analysis with rMATS. The only modification made to the standard rMATS protocol was setting the cstat parameter to 0.5, which helped reduce the number of false positives and exclude events with minimal PSI variations.

For the analysis with MAJIQ, we also used the alignment files generated by STAR. Several modifications were made to the standard protocol to align with the specific objectives of our study. Specifically, we disabled the search for novel splicing variants, as this aspect was not relevant to our research. We also set the minreads parameter to 5 to ensure that only splicing events supported by a minimum of five reads were considered. Furthermore, we enabled the quantification of redundant splicing events to ensure a comprehensive assessment, and activated the output of all splicing events for a more thorough data collection for subsequent analyses.

SUPPA was employed after transcripts quantification with Salmon [16], an isoform-based tool, from which splicing events were subsequently quantified. For SUPPA, we set the lower-bound parameter to 0.1, in order to not consider PSI differences under 0.1 to test for significance.

2.4 Performance Measures

To assess the tool performance, we employed the Area under the Receiver Operating Characteristic curve (AUC) [17]. The Receiver Operating Characteristic (ROC) curve shows the plot of the true positive rate (TPR, sensitivity) against the false positive rate (FPR, 1-specificity) at each threshold setting of the tool outcome score. ROC curves demonstrate the trade-off between sensitivity and specificity as the decision threshold varies, while AUC quantifies the overall ability of a classification model to discriminate between the positive and negative classes. An AUC of 1 indicates perfect discrimination, while an AUC of 0.5 suggests that the model is performing no better than random guessing.

3   Results

The results from the baseline experiment, which considers all events from all genes (Fig. 2A), show a clear distinction in terms of AUC performance among the analyzed tools, with SUPPA outperforming the other two. SUPPA achieves a higher overall performance with an AUC of 0.871 ± 0.001, in contrast to rMATS and MAJIQ, which perform more similarly to each other, reaching AUC values of 0.789 ± 0.003 and 0.784 ± 0.002, respectively. At low FPR rates, ROC curves show that MAJIQ exhibit a better sensitivity (TPR) with respect to others.

Analyzing individual splicing events (Fig. 2B), it becomes evident that SUPPA consistently outperforms the other tools for each event, with much more stable performance: we found AUCs of 0.869 ± 0.001 for SE, 0.950 ± 0.005 for RI, 0.864 ± 0.002 for A3SS, 0.853 ± 0.007 for A5SS, and 0.822 ± 0.018 for MXE.

In contrast, rMATS and MAJIQ exhibit more variable performance across different events. Specifically, MAJIQ outperforms rMATS in detecting A3SS events, with an AUC of 0.812 ± 0.001 compared to 0.733 ± 0.001, and for A5SS events, reaching an AUC of 0.827 ± 0.008 versus 0.768 ± 0.010. Conversely, rMATS excels over MAJIQ for SE events (0.809 ± 0.003 vs. 0.760 ± 0.003), RI events (0.901 ± 0.010 vs. 0.840 ± 0.006), and MXE events (0.724 ± 0.009 vs. 0.592 ± 0.001). Notably, MAJIQ performs particularly poorly on the MXE event.

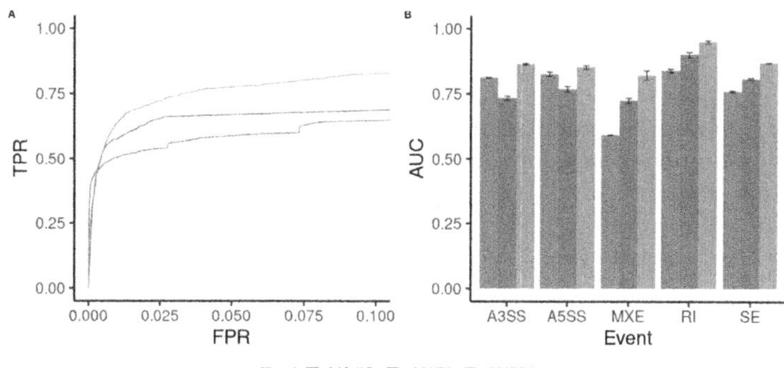

Fig. 2. Performance obtained from the baseline experiment, taking into account all genes. SUPPA outperforms both rMATS and MAJIQ globally, with the latter two showing comparable overall performance (A). When analyzing individual splicing events, SUPPA demonstrates much greater stability, while rMATS and MAJIQ exhibit higher variability. MAJIQ outperforms rMATS for A3SS and A5SS events, whereas rMATS shows superior performance for SE, RI, and MXE events (B).

When focusing solely on genes in the low expression group and calculating performance for the events of these genes (Fig. 3A), it emerges that SUPPA maintains remarkable stability (AUC of 0.873 ± 0.002), while rMATS and MAJIQ show a decline in performance (AUC values of 0.728 ± 0.01 and 0.734 ± 0.002, respectively).

Looking at individual splicing events (Fig. 3B), it becomes clear that the performance drop for rMATS and MAJIQ affects all splicing events equally. Thus, the relationship between the two tools remains essentially the same.

By varying the library size (Fig. 4A), SUPPA's performance, considering splicing events across all genes, remains relatively stable, with an AUC 0.867 $\pm$ 0.001 at 10 million reads and 0.871 ± 0.001 at 100 million reads. In contrast, rMATS shows more stable performance between 20 million reads, with an AUC of 0.775 ± 0.004, and 100 million reads, where the AUC results equal to 0.807 $\pm$ 0.001, but experiences a higher decline at 10 million reads, where the AUC is 0.731 ± 0.002. A similar pattern is observed for MAJIQ: it shows AUCs of 0.763 $\pm$ 0.003 and 0.805 ± 0.003 for 20 million and 100 million reads respectively, but at 10 million reads the AUC is equal to 0.680 ± 0.003. So, for MAJIQ the performance drop at 10 million reads is higher than that of rMATS.

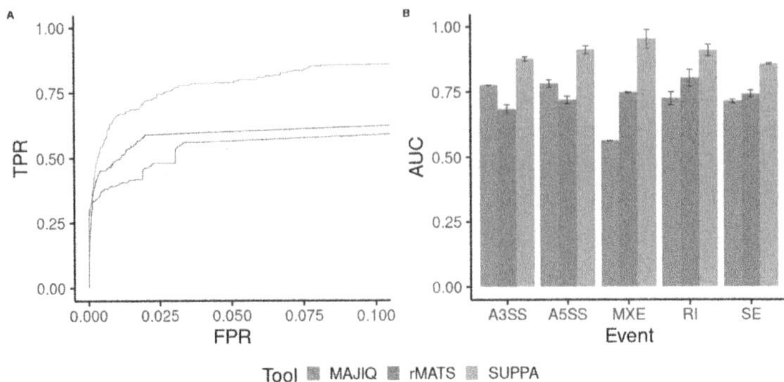

Fig. 3. Performances calculated from the baseline experiment, considering only low-expression genes. SUPPA's performance remains consistent with the results obtained when considering all events, while rMATS and MAJIQ tend to show a decline in performance (A). The performance drop for rMATS and MAJIQ affects all splicing events equally (B).

The different performance drop of rMATS and MAJIQ at 10 million reads is more evident for A3SS and A5SS events, which were better captured by MAJIQ at higher library sizes (Fig. 4C), with an AUC of 0.829 ± 0.007 against 0.751 ± 0.002 for A3SS and 0.846 ± 0.04 against 0.807 ± 0.008 for A5SS, but show comparable performance at 10 million reads (Fig. 4B), with an AUC of 0.697 ± 0.005 against 0.686 ± 0.002 for A3SS and 0.716 ± 0.002 against 0.698 ± 0.011 for A5SS.

The analysis of tool performance across varying library sizes, considering only low-expressed genes, reveals interesting characteristics. (Figure 5A). SUPPA continues to perform well (AUC of 0.858 ± 0.011 at 10 million reads and 0.872 ± 0.005 at 100 million reads). At 10 million reads, MAJIQ's performance is notably lower than that of rMATS (0.636 ± 0.001 vs. 0.514 ± 0.003); however, it improves progressively, surpassing rMATS's performance at 100 million reads (0.807 ± 0.004 vs. 0.759 ± 0.001). Nevertheless, it is evident that MXE events are still better detected by rMATS (Fig. 5B and 5C), even at 100 million reads (0.561 ± 0.001 vs. 0.768 ± 0.036).

Regarding the variation in fragment length, all tools tend to exhibit a progressive decline in performance as the length of the sequenced fragments increases (Fig. 6A). Considering the overall performance of SUPPA, the AUC decreases from 0.889 ± 0.003 for 300 bp fragments to 0.784 ± 0.003 for 900 bp fragments. In comparison, rMATS decreases from 0.819 ± 0.003 to 0.710 ± 0.003, while MAJIQ drops from 0.795 ± 0.004 to 0.722 ± 0.002. Performance tends to decline similarly across all types of events, as shown in Fig. 6B and 6C.

Considering splicing events involving genes with low expression levels, the trend remains essentially the same for both global events (Fig. 7A) and individual events when comparing 300 bp (Fig. 7B) and 900 bp fragments (Fig. 7C).

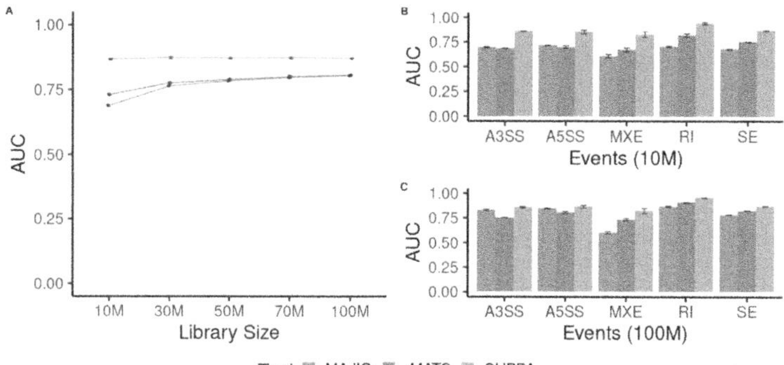

Fig. 4. Performance of the tools across different library sizes, considering all splicing events.. SUPPA's performance remains stable despite changes in this parameter, whereas rMATS and MAJIQ show a decline in performance below 20 million reads (A). The reduction of MAJIQ's performance can be seen in particular for A3SS and A5SS events, where, at 100 million reads, MAJIQ outperforms rMATS (C), whereas at 10 million reads, the two tools exhibit comparable performance (B).

Regarding the change in read length, if we consider all splicing events (Fig. 8A), this parameter does not affect SUPPA's performance, which shows an AUC of 0.864 ± 0.002 for 50 bp reads and 0.873 ± 0.002 for 150 bp reads. However, this parameter slightly impacts the performance of rMATS and MAJIQ. rMATS shows performance of 0.759 ± 0.003 and 0.806 ± 0.003, respectively, while MAJIQ exhibits performance of 0.735 ± 0.001 and 0.803 ± 0.002, respectively. When looking at individual events, with 50 bp reads there is a larger performance gap between the two tools and SUPPA (Fig. 8B), which narrows with 150 bp reads (Fig. 8C).

The performance of rMATS and MAJIQ becomes more interesting when considering low-expression genes. With 50 bp reads, rMATS outperforms MAJIQ with an AUC of 0.690 ± 0.005 versus 0.604 ± 0.003, but with 150 bp reads, the situation reverses, and MAJIQ surpasses rMATS with an AUC of 0.784 ± 0.004 versus 0.752 ± 0.004 (Fig. 9A). Looking at individual events, with 50 bp reads (Fig. 9B), rMATS shows better performance than MAJIQ across all events. However, with 150 bp reads (Fig. 9C), MAJIQ outperforms rMATS for A5SS (0.813 ± 0.009 vs. 0.765 ± 0.007) and A3SS (0.818 ± 0.006 vs. 0.703 ± 0.016), the two tools are comparable for SE (0.770 ± 0.007 vs. 0.768 ± 0.002) and RI (0.792 ± 0.026 vs. 0.803 ± 0.012), while rMATS remains superior for MXE (0.561 ± 0.001 vs. 0.746 ± 0.001).

Regarding the change from single-end to paired-end sequencing, considering all genes, no significant performance differences are observed, whether considering events globally (Fig. 10A) or individually (Fig. 10B and 10C). This suggests that the type of sequencing approach employed does not markedly influence the detection accuracy of splicing events in the analyzed datasets. Both single-end

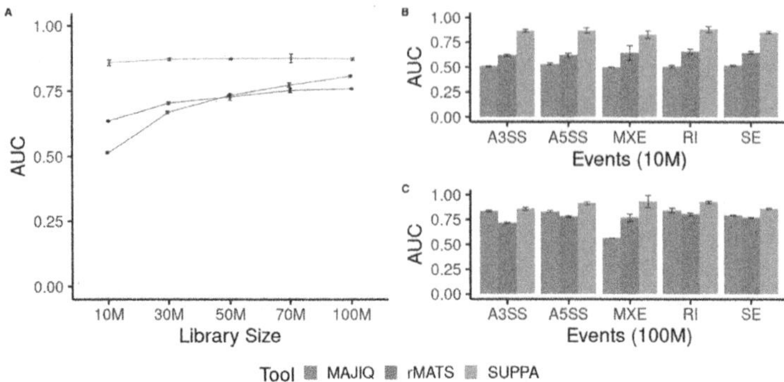

Fig. 5. **Performance assessed while varying the library size, focusing only on low expressed genes..** SUPPA continues to demonstrate good performance, while MAJIQ exhibits extremely low performance at 10 million reads, which improves as the library size increases, surpassing rMATS at 100 million reads (A). When considering individual events, at 10 million reads, MAJIQ shows poor performance across all events (B), whereas at 100 million reads, its performance exceeds that of rMATS for all events except for the MXE (C).

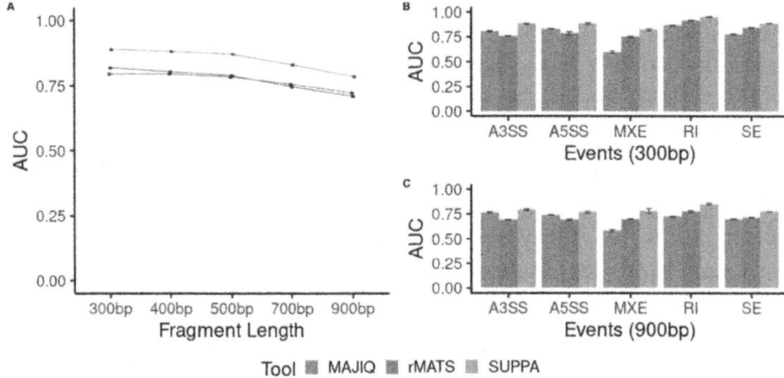

Fig. 6. **Performances obtained for different fragment lengths and considering all genes**. All tools exhibit a decrease in performance as the fragment length increases (A). This decline affects all splicing events when comparing 300 bp fragments (B) to 900 bp fragments (C).

and paired-end methods provide comparable sensitivity and specificity in identifying splicing events.

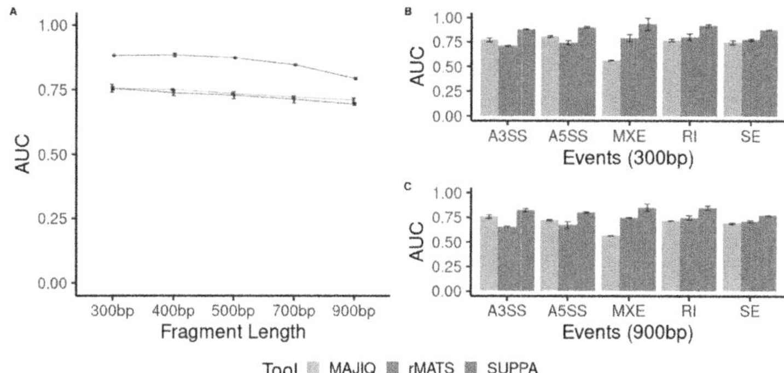

Fig. 7. Performances obtained for low expressed genes by fragment lengths changing. All tools exhibit a decrease in performance as the fragment length increases (A). This decline affects all splicing events when comparing 300 bp fragments (B) to 900 bp fragments (C).

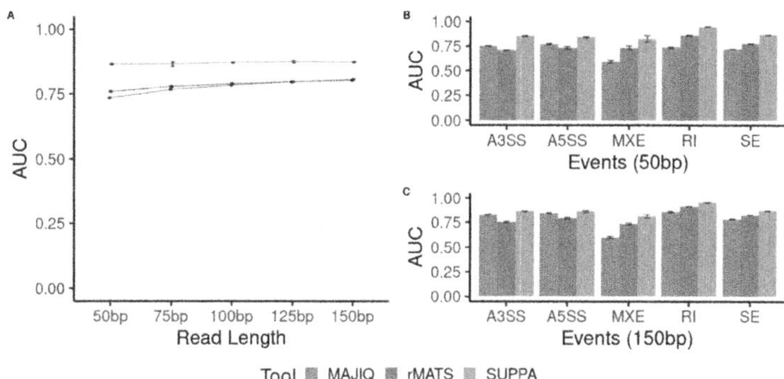

Fig. 8. Performances of tools considering all events with read lengths changing. SUPPA shows consistently stable performance, while rMATS and MAJIQ tend to experience a slight decline in performance as the read length decreases (A). Considering individual events, there is a larger performance gap between the two tools and SUPPA with 50 bp reads (B), which narrows with 150 bp reads (C).

When focusing on low-expression genes, a slight difference in performance emerges (Fig. 11A), with MAJIQ showing a greater decline compared to rMATS in single-end sequencing (AUC of 0.694 ± 0.010 vs. 0.645 ± 0.005). Considering individual events, the most notable difference is in A3SS and A5SS events, which are better captured by MAJIQ in paired-end sequencing (Fig. 11B), with an

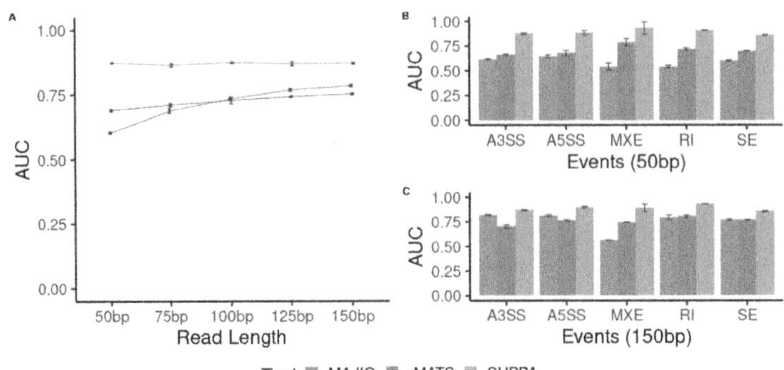

Fig. 9. Performances of the three tools considering only low expressed genes with read lengths changing.. MAJIQ's performance is lower than rMATS with shorter reads, but it progressively improves, surpassing rMATS with longer reads (A). Initially, rMATS outperforms MAJIQ across all events (B), but as read length increases, MAJIQ becomes superior to rMATS for A3SS and A5SS events, comparable for SE and RI events, while still performing worse for MXE events (C).

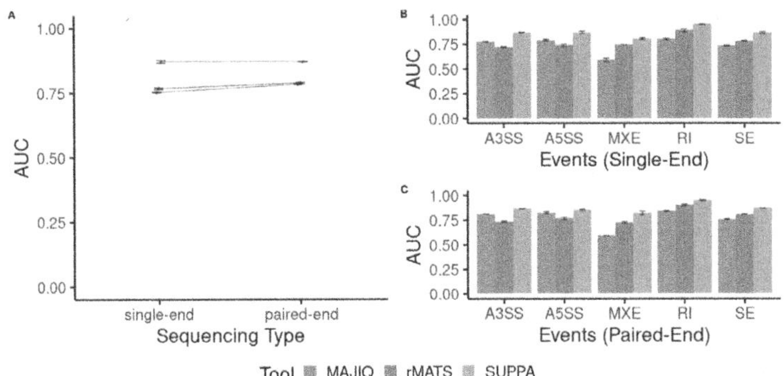

Fig. 10. Tools performances for all events considering sequencing type changing. No significant performance differences are observed, whether considering events globally (A) or individually (B and C).

AUC of 0.776 ± 0.001 against 0.684 ± 0.016 for A3SS and 0.781 ± 0.013 versus 0.718 ± 0.014 for A5SS, while in single-end sequencing, the performance of the two tools becomes comparable (Fig. 11C), with an AUC of 0.678 ± 0.013 versus 0.658 ± 0.023 for A3SS and 0.696 ± 0.009 against 0.699 ± 0.044 for A5SS.

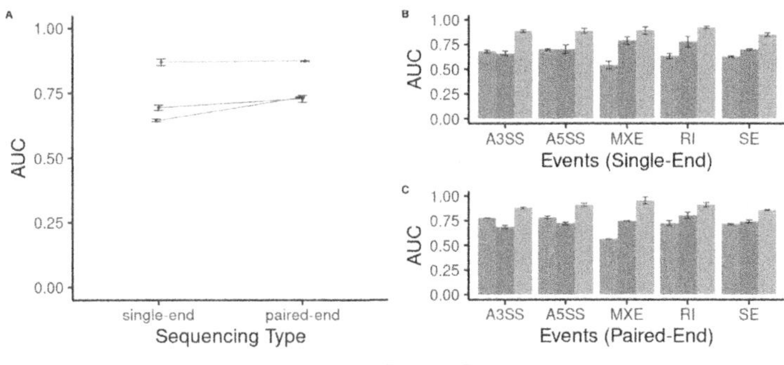

Fig. 11. Performance of the tools with low expression genes by sequencing type changing. It emerges only a greater decline of MAJIQ compared to rMATS in single-end sequencing (A). With individual events, the most notable difference is in A3SS and A5SS events, which are better captured by MAJIQ in paired-end sequencing (B), while in single-end sequencing, the performance of the two tools becomes comparable (C).

4   Conclusion

In this study, we evaluated the performance of three widely used event-based tools for analyzing AS events: SUPPA, rMATS, and MAJIQ. By utilizing simulated RNA-seq data, we discovered that SUPPA consistently outperformed both rMATS and MAJIQ for each condition, event and gene expression level, establishing its position as the most reliable tool among the three.

Our comprehensive assessment revealed that performance of all tools can change when subjected to different sequencing parameters. Specifically, SUPPA is less sensitive to parameter changes, while the performance of rMATS and MAJIQ tends to vary more; this difference is particularly evident in genes with low expression levels, as events associated with these genes are more challenging to detect.

Furthermore, when we delved into the analysis of specific splicing events, SUPPA also demonstrated remarkably stable results, reinforcing its reliability for detecting differential splicing. In contrast, rMATS and MAJIQ displayed notable variability in their performance, with MAJIQ exhibiting the highest degree of variability across various splicing events, suggesting that its performance may be more context-dependent.

Overall, our findings strongly indicate that SUPPA is the most effective tool for identifying differential AS events, providing researchers with a robust option for their analyses. SUPPA's superiority may be attributed to its significantly different analytical approach compared to rMATS and MAJIQ. Specifically, SUPPA quantifies splicing events based on transcripts quantified through SALMON, rather than directly using reads aligned on the reference genome. This approach, which involves a preliminary quantification of transcripts followed by splicing event quantification, appears to be a more effective strategy.

These results offer valuable insights and guidance for selecting the most appropriate tool for RNA-seq analysis, tailored to meet specific experimental requirements.

References

1. Wahl, M.C., Will, C.L., Lührmann, R.: The spliceosome: design principles of a dynamic RNP machine. Cell **136**(4), 701–718 (2009)
2. Wang, Y., et al.: Mechanism of alternative splicing and its regulation. Biomed. Rep. **3**(2), 152–158 (2015)
3. Mehmood, A., Laiho, A., Venäläinen, M.S., McGlinchey, A.J., Wang, N., Elo, L.L.: Systematic evaluation of differential splicing tools for RNA-seq studies. Briefings Bioinform. **21**(6), 2052–2065 (2020)
4. Schafer, S., Miao, K., Benson, C.C., Heinig, M., Cook, S.A., Hubner, N.: Alternative splicing signatures in RNS-seq data: percent spliced in (PSI). Curr. Protoc. Hum. Genet. **87**(1), 11–16 (2015)
5. Muller, I.B., et al.: Computational comparison of common event-based differential splicing tools: practical considerations for laboratory researchers. BMC Bioinform. **22**(1), 347 (2021)
6. Jiang, M., Zhang, S., Yin, H., Zhuo, Z., Meng, G.: A comprehensive benchmarking of differential splicing tools for RNA-seq analysis at the event level. Briefings Bioinform. **24**(3), bbad121 (2023)
7. Shen, S., et al.: rMats: robust and flexible detection of differential alternative splicing from replicate RNA-seq data. Proc. Nat. Acad. Sci. **111**(51), E5593–E5601 (2014)
8. Trincado, J.L., et al.: Suppa2: fast, accurate, and uncertainty-aware differential splicing analysis across multiple conditions. Genome Biol. **19**, 1–11 (2018)
9. Vaquero-Garcia, J., et al.: A new view of transcriptome complexity and regulation through the lens of local splicing variations. Elife **5**, e11752 (2016)
10. Li, B., Dewey, C.N.: RSEM: accurate transcript quantification from RNA-seq data with or without a reference genome. BMC Bioinform. **12**, 1–16 (2011)
11. Pickrell, J.K., et al.: Understanding mechanisms underlying human gene expression variation with RNA sequencing. Nature **464**(7289), 768–772 (2010)
12. Cheung, V.G., et al.: Polymorphic cis-and trans-regulation of human gene expression. PLoS Biol. **8**(9), e1000480 (2010)
13. Foissac, S., Sammeth, M.: Astalavista: dynamic and flexible analysis of alternative splicing events in custom gene datasets. Nucleic Acids Res. **35**(suppl 2), W297–W299 (2007)
14. Liao, Y., Shi W., Smyth, G.K., Dai, J.: Mapping, quantification and variant analysis of sequencing data (2024). http://bioconductor.org/packages/Rsubread
15. Dobin, A., et al.: Star: ultrafast universal RNA-seq aligner. Bioinformatics **29**(1), 15–21 (2013)
16. Patro, R., Duggal, G., Love, M.I., Irizarry, R.A., Kingsford, C.: Salmon provides fast and bias-aware quantification of transcript expression. Nat. Methods **14**(4), 417–419 (2017)
17. Sokolova, M., Japkowicz, N., Szpakowicz, S.: Beyond accuracy, F-score and ROC: a family of discriminant measures for performance evaluation. In: Australasian Joint Conference on Artificial Intelligence, pp. 1015–1021. Springer (2006)

Methods and Tools to Facilitate RE:IN Modeling and Analysis of GRNs

Daniel Grimland[1,2] , Eitan Tannenbaum[2] , and Hillel Kugler[2(✉)]

[1] The Future Scientists Center Alpha Program, Bar-Ilan University,
Ramat Gan, Israel
`daniel.grimland@biu.ac.il`
[2] Faculty of Engineering, Bar-Ilan University, Ramat Gan, Israel
{`eitan.tannenbaum,hillelk`}`@biu.ac.il`

Abstract. Stem cells play a central role in the development of organisms; hence, studying their gene regulatory networks (GRNs) is of great importance. The Reasoning Engine for Interaction Networks (RE:IN) is a toolset that supports modeling of GRNs to investigate their dynamics systematically and efficiently and make new predictions. Here we constructed a RE:IN model of the GRN which describes the regulation of gene expression in purple sea urchin stem cells. It consists of a constrained abstract Boolean network - a collection of Boolean networks, each corresponding to a possible structure and logic of the GRN consistent with experiments. We examined the model's compatibility with observed behavior and explored its robustness. To this end, we developed several new methods for modeling GRNs in RE:IN. These include methods for handling cases where models don't behave in accordance with observed behavior, tools for fast and efficient RE:IN modeling, and tools for synthesizing complex conditions in which models can be tested. Our results show that the current model cannot behave according to the entirety of the expected behavior. Moreover, we show that the model is robust to perturbations in a subset of key genes in the network. These results suggest that there is still work to be done to better capture the intricacies of this GRN in RE:IN. Furthermore, the tools we developed proved to be useful and may serve future research on GRNs within the RE:IN framework.

Keywords: Gene regulatory networks · Formal verification · Computational modeling

1 Introduction

Stem cells are cells with the ability to proliferate indefinitely and differentiate into cells with more specific functions. These characteristics give them an important role in the homeostasis and proper development of organisms. Therefore,

---

D. Grimland and E. Tannenbaum—These authors contributed equally to this work.

L. Cerulo et al. (Eds.): CIBB 2024, LNBI 15276, pp. 43–57, 2025.
https://doi.org/10.1007/978-3-031-89704-7_4

understanding the processes that control their dynamics and actions is highly important [9,12]. Gene Regulatory Networks (GRNs) control the dynamics of gene expression and determine cellular decision-making. These networks consist of components in the cell responsible for gene expression regulation and describe the interactions of activation and repression between them [13].

One method of studying GRNs is computational Boolean modeling. In this approach, we describe the state of a GRN in discrete time intervals by defining a Boolean variable for each network component and specifying update functions for each component according to the interactions and cis-regulatory logic of the GRN [7,8,10].

However, traditional methods ignore the gap of knowledge on the exact structure and cis-regulatory logic of networks and therefore assume some specific structure and cis-regulatory logic of the network, which is consistent with experiments, to be correct, thus only examining a specific network. Several approaches and tools [1,2,14,16] including the Reasoning Engine for Interaction Networks (RE:IN) [17,18] aim to mitigate this issue.

RE:IN modeling has potential to generate new biological knowledge, support biological research on a myriad of systems, and particularly in GRNs of developmental and stem cell systems. Hence, this study aimed to create tools and methods to facilitate the process of modeling in RE:IN, namely, the ABN Generator and Experiment Generator. Here we illuminate these methods, and illustrate the potential and possible use of these tools through the lens of a particular example, that of the endomesoderm GRN of the embryo of purple sea urchins. We developed a model of this GRN, and examined its accuracy, that is how well it captures the behaviors of the GRN (by the cardinality of the maximal satisfiable combinations of the observations of hours during development), and how robust it is to genetic perturbations of knockouts and forced expressions (by perturbing, in groups, key GRN components).

The purple sea urchin endomesoderm GRN describes the gene expression of stem cell domains in the endomesoderm and has extensive prior experimental knowledge on it [4,6]. The endomesoderm is the combination of two germ layers of the embryo: the endoderm and the mesoderm. These layers appear during the gastrulation phase of the embryo's development. During development, embryonic stem cells go from pluripotency to multipotency by differentiating into cells of one of the three germ layers and specific domains inside them. Hence, the endomesoderm GRN system describes the interactions of gene expression regulators with each other and how these regulators influence the differentiation of embryonic stem cells to cells of more specific domains, such as vegetal 1 endoderm cells, oral NSM cells, PMC cells, etc., via the alteration of gene expression patterns [6].

Owing to the research done on the endomesoderm GRN system in the purple sea urchin, its topology (i.e., the components and interactions of the system) and behavior (that is, the states of the GRN over different hours in development) are relatively well characterized. However, the exact mechanisms and interactions that correspond to specific developmental outputs are not entirely known [15].

2   Background

A Boolean variable is a variable σ ranging over the set of states $\mathbb{B}$, where the state 0 corresponds to the truth-value of FALSE, and the state 1 to the truth-value of TRUE. A Boolean function B of k variables is a function $B : \mathbb{B}^k \to \mathbb{B}$.

A Boolean network of a biological system effectively captures the components of the system, and the way components change value. Formally, A Boolean network $\mathcal{N}$ is a 2-tuple (V, B) where

$$V = \{\sigma_1, \ldots, \sigma_n\}$$

is the set of Boolean variables of the network, and

$$B = \{B_1, \ldots, B_n\}$$

is a set of Boolean update functions, where $\forall i,\ 1 \leq i \leq n : B_i : \mathbb{B}^n \to \mathbb{B}$, each corresponding to the variable with the same index.

For Boolean networks to model biological systems dynamically, they need to have a specified way one state leads to the next. This formalism is given by the synchronous and asynchronous Boolean models corresponding to a Boolean network. Given a Boolean network $\mathcal{N}$ of size n:

1. The synchronous Boolean model, $\mathcal{S}_{\mathcal{N}}$ is a 2-tuple $(Q, \to)$ where $Q := \mathbb{B}^n$ is the set of states of the model, and $\to$ is a relation over Q describing which states lead to each other, defined by

$$q_1 \to q_2 \iff \forall i, 1 \leq i \leq n : B_i(q_1) = q_2(\sigma_i)$$

2. The asynchronous Boolean model, $\mathcal{AS}_N$ is a 2-tuple $(Q, \to)$ where $Q := \mathbb{B}^n$ is the set of states of the model, and $\to$ is a relation over Q describing which states lead to each other, defined by

$$q_1 \to q_2 \iff \exists i, 1 \leq i \leq n \forall j, 1 \leq j \neq i \leq n :$$
$$(B_i(q_1) = q_2(\sigma_i)) \wedge (q_1(\sigma_j) = q_2(\sigma_j))$$

where for each variable σ of $\mathcal{N}$, and state $q \in Q$, $q(\sigma)$ is the state of σ in the state q. The term Boolean model will refer to a model of a Boolean network that is either synchronous or asynchronous. We will use $\mathcal{M}_{\mathcal{N}}$ to denote a Boolean model of a Boolean network $\mathcal{N}$. We construct Boolean models to uncover their behavior, that is their progression through states over their executions. Formally, given a Boolean model $\mathcal{M}_{\mathcal{N}}$, a trajectory $\mathcal{T}$ of length k is a tuple of states $(q_1, \ldots, q_k) \in Q^k$ satisfying

$$\forall i, 1 \leq i \leq k - 1 : q_i \to q_{i+1}$$

As previously stated, in Boolean modeling it is common to assume some specific network structure and logic to be correct, and hence only examine one specific possible network. RE:IN is a method that can mitigate this issue, and that allows us to make predictions and analyze the behavior and structure of biological systems, which are partially known, from a perspective that would have been hard to achieve otherwise. This is achieved by describing systems as Abstract Boolean Networks (ABNs), which are essentially collections of Boolean networks of a GRN that together describe all possible structures and update logic. The Boolean networks comprising the RE:IN model of a system differ from one another by:

1. The interactions present within them: The networks of an ABN are defined in part by a set of definite interactions, which are present in all networks, and a set of optional interactions, such that each network in the ABN has a different combination of interactions [18].
2. Their update logic: The ABN defines for each component a set of prebuilt general update functions called Regulation Conditions (RCs), from a group of 20 functions [18]. The reasoning behind allowing these 20 specific functions is explained in Sect. 6, alongside formal definitions.

The networks composing an ABN are all combinations of optional interactions, together with all combinations of RCs for each component, from their respective groups. Formally, an ABN $\mathcal{A}$ is a 5-tuple $(C, I, I^?, R, s)$ where:

1. C is the set of components of the ABN.
2. $I \subseteq C^2 \times \mathbb{B}$ is the set of definite interactions of the ABN.
3. $I^? \subseteq C^2 \times \mathbb{B}$ is the set of optional interactions of the ABN, and satisfies $I \cap I^? = \emptyset$.
4. $R : C \to \mathcal{P}(\{0, \ldots, 19\})$ is the function that assigns RC numbers for each component of the ABN.
5. $s \in \mathbb{B}$ and is 1 if the ABN updates synchronously, or 0 if it updates asynchronously.

ABNs may be analyzed for the satisfiability of certain behaviors or properties, specified by researchers, by performing a process similar to bounded model checking and utilizing an SMT solver [11] to check which models of the ABN satisfy the required behaviors. To be more exact, RE:IN allows to define behaviors checked for satisfiability as constructs termed constraints. These in turn are built of two parts:

1. A Boolean formula called a constraint expression, built of atoms called observations, which specify restrictions on the states of components over time points during multiple executions of our models, corresponding to experiment labels specified in the observations.
2. Perturbations over model components specifying that certain components are inactive or active (knocked out (KO) or forcefully expressed (FE), respectively) indefinitely for certain executions [18].

With that, an ABN, or RE:IN model satisfies certain constraints if and only if there exists a model and executions corresponding to experiments that make the constraint expression true [18]. The number of models of the ABN that satisfy the constraints is called the number of solutions of the ABN. Other formal definitions of terms and concepts in RE:IN modeling can be found in Sect. 6.

3 Data and Methods

Data and Models. Through the course of this work, we designed two ABNs to computationally model the purple sea urchin embryo endomesoderm, using data discovered and integrated in prior research [3]. Of relevance to the tools made in this work and the experiments discussed here is the domain-divided model (DDM). The DDM is composed of Boolean networks, each containing multiple copies of GRN components, one for each cell domain the component is present in. The interactions between components of the DDM were all set to be optional, and the regulation conditions for each component were 1 to 17. A graphical representation of the DDM is shown in Fig. 1.

To measure the accuracy and robustness of the DDM, we created a collection of 10 sets of observations, where every set corresponds to one hour of development and hence contains the observations regarding the states of components in that hour, and all observations belong to the same experiment. The collection of observations was created using data from prior studies [3].

Software and Modeling Methods Developed Through the Course of the Work. To enable the synthesis and preprocessing of biological data into RE:IN models, we developed the ABN generator, which is a Python script external to RE:IN that takes a text file as input that includes in it formatted strings corresponding to interactions, components, and possible update logic of a system and outputs a corresponding RE:IN file defining a RE:IN model. It also grants us a framework to alter the result by programming the required changes.

Constructing a RE:IN file with many complex gene interactions and experimental constraints can be tedious and prone to mistakes if done manually. To enable the synthesis of complex and diverse constraints in a standard format with minimal effort we developed the Experiment Generator. The generator is another Python script external to RE:IN that takes as input a RE:IN file defining an ABN and a text file describing a collection of sets of observations. We then need to define for each set of observations a set of timestamps at which it can be satisfied. Secondly, we define for the sets of every experiment the order in which the constraints must be satisfied. With the aforementioned input the generator creates a new RE:IN file by concatenating the ABN file with generated RE:IN constraints that are satisfiable if and only if each set is satisfied by a timestamp in its set and if the sets are satisfied in the correct order.

One common way we used the Experiment Generator was to allow us to, in a simple and generally applicable manner, make collections of sets of observations more flexible temporally by providing time intervals rather than time points

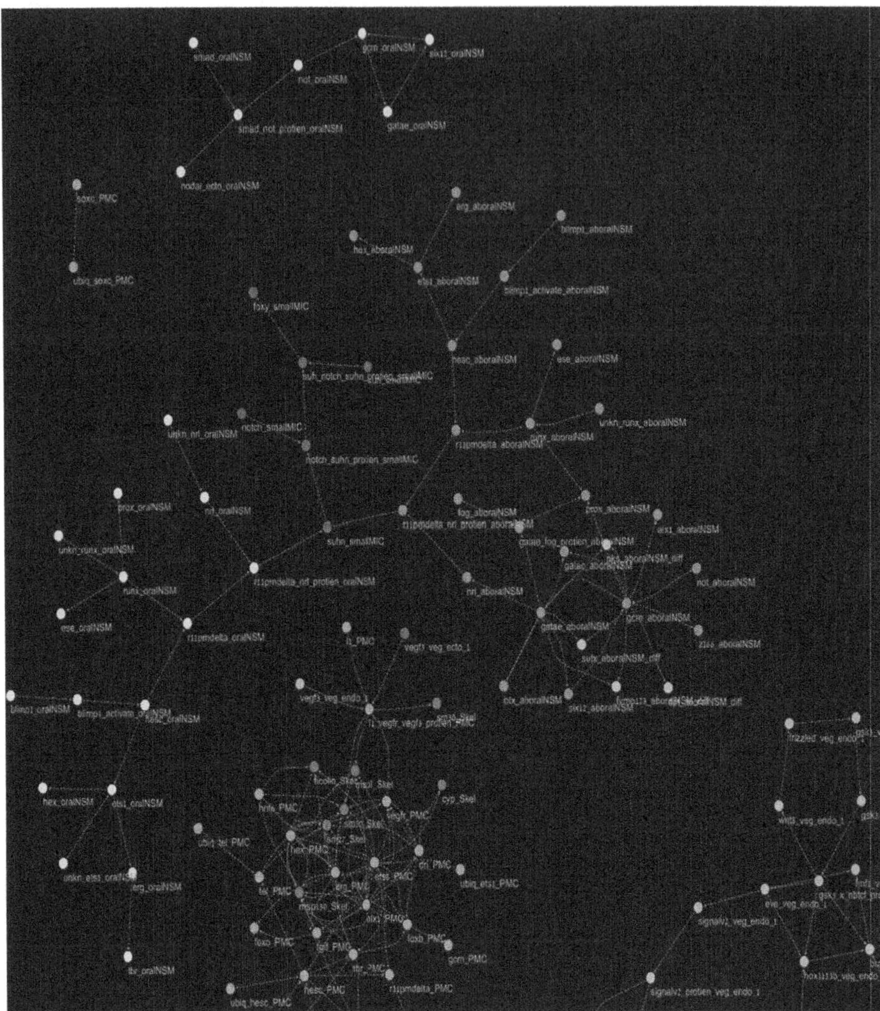

Fig. 1. In the diagram above, a representation of part of the structure of the DDM is shown. The components and interactions of a particular cellular domain are marked with specific colors. Dark-blue components represent small micromere cells, blue components represent primary mesenchyme cells, light-blue and orange components represent aboral non-skeletogenic mesoderm cells, yellow components represent oral non-skeletogenic mesoderm cells, pink components are of vegetal 1 endoderm cells, purple components are of skeletogenic cells, and red cells are of vegetal 1 ectoderm cells. The interactions between components are colored green for activation and red for inhibition. Finally, all interactions are drawn as dashed lines, as they are all optional in the model. (Color figure online)

for satisfying sets of observations (that is allowing observations to be satisfied in one time point out of a range of points) and having temporal distances between intervals. One way to implement this using the generator was using a method we termed FZST, which takes collections of sets of observations, where each set is of one experiment and the sets have initial time points, and defines the timestamps for each set according to two numbers: (1) The number of time stamps around each set may be satisfied (e.g., a number of 0 would mean the set may only be satisfied at its original time point. A number of 1 would mean it may be satisfied in its original time point, or a time point 1 less than or more than it). This number is termed the "fuzzy number" (FZ) setting of the generated constraints. (2) By how many time points to amplify the number of time points between consecutive sets of observations (e.g., a number of 2 would mean that between two sets of observations with an initial distance of 3, the distance would now be 6). This number is called the "stranding number" (ST) setting of the generated constraints.

In this work we are going to consider cases of FZST settings where if the fuzzy and stranding numbers of one model are less than or equal to the other this implies that the first model will be more restrictive than the latter. Furthermore, we are not going to consider FZST settings where there are intersections between the neighborhoods around which sets may be satisfied.

4   Results

Testing the Constraints of Hours 21-30 With and Without Temporal Flexibility. To test accuracy, we tested whether the entire collection of sets of observations we developed could be satisfied by the DDM. To accomplish this, we created RE:IN constraints using the FZST method with settings of FZ = 0, ST = 0, and FZ = 5, ST = 10, and tested them for satisfaction by the DDM. The first settings were chosen to see if no flexibility would suffice, and the second were chosen to check whether large FZST settings would. The exact reasoning behind choosing the second pair of numbers will be explained later. We found the constraints for both settings to be unsatisfiable using RE:IN under the DDM model. Hence, there are scenarios too restrictive for the FZST method to suffice.

Finding the Cardinality of the Largest Combinations of Sets of Observations of Hours During Development that are Satisfiable. The prior result suggested that the DDM is not consistent with large subsets of the collection of sets of observations. However, further analysis is needed to determine the maximal subsets of sets of observations that are satisfiable. As such, we choose subsets of the entire collection formed by choices of three sets of observations from it. We generated constraints for each such tuple with the settings FZ = 5 and ST = 10. Analyzing all combinations for solutions yielded no solutions for almost all combinations of sets, besides those of the hours 21-23, 21, 25, 26, and 21, 28, 29. Using these results, we can prove that the cardinality is 3. We decided to examine how taking away temporal flexibility would affect the satisfiability of the maximal combinations found. Hence, we tested the satisfiability of the constraints of hours 21-23, 21, 25, and 26, and 21, 28, and 29, under no temporal flexibility, that is, with a fuzzy number of 0 and a stranding number of 0. The constraints of hours 21, 25, and 26, and 21, 28, and 29 were still satisfiable after the change. However, hours 21-23 were found to be unsatisfiable without temporal flexibility, suggesting the method can indeed help specify constraints in a flexible and friendlier language.

Analyzing the Constraints of Hours 21-23 Under Different Flexibility Settings. Since we couldn't use the entirety of constraints to check robustness, we looked at the next best constraints in terms of challenge and similarity of constraints: one of the maximal satisfiable combinations of hours with minimal FZST settings. We chose the constraints of hours 21-23 for this, being the only ones not satisfiable by $FZ = ST = 0$. We examined all combinations of fuzzy and stranding numbers lower than 5 and 10, respectively, on these constraints where there is no overlap between environments. The set of possible stranding numbers without overlap given a certain fuzzy number is $\{k \in \mathbb{N}_0 \mid 2n \leq k\}$, hence we examined every possible fuzzy number from 0 to 5, with every possible stranding number in the shown bounds and below 10. The satisfiability results are presented in Table 1.

Table 1. Every cell of the table denotes constraints derived from the sets of observations of hours 21-23 when the fuzzy and stranding number settings are set to the respective values as shown in the column and row of numbers. White cells denote FZST settings not considered, red cells denote considered settings that turned out unsatisfiable, and blue cells denote considered settings that turned out satisfiable. These results show why earlier we considered the FZST numbers $FZ = 5$, $ST = 10$, since here it was suggested that for $FZ = 0$ or 1 the ST chosen could affect satisfiability, so to ensure proper temporal flexibility we had to choose a higher FZ number. To be safe, we chose $FZ = 5$ and chose the ST number corresponding to it for which the neighborhoods are adjacent.

FZ \ ST	0	1	2	3	4	5	6	7	8	9	10
0											
1											
2											
3											
4											
5											

These results allowed us to test robustness in a meaningful manner by employing perturbations on a subset of genes in the most restrictive and satisfying subset of the constraints. Hence, we can use the FZST method effectively when studying robustness.

5 Conclusions

From our results on the satisfiability of the full constraints on the endomesoderm GRN with and without temporal flexibility, it can be inferred that while our methods are limited in their ability to lessen the restrictiveness of some constraints, namely constraints that are already very restrictive and unsatisfiable, they do display some promise in this regard. This can be inferred from the constraints of hours 21-23 being unsatisfiable without the usage of our tools and being satisfiable when incorporating our methods for temporal flexibility in models. Moreover, we can conclude from the entirety of our research on the endomesoderm GRN in RE:IN using the tools detailed here that those are adequate for use in large models, like the one studied here.

The new tools detailed here have been a major help in conducting our research on the GRN by speeding up modeling, allowing us to generate complex experiments automatically, and allowing us to perform robustness research with greater ease. However, it is important to stress that the conclusions of this work require further research on a wider array of biological networks to establish the methodology detailed here.

In this study, we have developed important tools and methods for future research into the endomesoderm GRN using the RE:IN tool and RE:IN modeling in general. We have constructed useful tools for the synthesis of constraints over RE:IN programs and the synthesis of RE:IN ABNs, allowing for the automation of a large part of the modeling process and for experimentation on models to be done with ease. Moreover, through our attempt to cope with the difficulty of RE:IN models satisfying constraints, we developed useful strategies for RE:IN modeling for dealing with situations where satisfactory models for tested constraints do not exists, namely the idea of FZST.

6 Formal Definitions

In this section we present formal definitions for the main concepts related to abstract Boolean networks and the Reasoning Engine.

Definition 1. *A Boolean variable is a variable σ ranging over the set of states $\mathbb{B}$. The state 0 corresponds to the truth-value of FALSE, and the state 1 to the truth-value of TRUE. A Boolean function B of k variables is a function $B : \mathbb{B}^k \to \mathbb{B}$.*

A Boolean network of a biological system effectively captures the components of the system, and the way components change value as expected from Boolean models. Formally, we have the following definition.

Definition 2. *A Boolean network $\mathcal{N}$ is a 2-tuple (V, B) where*

$$V = \{\sigma_1, \ldots, \sigma_n\}$$

is the set of Boolean variables of the network, and

$$B = \{B_1, \ldots, B_n\}$$

is the set of Boolean update functions of the network, each corresponding to the variable with the same index.

For Boolean networks to model biological systems dynamically, they need to have a specified way one state leads to the next. This formalism is given by the synchronous and asynchronous Boolean models corresponding to a Boolean network.

Definition 3. *Given a Boolean network $\mathcal{N}$ of size n:*

1. *The synchronous Boolean model, $\mathcal{S}_{\mathcal{N}}$ is a 2-tuple $(Q, \rightarrow)$ where $Q := \mathbb{B}^n$ is the set of states of the model, and $\rightarrow$ is a relation over Q describing which states lead to each other, defined by*

$$q_1 \rightarrow q_2 \iff \forall i, 1 \leq i \leq n : B_i(q_1) = q_2(\sigma_i)$$

2. *The asynchronous Boolean model, $\mathcal{AS}_N$ is a 2-tuple $(Q, \rightarrow)$ where $Q := \mathbb{B}^n$ is the set of states of the model, and $\rightarrow$ is a relation over Q describing which states lead to each other, defined by*

$$q_1 \rightarrow q_2 \iff \exists i, 1 \leq i \leq n \, \forall j, 1 \leq j \neq i \leq n :$$
$$(B_i(q_1) = q_2(\sigma_i)) \wedge (q_1(\sigma_j) = q_2(\sigma_j))$$

where for each variable σ of $\mathcal{N}$, and state $q \in Q$, $q(\sigma)$ is the state of σ in the state q.

Remark 1. The term Boolean model will refer to a model of a Boolean network that is either synchronous or asynchronous. We will use $\mathcal{M}_{\mathcal{N}}$ to denote a Boolean model of a Boolean network $\mathcal{N}$.

Definition 4. *Given a Boolean Model $\mathcal{M}_{\mathcal{N}}$, a trajectory $\mathcal{T}$ of length k is a tuple of states $(q_1, \ldots, q_k) \in Q^k$ satisfying*

$$\forall i, 1 \leq i \leq k - 1 : q_i \rightarrow q_{i+1}$$

RE:IN allows for the specification of GRNs as abstract Boolean networks (ABNs), which are essentially sets of Boolean networks of a GRN that together describe all possible structures and update logic, as found in experiments. The Boolean networks comprising the RE:IN model of a system differ from one another by:

1. The interactions present within them: The networks of an ABN are defined in part by a set of definite interactions, which are present in all networks, and a set of optional interactions, such that each network in the ABN has a different combination of interactions.
2. Their update logic: The ABN defines for each component a set of prebuilt general update functions called Regulation Conditions (RCs), from a group of 20 functions. The reasoning behind allowing these 20 specific functions is explained later.

The networks composing an ABN are all combinations of optional interactions, together with all combinations of RCs for each component, from their respective groups.

Definition 5. *An ABN $\mathcal{A}$ is a 5-tuple $(C, I, I^?, R, s)$ where:*

1. *C is the set of components of the ABN.*
2. *$I \subseteq C^2 \times \mathbb{B}$ is the set of definite interactions of the ABN.*
3. *$I^? \subseteq C^2 \times \mathbb{B}$ is the set of optional interactions of the ABN, and satisfies $I \cap I^? = \emptyset$.*
4. *$R : C \to \mathcal{P}(\{0, \ldots, 19\})$ is the function that assigns RC numbers for each component of the ABN.*
5. *$s \in \mathbb{B}$ and is 1 if the ABN updates synchronously, or 0 if it updates asynchronously.*

Remark 2. An interaction (c, c', s) specifies an interaction from c to c' which is activatory if $s = 1$ and inhibitory if $s = 0$.

We would like to talk about ABNs as sets of Boolean networks, and to do so we need to have formal definitions of RCs. To this end, we have the notion of a network topology, which describes a possible structure of interactions in a network arising from an ABN.

Definition 6. *Given an ABN $\mathcal{A}$, a topology is a set $I \subseteq \mathcal{S} \subseteq I \cup I^?$. The set of states of the ABN in a given topology is defined as $\mathbb{B}^n$*

As stated, RCs are general update functions, and to that end, instead of accepting as input the detailed states of the regulators of a component, they input a general state of the activators and repressors of the component.

Definition 7. *Given a topology $\mathcal{S}$ of $\mathcal{A}$ of size n, and a component $c \in C$:*

1. *Its set of activators is defined as $A_c := \{c' \in C \mid (c', c, 1) \in \mathcal{S}\}$. The state of the activators in a given state q is defined as*

$$a(c, q) := \begin{cases} 1, & \forall c' \in A_c : q(c') = 1 \\ 0, & \exists c_1', c_2' \in A_c : (q(c_1') = 1) \wedge (q(c_2') = 0) \\ -1, & \forall c' \in A_c : q(c') = 0 \\ -2, & A_c = \emptyset \end{cases}$$

2. *Its set of inhibitors is defined as $R_c := \{c' \in C \mid (c', c, 0) \in \mathcal{S}\}$. The state of the inhibitors in a given state q is defined as*

$$i(c, q) := \begin{cases} 1, & \forall c' \in R_c : q(c') = 1 \\ 0, & \exists c'_1, c'_2 \in R_c : (q(c'_1) = 1) \wedge (q(c'_2) = 0) \\ -1, & \forall c' \in R_c : q(c') = 0 \\ -2, & R_c = \emptyset \end{cases}$$

In the prior definition, a state of 1 of the set of activators (or repressors) implies all components of the activators are active. A state of 0 implies some activators (repressors) are active, while others are inactive, a state of -1 implies all activators (repressors) are inactive, and a state of -2 implies there are no activators (repressors).

The RCs of RE:IN are functions from components and states to the possible states of the updated components. The set of RCs of a component in an ABN comprises its set of possible update functions. RCs are divided into two groups. The first group is defined in terms of regulation functions. These are functions from the states of the activators and repressors of components which satisfy two main conditions: initialization, which means that if all activators of a component are active, and no repressors are active, the regulation condition needs to activate it, and monotonicity. The latter means, in rough terms, that in biological situations which entail "greater activation" than others (i.e. all activators are active and no repressors, compared to all activators being active and some repressors being active), regulations conditions cannot return a "lesser output".

Definition 8. *A regulation function is a function $r : \{-1, 0, 1\}^2 \to \mathbb{B}$ which satisfies:*

1. *Initialization:*
$$r(1, -1) = 1, r(-1, 1) = 0$$

2. *Monotonicity:*

$$\forall a, \forall b, \forall c \in \{1, 0, -1\}, b < c : (r(a, b) \geq r(a, c)) \wedge (r(b, a) \leq r(c, a))$$

Remark 3. In the previous definition we did not consider cases where no inhibitors or activators are present, although these are possible cases [18]. We ignore these edge cases in the current work.

Remark 4. In the previous definition the input to the regulation functions is a tuple of the form $(a(c, q), i(c, q))$, meaning their first input corresponds to the state of the activators of the updated component and the second input to the state of inhibitors.

The second group of RCs is comprised of two RCs termed the "threshold RCs". These update the state of a component depending on its balance of activators and regulators, where more activators lead to activation, and more repressors to repression.

Definition 9. *Given a topology $\mathcal{S}$ of an ABN $\mathcal{A}$ of size n, the two threshold regulation conditions on it are functions $r_t^{\mathcal{S}}, r_{td}^{\mathcal{S}} : C \times Q \to \mathbb{B}$ defined by*

$$r_t^{\mathcal{S}}(c, q) := \begin{cases} 1, & |\{c' \in A_c \mid q(c') = 1\}| \geq |\{c' \in R_c \mid q(c') = 1\}| \\ 0, & Otherwise \end{cases}$$

$$r_{td}^{\mathcal{S}}(c, q) := \begin{cases} 1, & |\{c' \in A_c \mid q(c') = 1\}| > |\{c' \in R_c \mid q(c') = 1\}| \\ 0, & Otherwise \end{cases}$$

With definitions 9 and 8 in mind, we are finally ready to define the general concept of an RC.

Definition 10. *Given a topology $\mathcal{S}$ of an ABN $\mathcal{A}$, a regulation condition over it is a function $r^{\mathcal{S}} : C \times Q \to \mathbb{B}$ that is either one of the threshold regulation conditions over it, or is defined by $r(c, q) = r'(a(c, q), i(c, q))$ where r' is a regulation function.*

Lemma 1. *There are exactly 18 unique regulation functions.*

Proof. Since regulation functions are defined on a very small set of states, we can use software to enumerate and check for all possible functions $\{-1, 0, 1\}^2 \to \mathbb{B}$ which are regulation functions and which are not. We have made a script to do this, presented at [5]. Using this program we can prove there are exactly 18 regulation functions, and find how they are defined. $\qquad\square$

Corollary 1. *Using lemma 1 we get that the set of possible RCs of each component in an ABN has 20 functions, 2 threshold RCs, and 18 regular RCs. We will denote them for a given topology $\mathcal{S}$ as $r_0^{\mathcal{S}}, \ldots, r_{19}^{\mathcal{S}}$, where the first 18 functions are the regular RCs, $r_{18}^{\mathcal{S}}$ is $r_t^{\mathcal{S}}$ and $r_{19}^{\mathcal{S}}$ is $r_{td}^{\mathcal{S}}$. This also explain why we defined the function R assigning RCs to components as we did in definition 6.*

Definition 11. *Given an ABN $\mathcal{A}$ of size n:*

1. *The topology set is defined as $\mathcal{S}(\mathcal{A}) := \{I \cup I' \mid I' \subseteq I^?\}$.*
2. *The RC set given a certain topology $\mathcal{S}$ is defined as*

$$\mathcal{RC}(\mathcal{S}) := R(c_1) \times \cdots \times R(c_n)$$

Recall that an ABN is a collection of Boolean networks defined by its optional interactions and the RCs of each component. This notion is captured in the network set of an ABN.

Definition 12. *Given an ABN $\mathcal{A}$ of size n:*

1. The network set of the ABN is defined as

$$\mathcal{N}(\mathcal{A}) := \{(\{\sigma_1,\ldots,\sigma_n\},\{f_1,\ldots,f_n\}) \mid \exists \mathcal{S} \in \mathcal{S}(\mathcal{A})$$
$$\exists (a_1,\ldots,a_n) \in \mathcal{RC}(\mathcal{A}) \forall 1 \le i \le n \in \mathbb{N} : f_i = r_{a_i}^{\mathcal{S}} \}$$

2. The model set of the ABN is defined as

$$\mathcal{M}(\mathcal{A}) := \begin{cases} \{\mathcal{S}_{\mathcal{N}} \mid \mathcal{N} \in \mathcal{N}(\mathcal{A})\}, & s = 1 \\ \{\mathcal{AS}_{\mathcal{N}} \mid \mathcal{N} \in \mathcal{N}(\mathcal{A})\}, & s = 0 \end{cases}$$

Acknowledgments. The research was supported by the ISRAEL SCIENCE FOUNDATION (grant No.190/19), and by Bar-Ilan University's DATA SCIENCE INSTITUTE.

Disclosure of Interests. The authors declare no conflicts of interests.

Availability of Data and Software Code. Our models, data used to generate them and the experiments, programs, and experiments, are available at the following URL: https://github.com/danielgrimland/Sea_Urchin_Endomesoderm_GRN.

References

1. Brim, L., Pastva, S., Šafránek, D., Šmijáková, E.: Temporary and permanent control of partially specified boolean networks. Biosystems **223**, 104795 (2023)
2. Chevalier, S., Noël, V., Calzone, L., Zinovyev, A., Paulevé, L.: Synthesis and simulation of ensembles of boolean networks for cell fate decision. In: Abate, A., Petrov, T., Wolf, V. (eds.) CMSB 2020. LNCS, vol. 12314, pp. 193–209. Springer, Cham (2020). https://doi.org/10.1007/978-3-030-60327-4_11
3. Davidson, E.: Endomesoderm GRN model. https://grns.systemsbiology.net/SpEndomes/
4. Davidson, E.H., Hough-Evans, B.R., Britten, R.J.: Molecular biology of the sea urchin embryo. Science **217**(4554), 17–26 (1982)
5. Grimland, D., Tannenbaum, E., Kugler, H.: Regulation conditions calculator (2024). https://github.com/danielgrimland/Sea_Urchin_Endomesoderm_GRN/tree/main/Regulation%20Conditions%20Calculator
6. Hinman, V.F., Jarvela, A.: Developmental gene regulatory network evolution: insights from comparative studies in echinoderms. Genesis **52**, 193–207 (2014)
7. Ji, Z., Yan, K., Li, W., Hu, H., Zhu, X.: Mathematical and computational modeling in complex biological systems. BioMed Res. Int. **2017** (2017)
8. Kauffman, S.: Metabolic stability and epigenesis in randomly constructed genetic nets. J. Theoret. Biol. **22**(3), 437–467 (1969)
9. Kwon, S.G., Kwon, Y.W., Lee, T.W., Park, G.T., Kim, J.H.: Recent advances in stem cell therapeutics and tissue engineering strategies. Biomater. Res. **22** (12 2018)

10. Morris, M.K., Saez-Rodriguez, J., Sorger, P.K., Lauffenburger, D.A.: Logic-based models for the analysis of cell signaling networks. Biochemistry **49**, 3216–3224 (2010)
11. de Moura, L., Bjørner, N.: Z3: an efficient SMT solver. In: Proceedings of the 14th International Conference on Tools and Algorithms for the Construction and Analysis of Systems (TACAS'08), pp. 337–340 (2008)
12. Nichols, J., Smith, A.: Pluripotency in the embryo and in culture. Cold Spring Harb. Perspect. Biol. **4**(8), a008128 (2012)
13. Okawa, S., Nicklas, S., Zickenrott, S., Schwamborn, J.C., del Sol, A.: A generalized gene-regulatory network model of stem cell differentiation for predicting lineage specifiers. Stem Cell Rep. **7**, 307–315 (2016)
14. Paoletti, N., Yordanov, B., Hamadi, Y., Wintersteiger, C.M., Kugler, H.: Analyzing and synthesizing genomic logic functions. In: Proceedings of the 26th International Conference on Computer Aided Verification (CAV'14), pp. 343–357 (2014)
15. Peter, I.S., Davidson, E.H.: Assessing regulatory information in developmental gene regulatory networks. Proc. Nat. Acad. Sci. USA **114**, 5862–5869 (2017)
16. Videla, S., et al.: Revisiting the training of logic models of protein signaling networks with ASP. In: International Conference on Computational Methods in Systems Biology, pp. 342–361. Springer (2012)
17. Yordanov, B., Dunn, S.J., Gravill, C., Arora, H., Kugler, H., Wintersteiger, C.M.: The reasoning engine: a satisfiability modulo theories-based framework for reasoning about discrete biological models. J. Comput. Biol. **30**(9), 1046–1058 (2023)
18. Yordanov, B., Dunn, S.J., Kugler, H., Smith, A., Martello, G., Emmott, S.: A method to identify and analyze biological programs through automated reasoning. npj Syst. Biol. Appl. **2** (2016)

Gene Set-Focused Analysis of RNA-Seq Data with MIEP (Make-It-Easy-Pipeline)

Alberto Corradin[1]([✉]) [iD], Francesco Ciccarese[1] [iD], Vittoria Raimondi[1] [iD], Loredana Urso[1] [iD], Micol Silic-Benussi[1] [iD], and Vincenzo Ciminale[1,2]([✉]) [iD]

[1] Veneto Institute of Oncology IOV – IRCCS, Padova, Italy
alberto.corradin@iov.veneto.it, v.ciminale@unipd.it
[2] Department of Surgery, Oncology and Gastroenterology, University of Padova, Padova, Italy

Abstract. Target identification and selection is a critical rate-limiting step in drug discovery, especially the context of cancer research, following the identification of many driver mutations and tumor-associated transcriptional signatures. To this end, the development of refined bioinformatic tools is essential to gain insights from data deriving from high-throughput technologies. We have developed the MIEP (make-it-easy-pipeline), a user-friendly R package implementing an integrated pipeline for RNA-seq data analysis.

MIEP represents an easy-to-use tool that performs statistical testing, annotates sequences, reduces biases, and summarizes results in HTML tables, volcano plots, and heat maps. MIEP performs dimensionality reduction of data, tests the enrichment of Gene Ontology terms and includes machine learning algorithms for sample classification. In our hands, the use of MIEP in the cancer research field facilitated the identification of phenotype-linked signal transduction pathways whose biological relevance was experimentally verified.

Calling up a single function, a series of integrated analyses is launched in sequence. Settings can be changed while running the analysis thanks to user-friendly, interactive apps. Automatic procedures optimize the selection of the hyperparameters affecting the different algorithms. Together with careful handling of exceptions, these characteristics make MIEP a handy tool for researchers with only basic knowledge of programming.

MIEP also allows editing new gene sets based on results of multiple analyses and provides gene set-focused data representations. This can accelerate the understanding of pathogenic mechanisms as well as the development of new drugs because the gene set-focused approach greatly facilitates the identification of functional hubs and new potential pharmacological targets.

Keywords: RNA-seq data · pipeline · gene set · graphical supports

Supplementary Information The online version contains supplementary material available at https://doi.org/10.1007/978-3-031-89704-7_5.

1   Introduction

Simplicity of the message and clear supporting images are key elements of communication. This principle is obvious but is frequently undervalued in research laboratories where the complexity and importance of the object of study may overshadow other aspects. Paradoxically, when information is overwhelming, simplicity and immediacy are most needed. The increasing application of Next Generation Sequencing (NGS) to identify alterations in biological pathways has necessitated the development of bioinformatic pipelines of increasing complexity to deal with large datasets. Bioinformatic researchers have added details and potentialities to their creations to address ever-emerging new problems posed by biologists.

This has helped the development of scientific research at the price of an increased complexity in tools [1–6]. The integration of wet and dry labs can be an effective approach to address both biological and computational issues. However, this is not always the case because of limitations in resources or personnel. As an alternative approach, we propose here an intuitive instrument that enables researchers in the biomedical field to perform preliminary analyses of their RNA-seq data without the assistance of specialized bioinformaticians.

2   Methods

Our approach was based on two assumptions: (i) that raw gene counts derived from sequence reads are available to researchers (NGS is often performed by service companies nowadays); (ii) that biologists are searching for gene targets for a subsequent in-depth analysis and wet lab testing. In the particular case of biomedical research, the effort to identify novel drug targets is increasing day by day [7–11].

Deep understanding of pathways and metabolic processes is required to identify relevant targets among a large number of differentially expressed genes and to reveal the causal links between gene products upstream mechanisms and functional pathway dysregulation. Graphical representations able to convey the informative contents of RNA-seq data with clarity and immediacy can greatly facilitate the identification of potential gene targets.

Our MIEP (make-it-easy-pipeline) toolbox generates a wide range of graphical outputs in a user-friendly manner with a high level of automation. This new R package aggregates well-established algorithms in a unique pipeline and enables biologists with basic knowledge of programming to launch their analyses with only one R command. In addition, MIEP develops and edits gene sets based on the results of gene ontology enrichment and machine learning. By focusing on gene sets involved in different pathways, the generation of testable hypotheses can be greatly accelerated. In fact, alterations in key signal transduction pathways play a critical role in pathogenesis of cancer and other diseases.

3   Results

MIEP (make-it-easy-pipeline) aggregates a diversity of tools for the analysis of RNA-seq data resulting in an integrated, interactive, and user-friendly R package.

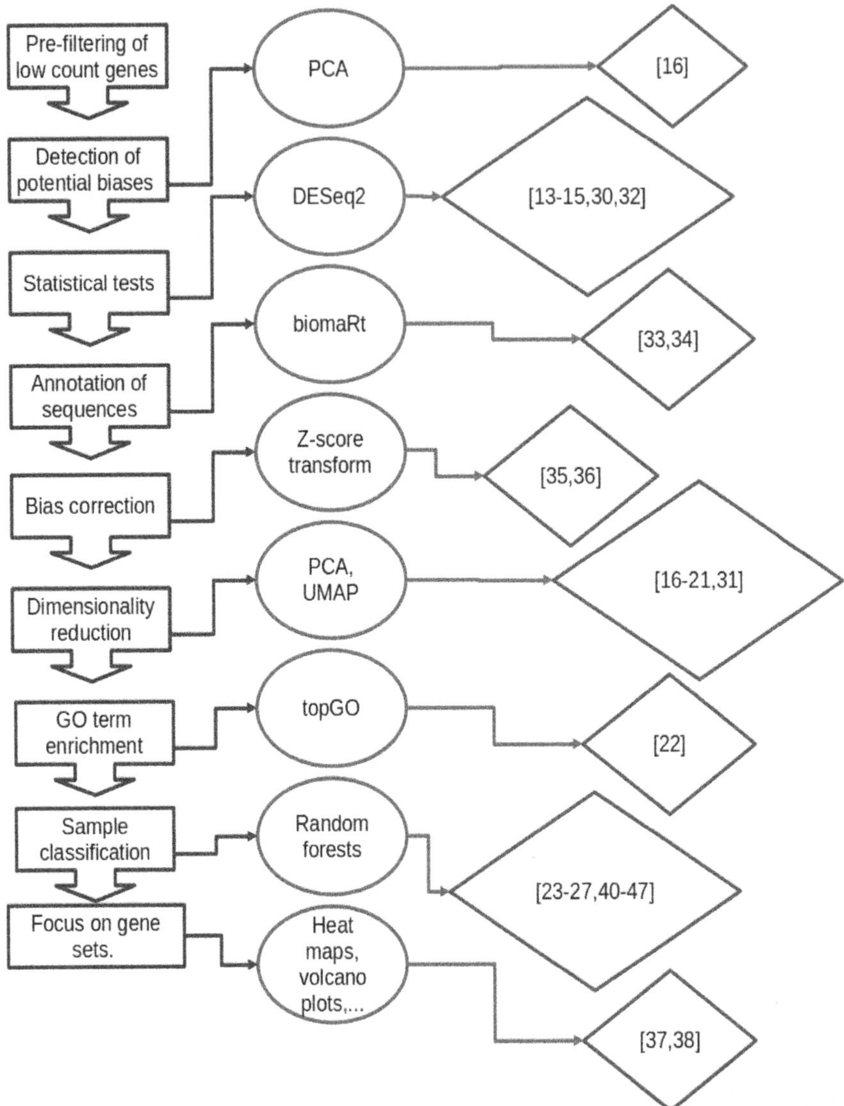

Fig. 1. The flow chart depicts the steps of data elaboration performed by MIEP (left). Main algorithms, or R packages, called to elaborate data are shown in the center. References to exploited algorithms/packages are reported on the right.

When supplied with raw gene counts, MIEP performs statistical testing, annotates sequences with gene ID names based on Hugo Gene nomenclature (https://www.genena mes.org/), reduces potential biases due to sample characteristics, and summarizes the main results by means of tables in HTML format, heat maps, and dendrograms based on hierarchical clustering. Heat maps and other graphical depictions can be customized to better highlight differences by applying specifically designed color palettes.

Threshold selection to identify differentially expressed features and data shrinkage can be defined interactively, thanks to *Shiny* apps [12] which provide a well-known user-friendly interface. Results of *DESeq2* [13] are followed by graphical supports, including mean-average (MA) plots, volcano plots, and visual comparisons of the effects of different shrinkage methods, including *apeglm* [14] and *ashr* [15].

Dimensionality reduction is carried out by means of principal component analysis (PCA) [16] and uniform manifold approximation and projection (UMAP) [17], supplying both 2D and 3D representations. T-distributed stochastic neighbor embedding (t-SNE) [18–20] and support vector machines (SVM) [21] are available to the user for additional investigations of data. They exceed the scope of the basic pipeline, and therefore are not included in the flow chart in Fig. 1, which summarizes the main steps of data elaboration. The enrichment of Gene Ontology (GO) terms can be accomplished on the basis of multiple, different data sets. Differentially expressed features are obvious candidates to constitute the significant subset of genes to compare against the gene universe of all the detected transcripts. Furthermore, alternative subsets of features can be extracted based on the ranking of importance in composing principal components. For example, if PCA separates samples based on the treatment, genes participating with higher weights to construct the first principal component are likely involved in the functional mechanisms of interest. After identifying the most enriched GO terms thanks to the *topGO* R package [22], corresponding sets of differentially expressed genes can be edited.

Conditional random forests [23–25] can classify samples depending on treatments, allowing the subsequent calculation of variables importance [26, 27], i.e. features' usefulness in separating classes. Genes characterized by evident importance are candidates to edit new gene sets deserving further investigation.

Gene sets of interest that merit in-depth investigation may be supplied either by downloading corresponding Gene Matrix Transposed files (*.gmt extension) from the GSEA [28] web site (www.gsea-msigdb.org/gsea/index.jsp), or through their automatic generation, following tests of GO term enrichment or sample classification. In both cases, MIEP will provide heat maps focused on gene sets, volcano plots, and tables reporting results of statistical tests in HTML format. Detailed graphical representations of fold-change values are produced, in order to facilitate the development of testable hypotheses and identification of potential targets for subsequent wet lab testing.

4 Practice Dataset and Analysis

To depict how the MIEP integrated pipeline develops, a practice dataset is reported in the vignette [29] that accompanies the R package. Vignettes can be called from the R shell with the R command:

```
> browseVignettes(package="MIEP")
```

MIEP's vignette details the potentialities of the pipeline and provides a guide to change default settings at run time. For the reported practice analysis, *in silico* data were generated based on real data downloaded from the GEO database (GSE198518). However, raw gene counts were combined to increase the number of samples with simulated replicates, and gaussian noise was added. Therefore no medical implication can be derived from this example because of distortions applied to the original data. The final database is included in the MIEP R package (folder inst/extdata [29]), and aims exclusively to supply an example to users.

4.1 Quick Start

MIEP can be launched with a single command:

```
>      MIEP(filePath1=".../projectName/pipeline",      filePath2=
".../gmtFiles")
```

where parameter "*filePath1*" represents the path to the folder where MIEP will store the results of the analyses for research project "*projectName*". Parameter "*filePath2*" represents the path to the folder containing the gene sets of interest to the user (i.e. downloaded from GSEA web site or customized). Alternatively, the pipeline can be launched by running the user-friendly and intuitive R command:

```
> MIEP()
```

This assumes required paths were previously listed in a text file to be selected interactively (see MIEP's vignette for details).

4.2 Pre-filtering of Low Gene Counts

The *DESeq2* R package is used to normalize raw gene counts [30] and to perform statistical tests that highlight differences in samples. Love's algorithm [13] requires two input files: the raw gene counts and a description of samples to compare (details in the vignette).

Pre-filtering of low-count genes is recommended by *DESeq2*'s authors. MIEP is endowed with two filters that we informally called "*pheno-filter*" and "*batch-filter*", in accord with the intuitive approach characterizing MIEP. The first considers read counts in every combination of conditions (e.g. treatment) and analyzed samples (e.g. cell lines, patient samples). Only the features that show average copy numbers higher than the imputed threshold in every combination will be retained. On the other side, "*batch-filter*" evaluates the total number of reads per feature in the whole batch. This approach gains interest when some transcripts are expected to appear only in a specific subset of samples, and are rare or absent elsewhere. Since the above-described situation represents a particular case, the default threshold is set to zero (no application). Shiny apps allow the user to change MIEP's settings at run time.

4.3 Detection of Potential Biases in Data

The presence of biases due to a sample's particular characteristics is next verified by MIEP R package with PCA [31]. If detected (as in Fig. 2), treated samples are compared against corresponding controls with a single-factor approach (Wald statistical test [32], $p < 0.05$ with Benjamini-Hochberg multiple testing correction) for every subset of consistent (unbiased) samples (e.g. data derived from the same patient samples as in the practice analysis, or belonging to the same cell line), separately. This approach was preferred to multi-factor tests to avoid potential misinterpretation of systematic biases as factors of variability in data.

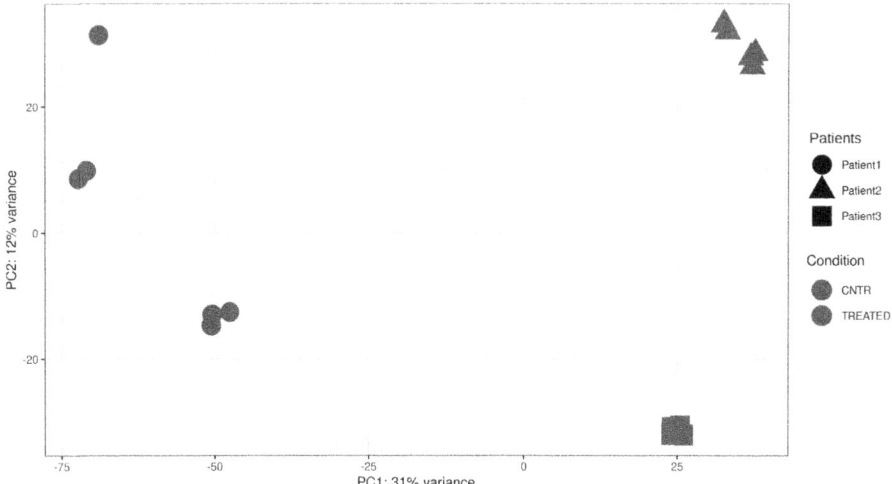

Fig. 2. PCA performed on normalized gene counts suggests data are strongly affected by samples' specific characteristics.

4.4 Differentially Expressed Features

Biologists are usually interested in differentially expressed transcripts. These are most often defined as statistically significant features showing relevant fold-change (FC) values when compared to controls. MIEP's default threshold is FC > 2, a cut-off widely reported in the literature. However, an alternative is proposed to address unusual situations.

For example, in time series logging kinetics of gene expression, the change in expression levels may be subtle at initial time points and require a lower threshold for detection. To avoid arbitrary choices, MIEP evaluates the distribution of FC values (more precisely the distribution of absolute values of logarithmic FCs). The threshold is then set to the 95th percentile. This means that only statistically significant sequences that show fold changes belonging to the top 5% of (|logFC|) values are classified as differentially expressed.

The released version of MIEP R package was developed to investigate coding genes, which were the focus of our research. Therefore, subsequently to annotation of sequences [33, 34] on the basis of Hugo Gene nomenclature (HGCN, https://www.genenames.org), the following categories are automatically discarded: long intergenic non-coding RNA (lncRNA), uncharacterized open reading frames, families with sequence similarity, small nucleolar RNA, pseudogenes, divergent transcripts, novel proteins, intronic transcripts, competing endogenous lncRNA, and overlapping transcripts. MicroRNAs sequences can be saved by changing default settings. An expert user can adapt R scripts to different research necessities.

4.5 Z-score Transformation

Z-score transform is widely applied to preprocess the data to input machine learning algorithms [35], such as the random forests included in the pipeline. Figure 3 shows a PCA of data that had been subjected to Z-score transformation [36] to reduce biases due to sample-related specificities; in the resulting graph, treated samples are clearly separated from controls. This is confirmed by heat maps shown in Supplementary Figures S1 and S2. The hierarchical clustering in Figure S1 gathers data based on patients whereas Figure S2 highlights two main clusters: treated samples are separated from controls.

We acknowledge that this procedure (preliminary check of biases followed by Z-score transformation of data) should not be excessively generalized and that some problems may require input by a bioinformatics expert. However, the proposed approach is consistent with our intent to provide biologists with an automatic tool for preliminary analysis of RNA-seq data. Importantly, it avoids the naïvety of mixing data generated by different researchers in different settings without checking if they are affected by systematic biases. Particular cases should be studied with more attention by skilled bioinformaticians. In fact, MIEP does not address the correction of so-called "batch effect" and does not perform any statistical test on Z-score-transformed data. *DESeq2*'s procedures ensure the reliability of the results because statistical tests are performed on subsets of unbiased data and by strictly following the indications of *DESeq2*'s authors. We again stress that MIEP is designed to supply only preliminary analyses of RNA-seq count data, and to highlight potential biases affecting the dataset. Users are encouraged to consult with expert bioinformaticians to discuss the results.

4.6 MIEP Edits New Gene Sets Following Gene Ontology Term Enrichment or Conditional Random Forests

The enrichment of GO terms is tested by calling up the *topGO* R package. Here we incorporated the results of Alexa's algorithm [22] in order to select only the enriched GO terms corresponding to a high number of differentially expressed genes. In particular, GO terms of interest were ranked based on a weight parameter. This was calculated by combining the p-value originated from the statistical test performed by *topGO* (and corrected for multiple testing with the Benjamini-Hochberg method) with the number of significant features related to the term under consideration. Only high-ranked GO terms are selected for further consideration. In other words, terms whose apparent enrichment was due to unexpected scarcity of significant transcripts in the tested set (low-ranked terms) are discarded. Finally, differentially expressed genes corresponding to selected GO terms are used to edit new gene sets.

Conditional random forests classify samples based on treatment; variables' importance is calculated subsequently. Values defined as outliers based on box plots (precisely, data points located above the higher whisker) are considered to be of interest: corresponding genes are used to compose new gene sets to be further investigated because of their evident importance for classification.

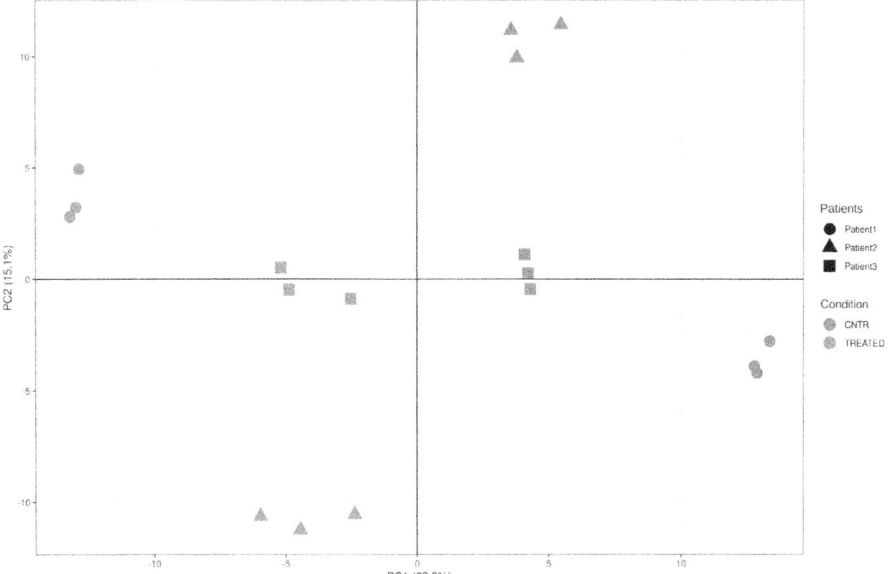

Fig. 3. PCA performed on Z-score-transformed data separates samples based on treatment. This facilitates the study of the effects of treatment on patients.

4.7 Focus on Gene Sets

MIEP produces tables with test results in HTML format, gene expression heat maps, volcano plots, heat maps and dot plots showing fold-change values, etc. for every gene set edited along the pipeline or downloaded from the GSEA web site (see Sect. 4.1). These graphical depictions [36] can provide indications of immediate understanding for biologists planning subsequent wet lab tests.

5 Discussion

MIEP represents an easy-to-use tool that facilitates a preliminary analysis of RNA-seq data thanks to extensive graphical support [37, 38]. MIEP's R functions directly available to users are enriched by *Roxygen* comments [29, 39] allowing the user to call up help functions for explanations about their use. This new R package is composed of more than 200 R scripts and characterized by very careful handling of exceptions (errors and warnings) that facilitates its use by biologists having basic programming skills.

Moreover, interesting technical features allow running the pipeline with limited computational resources. The settings of random forests' parameters are optimized by means of a grid search approach [40, 41], whose computational burden is eased by parallel computing. In addition, distinct implementations [42–44] are automatically tested before grid search to select the fastest algorithm. The influence of random seeds [45, 46] was considered by retrieving them from the global environment and a *hidden* environment of the package [47] was declared in the main script (then created) to facilitate the transmission of variables along the grid search or object construction. Since expanded swap memory is often used as a surrogate when RAM is limited, swappiness in Linux OS can be increased for machine learning: a Shiny app allows inserting a superuser password to invoke system commands if desired (a future version of the software will include equivalent features adapted for Windows OS). These features may appear incongruent with respect to the described proposal to provide an easy-to-use tool to medical researchers. However, these advanced technical devices allow MIEP to run on notebook computers. This feature enhances the potentialities of MIEP pipeline to be adopted for preliminary analysis of RNA-seq data by any biologist equipped with a computer and basics of programming.

This marks a fundamental difference with respect to previous tools, which require more extended computational resources and are targeted for in-depth analysis of data. Second, most existing pipelines start with *"fastq"* files or provide tools for sequence alignment. Instead, MIEP's input is represented by a small size, tab-separated-values file reporting raw sequence counts. This could appear as a limitation, but it is within reach of the researchers to whom MIEP is addressed and considering their computational resources. Consistently, MIEP can be launched with a single command in the R shell because all of the algorithms composing the pipeline are managed by automatic procedures. In contrast, other pipelines/tools require a higher level of expertise and programming skills from their users.

6 Conclusions

As highlighted, MIEP is targeted only for a preliminary analysis, which serves to as a bridge between RNA-seq data generation and more in-depth analyses. This is accomplished by means of a lightweight, easy-to-use R package, which was developed for data mining (with particular focus on pathways). We realize that some bioinformaticians may argue that the automatism characterizing MIEP could produce misleading results in the absence of expert supervision. However, when applied to our cancer research projects, MIEP provided fundamental insights that were experimentally verified (manuscript in preparation), and provided insights on signal transduction pathways and gene functions in cancer cells. In addition, the availability of source codes on github (https://github.com/AlbertoCorradinPhD/MIEP) permits the reusability of scripts and their further development. In summary, we believe MIEP R package represents a useful, intuitive tool, with features and targets that distinguish it from its predecessors.

Funding. MIEP and its developers were supported in part by funding from the Associazione Italiana per la Ricerca sul Cancro (AIRC IG#24935, to VC), the Italian Ministero dell' Università e della Ricerca (PRIN Prot. 2022HTJJYW, to VC), and by the Veneto Institute of Oncology IOV-IRCCS (HAEMOSTATIC (5x100 Bridge) cdc 099217, to MSB).

Disclosure of Interests. The authors have no competing interests to declare that are relevant to the content of this article.

Availability of Data and Software Code. Our software code is publicly available at the following URL: https://github.com/AlbertoCorradinPhD/MIEP.

References

1. Goecks, J., Nekrutenko, A., Taylor, J., Galaxy Team, T.: Galaxy: a comprehensive approach for supporting accessible, reproducible, and transparent computational research in the life sciences. Genome Biol. **11**, R86 (2010). https://doi.org/10.1186/gb-2010-11-8-r86
2. Wolfinger, M.T., Fallmann, J., Eggenhofer, F., Amman, F.: ViennaNGS: a toolbox for building efficient next- generation sequencing analysis pipelines. F1000 Res. **4**, 50 (2015). https://doi.org/10.12688/f1000research.6157.2
3. Schorderet, P.: NEAT: a framework for building fully automated NGS pipelines and analyses. BMC Bioinf. **17**, (2016). https://doi.org/10.1186/s12859-016-0902-3
4. Singer, J., et al.: NGS-pipe: a flexible, easily extendable and highly configurable framework for NGS analysis. Bioinformatics **34**, 107–108 (2017). https://doi.org/10.1093/bioinformatics/btx540
5. Causey, J.L., et al.: DNAp: a pipeline for DNA-seq data analysis. Sci. Rep. **8** (2018). https://doi.org/10.1038/s41598-018-25022-6
6. Morandi, E., et al.: HaTSPiL: a modular pipeline for high-throughput sequencing data analysis. PLoS ONE **14**, e0222512 (2019). https://doi.org/10.1371/journal.pone.0222512
7. Stock, J.K., Jones, N.P., Hammonds, T., Roffey, J., Dillon, C.: Addressing the right targets in oncology: challenges and alternative approaches. SLAS Discovery. **20**, 305–317 (2015). https://doi.org/10.1177/1087057114564349
8. Schenone, M., Dančík, V., Wagner, B.K., Clemons, P.A.: Target identification and mechanism of action in chemical biology and drug discovery. Nat. Chem. Biol. **9**, 232–240 (2013). https://doi.org/10.1038/nchembio.1199
9. Hahn, W.C., et al.: An expanded universe of cancer targets. Cell. **184**, 1142–1155 (2021). https://doi.org/10.1016/j.cell.2021.02.020
10. Chua, H.N., Roth, F.P.: Discovering the targets of drugs via computational systems biology. J. Biol. Chem. **286**, 23653–23658 (2011). https://doi.org/10.1074/jbc.r110.174797
11. Thafar, M.A., et al.: OncoRTT: Predicting novel oncology-related therapeutic targets using BERT embeddings and omics features. Front. Genet. **14** (2023). https://doi.org/10.3389/fgene.2023.1139626
12. Chang, W., et al.: shiny: web application framework for r (2024)
13. Love, M.I., Huber, W., Anders, S.: Moderated estimation of fold change and dispersion for RNA-seq data with DESeq2. Genome Biol. **15**, 550 (2014). https://doi.org/10.1186/s13059-014-0550-8
14. Zhu, A., Ibrahim, J.G., Love, M.I.: Heavy-tailed prior distributions for sequence count data: removing the noise and preserving large differences. Bioinformatics **35**, 2084–2092 (2018). https://doi.org/10.1093/bioinformatics/bty895
15. Stephens, M., et al.: ashr: methods for adaptive shrinkage, using empirical bayes (2023)
16. Kassambara, A.: Practical guide to principal component methods in R. Stdha.com, United States (2017)
17. Melville, J.: uwot: the uniform manifold approximation and projection (UMAP) method for dimensionality reduction (2024)
18. Krijthe, J.H.: Rtsne: T-distributed stochastic neighbor embedding using barnes-hut implementation (2015)
19. van der Maaten, L.J.P., Hinton, G.E.: Visualizing high-dimensional data using t-SNE. J. Mach. Learn. Res. **9**, 2579–2605 (2008)
20. van der Maaten, L.J.P: Accelerating t-SNE using tree-based algorithms. J. Mach. Learn. Res. **15**, 3221–3245 (2014)
21. Meyer, D., Dimitriadou, E., Hornik, K., Weingessel, A., Leisch, F.: e1071: Misc functions of the department of statistics, probability theory group (formerly: E1071), TU wien. (2023)

22. Alexa, A., Rahnenfuhrer, J.: topGO: enrichment analysis for gene ontology (2024)
23. Hothorn, T., Hornik, K., Zeileis, A.: Unbiased recursive partitioning: a conditional inference framework. J. Comput. Graph. Stat. **15**, 651–674 (2006). https://doi.org/10.1198/106186006 x133933
24. Zeileis, A., Hothorn, T., Hornik, K.: Model-based recursive partitioning. J. Comput. Graph. Stat. **17**, 492–514 (2008). https://doi.org/10.1198/106186008x319331
25. Hothorn, T.: Survival ensembles. Biostatistics **7**, 355–373 (2005). https://doi.org/10.1093/bio statistics/kxj011
26. Strobl, C., Boulesteix, A.-L., Zeileis, A., Hothorn, T.: Bias in random forest variable importance measures: Illustrations, sources and a solution. BMC Bioinformatics **8** (2007). https://doi.org/10.1186/1471-2105-8-25
27. Strobl, C., Boulesteix, A.-L., Kneib, T., Augustin, T., Zeileis, A.: Conditional variable importance for random forests. BMC Bioinf. **9** (2008). https://doi.org/10.1186/1471-2105-9-307
28. Subramanian, A., et al.: Gene set enrichment analysis: a knowledge-based approach for interpreting genome-wide expression profiles. Proc. Natl. Acad. Sci. **102**, 15545–15550 (2005). https://doi.org/10.1073/pnas.0506580102
29. Wickham, H.: R Packages. "O'Reilly Media, Inc." (2015)
30. Anders, S., Huber, W.: Differential expression analysis for sequence count data. Genome Biol. **11** (2010). https://doi.org/10.1186/gb-2010-11-10-r106
31. Kassambara, A., Mundt, F.: factoextra: extract and visualize the results of multivariate data analyses. R Package Version 1.0.7 (2020)
32. Agresti, A.: An introduction to categorical data analysis. Wiley (2007)
33. Durinck, S., Spellman, P.T., Birney, E., Huber, W.: Mapping identifiers for the integration of genomic datasets with the R/Bioconductor package biomaRt. Nat. Protoc. **4**, 1184–1191 (2009). https://doi.org/10.1038/nprot.2009.97
34. Durinck, S., et al.: BioMart and Bioconductor: a powerful link between biological databases and microarray data analysis. Bioinformatics **21**, 3439–3440 (2005)
35. Géron, A.: Hands-on machine learning with Scikit-Learn and TensorFlow concepts, tools, and techniques to build intelligent systems. O'Reilly Media, Inc. (2019)
36. Cheadle, C., Vawter, M.P., Freed, W.J., Becker, K.G.: Analysis of microarray data using z score transformation. J. Mol. Diagn. **5**, 73–81 (2003). https://doi.org/10.1016/s1525-157 8(10)60455-2
37. Wickham, H.: Ggplot2: Elegant Graphics For Data Analysis. Springer-Verlag, New York (2016)
38. Acker, D.: gg3D: 3D perspective plots for ggplot2 (2024)
39. Wickham, H., Danenberg, P., Gábor C., Eugster, M.: roxygen2: In-line documentation for R (2024)
40. Schumann, E.: Numerical methods and optimization in finance (NMOF) manual. Package version 2.8–0
41. Gilli, M., Maringer, D.G., Schumann, E.: Numerical methods and optimization in finance. Oxford Academic Press, London; San Diego; Cambridge. An Imprint of Elsevier (2019)
42. R Core Team: R: A language and environment for statistical computing (2024)
43. Microsoft Corporation, Weston, S.: doParallel: Foreach parallel adaptor for the "parallel" package (2022)
44. Solymos, P., Zawadzki, Z.: pbapply: adding progress bar to "*apply" functions. (2023)
45. Itai, U.: The Influence of random seed on random Forest - Uri Itai - Medium, https://medium.com/@uriitai/the-influence-of-random-seed-on-random-forest-080dffecb2f3
46. Louppe, G.: understanding random forests: from theory to practice. (2014). https://doi.org/10.48550/arxiv.1407.7502
47. Wickham, H.: Advanced R. Chapman & Hall/Crc, Boca Raton (2019)

Cross Sequencing Integration of Compositional Microbiome Data in Cancer

Diego Fernández-Edreira, Jose Liñares-Blanco ⓘ,
and Carlos Fernandez-Lozano$^{(\boxtimes)}$ ⓘ

Machine Learning in Life Sciences Laboratory Department of Computer Science and
Information Technologies, Universidade da Coruña (CITIC), A Coruña, Spain
{diego.fedreira,jose.linares,carlos.fernandez}@udc.es

Abstract. High-throughput sequencing has revolutionized our understanding of the human microbiome, providing detailed insights into microbial communities under various health and disease conditions. Among the most common strategies for studying the microbiome are 16S rRNA amplicon sequencing and whole genome shotgun sequencing (WGS), each with its own advantages and limitations. However, integrating and comparing results from data obtained through these two sequencing techniques presents a challenge due to the inherent differences in methods and discrepancies among datasets and their sources. This work evaluates batch effect removal (BER) methods for integrating microbiome composition data from different sequencing platforms. Using data from ten different cohorts, we applied BER methods such as Combat, Limma, FAbatch, MMUPHin, and Percentile-normalization. Our results demonstrate the effectiveness of these methods in reducing batch effects. However, it remains unclear whether the remaining biological signal is reliable, which is critical. Additionally, we compared GG2 with standard databases (SILVA for 16S and WoL for WGS), showing that GG2 enables more unified analysis (increasing the number of taxa shared among cohorts from 94 genera and 58 species to 215 and 210, respectively). In conclusion, our findings suggest that appropriate BER methods can harmonize microbiome data from diverse sequencing platforms, but further experiments are needed to reliably understand how the biological signal is modulated in the process.

Keywords: Bioinformatics · Batch effect · Machine Learning · 16S rRNA · WGS

1 Introduction

The advent of new high-throughput sequencing techniques has revolutionized our understanding of the human microbiome, providing detailed insights into microbial communities in specific environments. Among the most common strategies for studying the microbiome are 16S rRNA amplicon sequencing and whole

L. Cerulo et al. (Eds.): CIBB 2024, LNBI 15276, pp. 69–78, 2025.
https://doi.org/10.1007/978-3-031-89704-7_6

genome sequencing (WGS). Comparing the two, 16S rRNA sequencing is more cost-effective, as it focuses on hypervariable regions of a shared gene, allowing for rapid taxonomic identification. In contrast, WGS offers higher resolution by sequencing entire genomes present in the sample, including information on functional genes and genetic variability within species [6].

However, integrating and analyzing data obtained through these two strategies presents a significant challenge due to the inherent methodological differences already mentioned. Traditionally, 16S sequencing uses databases such as SILVA [11], while WGS aligns against databases like Web of Life (WoL) [13]. These differences can introduce variations in taxonomic classification, affecting data interpretation and the conclusions drawn.

Greengenes2 (GG2) emerges as a solution to unify the analysis of 16S and WGS data, providing a database to integrate both types of sequencing [10]. This integrated approach facilitates the direct comparison of biomarkers and results between studies using different sequencing strategies, overcoming the limitations of the individual approaches previously mentioned [3].

In this work, we have evaluated various batch effect removal (BER) methods to integrate compositional microbiome data from different sequencing platforms [8]. Specifically, we used the following BER methods: Combat, Limma, FAbatch, MMUPHin, and Percentile-Normalization, which have proven effective in various applications previously [5,7,9]. In addition to analyzing the BER capacity of these methods, we also focused on quantifying whether the biological signal is preserved when applying them, which is critical in this field of application. A secondary objective of the study was to compare different sequencing strategies and the data generated through 16S amplicon sequencing and WGS. Using GG2, we evaluated its ability to integrate 16S and WGS data, comparing it with standard approaches (SILVA for 16S and WoL for WGS).

2   Materials and Methods

The proposed workflow is shown in the Fig. 1. It consists of two phases: initially, taxonomic assignment is performed for the different data sets, both 16S rRNA and WGS, using GG2, WoL, and SILVA. Subsequently, benchmarking of the different BER techniques is conducted.

2.1   Cohorts

For this study, a total of ten cohorts were used, with an equal number from 16S sequencing and WGS. These cohorts were directly downloaded from the European Nucleotide Archive (PRJEB33634, PRJEB6070, DRA006684, PRJEB10878, PRJEB7774, and PRJNA447983) using their accession numbers, except for the cohort from Zackular et al. [12]. The studies PRJEB33634 and PRJNA447983 are divided into 3 and 2 cohorts respectively, thus comprising the 10 cohorts used in this work (see Table 1).

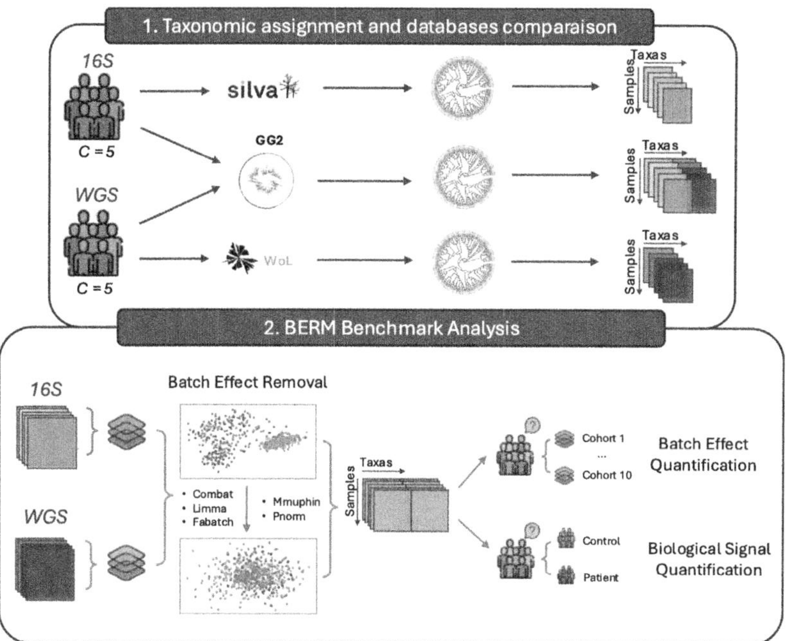

Fig. 1. Workflow. Proposed workflow with two phases-taxonomic assignment for 16S rRNA and WGS data sets using GG2, WoL, and SILVA, followed by benchmarking of various BER techniques.

As one of the main objectives is to quantify the biological signal after correcting for batch effects, only samples from control patients or those diagnosed with colorectal cancer were used, excluding adenoma samples.

Table 1. Cohorts Summary

Cohort	Samples	Controls	CRC	Sequencing
PRJEB33634_B1	15	8	7	16S
PRJEB33634_B2	64	37	27	16S
PRJEB33634_B3	80	28	52	16S
PRJEB6070_F	91	50	41	16S
Zackular	60	30	30	16S
DRA006684	80	40	40	WGS
PRJEB10878	128	54	74	WGS
PRJEB7774	109	63	46	WGS
PRJNA447983_1	53	24	29	WGS
PRJNA447983_2	60	28	32	WGS

2.2   Preprocessing of Sequences

For all 16S rRNA sequence data, the quality of the raw reads was visualized with FastQC v0.12.1 to ensure at least an average quality of 25. The reads were then imported into R v4.2.2 and assembled in amplicon sequence variants (ASVs) with the DADA2 package v1.26. To examine the length distribution of all sequences in aggregate terms a second quality check of the samples was performed using the quality plots provided by DADA2. For the processing of the sequences, the default parameters given by the author of DADA2 have been left unchanged (except trimleft in the case primers were presented). Non-chimeric assembled ASVs were taxonomically assigned (phylum to species) using the SILVA reference database v138.1 or GG2 reference database v2022.10 through qiime2 plugin.

For all WGS sequence data, the quality of the raw reads was first visualized using FastQC. Fastp v0.22.0 was then used for quality control. Subsequently, the samples were aligned using Bowtie2 v2.5.1 against the reference genome (WoL or WoL2 when using GG2). To produce the OGU table, the resulting alignments were classified using Woltka v0.1.5. Additionally, for the OGU table generated from alignments against WoL, taxonomic classification was performed using Woltka. For alignments against WoL2, the GG2 plugin for QIIME2 was used for taxonomic classification.

Once this was done, phyloseq objects (Phyloseq v1.42.0) were constructed using the ASV/OGU count table, the taxonomic table, and the clinical data from each cohort for downstream analysis. In the case of the phyloseq objects generated from SILVA and WoL, an attempt was made to unify the taxonomic names to a common format. After the construction of the phyloseqs, the next step was filtering out samples belonging to adenoma, followed by filtering out species that could not be identified. Next, agglomeration was performed at the species level, and species with zero counts were filtered out. Finally, given the compositional nature of microbiome data, each phyloseq object was transformed using centered log-ratio (CLR) transformation; relative abundance was also used for some batch-effect reduction methods that require it.

2.3   Batch Effect Removal Methods

The following BER methods have been used to correct the differences between the different platforms: Combat, Limma, FAbatch, MMUPHin, and Percentile-Normalization.

Combat combines an empirical Bayesian approach with effect estimation to adjust the data. Combat works by adjusting the mean and variance of each batch to match a global average. Key steps include adjusting the data for batch differences and covariates of interest and modeling the batch effect and adjusting it using empirical Bayes methods. This method is particularly effective when there are many batches and when batch variation is a significant problem.

Limma adjusts batch effects using a linear model with a design matrix that includes terms for both batch and experimental conditions. It then applies empirical Bayesian adjustments to stabilize variance estimates. It is versatile and allows modeling the effects of multiple factors besides the batch effect.

FAbatch is a method based on factor analysis theory. This method models batch effects as latent factors and corrects them by adjusting the factorial structure of the data. It uses Gaussian mixture models to identify and adjust latent batch effects in the data. It is suitable for correcting batch effects in complex studies and can be used for gene expression, metabolomics, and proteomics data. Its main advantage is adaptively identifying batch effects without the need to explicitly specify batch covariates.

MMUPHin employs mixed-effects models for correcting batch effects in multi-omics profiling data. This method adjusts batch effects by treating batches as random effects and biological variables of interest as fixed effects. It uses a hierarchical mixed model to separate biological signals from batch effects. It allows correction of both fixed and random effects, making it useful in studies with complex experimental designs. It is particularly effective when batch effects and biological effects are correlated.

Percentile-Normalization is a non-parametric normalization method that adjusts data to the corresponding percentile in a reference distribution, usually the average of all batches' percentiles. This method does not assume any specific distribution and can be used when data do not meet normality assumptions. It aligns the distribution of each batch so that each quantile has a similar distribution to the target distribution.

2.4   Machine Learning Methods

For the experiments the following ML algorithms have been used: Random Forest (RF) [1], Support Vector Machines (SVM) [2] and Generalized Linear Model (glmnet) [4].

Random Forest is an ensemble learning method that builds multiple decision trees during training and outputs the mode of their predictions (classification) or average (regression). It reduces overfitting and improves accuracy by aggregating the results of diverse models. It is robust to noise and effective for large datasets with many features. A search was conducted to determine the appropriate values for the hyperparameters: mtry (number of variables randomly sampled at each data division), nodesize (minimum size of the terminal nodes), and number of trees. The final configuration of the model was determined to be 1000 trees, an mtry of 4, and a nodesize of 5.

Support Vector Machines is a supervised learning algorithm used for classification and regression tasks. It works by finding the optimal hyperplane that separates data points of different classes with the maximum margin. SVMs are effective in high-dimensional spaces and are used for both linear and non-linear classification through kernel functions. For this study we used the kernel function RBF (radial Gaussian base) and we made a search of the appropriate values for C (penalty for misclassified observations) and sigma (standard deviation of Gaussian distribution). For this study, we used the RBF (Radial Basis Function) kernel, with a C parameter (penalty for misclassified observations) of 1 and a sigma (standard deviation of the Gaussian distribution) of 0.5.

Glmnet is an algorithm for fitting generalized linear models (GLMs) with elastic net regularization, which combines L1 (lasso) and L2 (ridge) penalties. It is particularly useful for high-dimensional data and variable selection, providing a balance between feature sparsity and model complexity. A search was conducted to determine the appropriate values for alpha (which controls the penalty of the elastic network) and lambda (which controls the total strength of the penalty). The final configuration of the model was an alpha of 0.15 and a lambda of 0.25.

2.5 Measures of Performance

Balanced Accuracy (B.Acc) is the average of sensitivity (true positive rate) and specificity (true negative rate). It is used to evaluate the performance of classification models, especially when dealing with imbalanced datasets.

Classification Error (CE) measures the proportion of incorrect predictions made by the model, calculated as the ratio of misclassified samples to the total number of samples.

Area Under the Receiver Operating Characteristic Curve (AUC-ROC) quantifies the overall ability of a model to discriminate between positive and negative classes. It represents the probability that a randomly chosen positive instance is ranked higher than a randomly chosen negative instance.

Precision-Recall Area Under the Curve (PRAUC) measures the area under the precision-recall curve, highlighting the trade-off between precision and recall, especially useful for imbalanced datasets where false negatives are more critical.

3 Results

The first step involved filtering out samples belonging to adenoma, followed by filtering out species that could not be identified. Next, agglomeration was performed at the species level, and species with zero counts were filtered out. Finally, to compare GG2 with SILVA/WoL, common taxa between both pairs of databases were matched. Given the compositional nature of microbiome data, each phyloseq object was transformed using centered log-ratio (CLR) transformation; relative abundance was also used for some batch-effect reduction methods that require it.

Subsequently, various BER methods were applied: Combat (default configuration), Combat (using cohort DRA006684 as reference), Limma, FAbatch, MMUPHin, and Percentile-normalization.

To assess batch effect correction, different Machine Learning (ML) algorithms were employed, including Random Forest (RF), Support Vector Machine (SVM), and Generalized Linear Model (GLMNET). Models were trained with these algorithms (predicting each sample's cohort membership using the first 15 principal components as predictors) on data from each BER method and compared with models trained on uncorrected data. The premise is that uncorrected data should enable models to differentiate each sample's cohort; a decrease in model performance relative to this benchmark suggests effective batch effect reduction

Additionally, the ability of these BER methods to maintain the biological signal (BS)-in this case, the ability to distinguish between controls and CRC samples-was assessed. Models were trained using the same algorithms (predicting the status of each sample, i.e., control or cancer, with the 210 species as predictors) on data from each BER method and on uncorrected data. In both experiments, 5-fold cross-validation with 10 repetitions was performed.

3.1 SILVA/WOL Versus GG2

As shown in Fig. 2.a, using GG2 results in a greater number of shared taxa across all cohorts. Thus, we transition from 94 genera and 58 species using a standard approach (WoL with WGS and Silva with 16S) to 215 and 210, respectively. Taking this into consideration, subsequent analyses were performed utilizing taxa acquired through GG2.

3.2 Quantification of BER and Maintenance of BS

Figure 2.b displays samples from all cohorts, with colors indicating their respective sequencing strategy. As seen in the PCA plot using the first two components, a dimensional discrepancy is evident between the two library types, implying the presence of a batch effect.

Figure 2.c illustrates the performance difference among different BER methods. In this figure, the balanced accuracy and classification error of each method are plotted. If the batch effect is not corrected, a mean balanced accuracy of 0.65 and a mean classification error of 0.27 are obtained. It appears that all methods are capable of clearly reducing the batch effect, with FAbatch being the best (achieving a balanced accuracy of 0.10 and a classification error of 0.85).

However, as illustrated in Fig. 2.d, certain methods appear effective in mitigating these differences at the expense of compromising BS. For example, while FAbatch appears to successfully correct the batch effect, it does not surpass uncorrected data in classifying between controls and cancer samples. Conversely, the method least effective in eliminating the batch effect (Percentile-normalization) appears to marginally enhance sample classification based on BS compared to uncorrected data. Lastly, employing Combat with DRA006684 as a reference, despite not being among the best methods in eliminating the batch effect (as observed in Fig. 2.c), demonstrates excellent performance in classifying samples by biological signal (achieving AUCROC and PRAUC values above 0.9).

In Fig. 2.e, a shift in the distribution of the two library types is evident compared to Fig. 2.b, with both now centered in the plot and almost aligned along both axes. This suggests that the previous disparity in composition between the two types has been rectified. However, when plotting PCAs with the first 4 components (Fig. 2.f), distinctions between controls and cancer patients are not discernible across any combinations. Coupled with the notably high results, this raises concerns regarding potential overfitting when employing Combat.

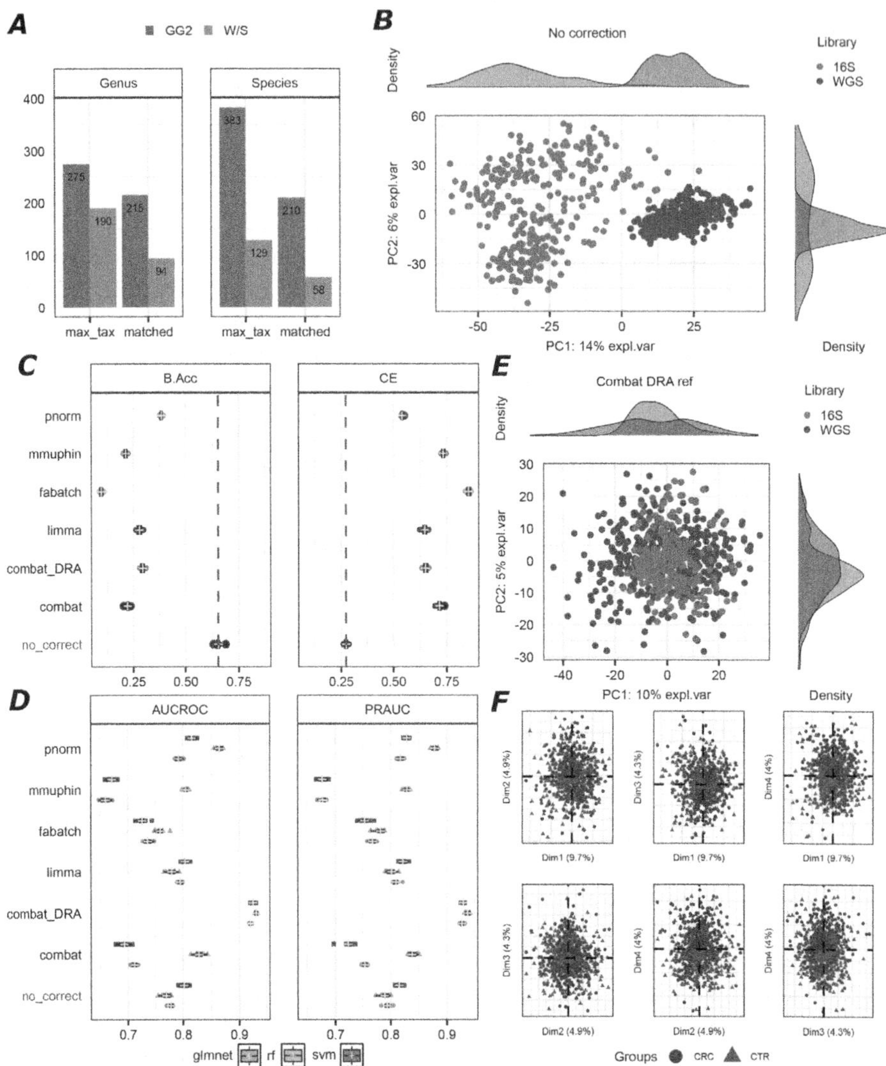

Fig. 2. Results. A) Databases comparison. **B)** PCA without correction. **C)** Batch-Effect quantification. **D)** Biological signal verification. **E)** PCA with data corrected with combat (using DRA006684 as reference). **F)** PCAs of first 4 components colored by status (CRC vs Controls).

4   Conclusions

This study has demonstrated that the use of GG2 can enhance reproducibility in microbiome studies by enabling a greater number of shared taxa between 16S rRNA sequencing and WGS data. Specifically, employing GG2 resulted in a significant increase in the number of identified genera and species compared to standard databases such as SILVA and WoL. Furthermore, the evaluated BER methods proved effective in reducing technical differences between 16S and WGS data. However, it was observed that some BER methods may alter the intrinsic biological signal of the samples, posing challenges for accurate interpretation of biological results. Therefore, this study underscores the importance of continuing research and optimizing batch effect correction methods that do not compromise biological signal. Future investigations should focus on developing and validating strategies that balance the removal of technical variations with the preservation of the biological integrity of the data, ensuring reliable and reproducible results in microbiome studies.

Acknowledgments. We thank the members of the Machine Learning in Live Sciences Lab for the helpful discussions. This work was supported by the Interreg Sudoe and the ERDF (S1/1.1/P0033).

Disclosure of Interests. The authors declare that they have no conflicts of interest related to this study.

References

1. Breiman, L.: Random forests. Mach. Learn. **45**(1), 5–32 (2001)
2. Cortes, C., Vapnik, V.: Support-vector networks. Mach. Learn. **20**(3), 273–297 (1995)
3. Deschasaux, M., et al.: Depicting the composition of gut microbiota in a population with varied ethnic origins but shared geography. Nat. Med. **24**(10), 1526–1531 (2018)
4. Friedman, J., Hastie, T., Tibshirani, R.: Regularization paths for generalized linear models via coordinate descent. J. Stat. Softw. **33**(1), 1 (2010)
5. Hornung, R., Boulesteix, A.L., Causeur, D.: Combining location-and-scale batch effect adjustment with data cleaning by latent factor adjustment. BMC Bioinf. **17**, 1–19 (2016)
6. Johnson, J.S., et al.: Evaluation of 16s rrna gene sequencing for species and strain-level microbiome analysis. Nat. Commun. **10**(1), 5029 (2019)
7. Johnson, W.E., Li, C., Rabinovic, A.: Adjusting batch effects in microarray expression data using empirical bayes methods. Biostatistics **8**(1), 118–127 (2007)
8. Leek, J., et al.: Tackling the widespread and critical impact of batch effects in high-throughput data. Nat. Rev. Genet. **11**(10), 733–739 (2010)
9. Ma, S., et al.: Population structure discovery in meta-analyzed microbial communities and inflammatory bowel disease using mmuphin. Genome Biol. **23**(1), 208 (2022)
10. McDonald, D., et al.: Greengenes2 unifies microbial data in a single reference tree. Nat. Biotech. 1–4 (2023)

11. Quast, C., et al.: The silva ribosomal rna gene database project: improved data processing and web-based tools. Nucleic Acids Res. **41**(D1), D590–D596 (2012)
12. Zackular, J.P., Rogers, M.A., Ruffin, M.T., IV., Schloss, P.D.: The human gut microbiome as a screening tool for colorectal cancer. Cancer Prev. Res. **7**(11), 1112–1121 (2014)
13. Zhu, Q., et al.: Phylogenomics of 10,575 genomes reveals evolutionary proximity between domains bacteria and archaea. Nat. Commun. **10**(1), 5477 (2019)

Medical Informatics

Private, Efficient and Scalable Kernel Learning for Medical Image Analysis

Anika Hannemann[1,2]($\boxtimes$), Arjhun Swaminathan[3,4], Ali Burak Ünal[3,4], and Mete Akgün[3,4]

[1] Department of Computer Science, Leipzig University, Leipzig, Germany
anika.hannemann@cs.uni-leipzig.de
[2] Center for Scalable Data Analytics and Artificial Intelligence (ScaDS.AI), Dresden/Leipzig, Germany
[3] Medical Data Privacy and Privacy-preserving Machine Learning (MDPPML), University of Tübingen, Tübingen, Germany
arjhun.swaminathan@uni-tuebingen.de
[4] Institute for Bioinformatics and Medical Informatics (IBMI), University of Tübingen, Tübingen, Germany

Abstract. Medical imaging is key in modern medicine. From magnetic resonance imaging (MRI) to microscopic imaging for blood cell detection, diagnostic medical imaging reveals vital insights into patient health. To predict diseases or provide individualized therapies, machine learning techniques like kernel methods have been widely used. Nevertheless, there are multiple challenges for implementing kernel methods. Medical image data often originates from various hospitals and cannot be combined due to privacy concerns, and the high dimensionality of image data presents another significant obstacle. While randomised encoding offers a promising direction, existing methods often struggle with a trade-off between accuracy and efficiency. Addressing the need for efficient privacy-preserving methods on distributed image data, we introduce OKRA (Orthonormal K-fRAmes), a novel randomized encoding-based approach for kernel-based machine learning. This technique, tailored for widely used kernel functions, significantly enhances scalability and speed compared to current state-of-the-art solutions. Through experiments conducted on various clinical image datasets, we evaluated model quality, computational performance, and resource overhead. Additionally, our method outperforms comparable approaches.

Keywords: Distributed Learning · Kernel Methods · Privacy · Machine Learning · Medical Images

1 Introduction

Medical imaging is crucial in modern medicine, offering insights into the human body and revolutionizing healthcare by aiding in disease detection, treatment

---

A. Hannemann and A. Swaminathan—These authors contributed equally to this work.

© The Author(s), under exclusive license to Springer Nature Switzerland AG 2025
L. Cerulo et al. (Eds.): CIBB 2024, LNBI 15276, pp. 81–95, 2025.
https://doi.org/10.1007/978-3-031-89704-7_7

planning, and patient monitoring. Kernel-based machine learning, or kernel learning, has become widely accepted in medical image analysis for identifying complex patterns by mapping data to higher-dimensional spaces [21,28]. Techniques like kernel-based Support Vector Machines (SVM), Principal Component Analysis (PCA), Canonical Correlation Analysis (CCA), Gaussian processes, and k-means are particularly effective for smaller [13], high-dimensional datasets and offer greater interpretability [28].

However, medical data is often distributed across various institutions, complicating the implementation of kernel learning due to privacy risks. Health data can re-identify individuals [14], and regulations like GDPR prohibit sharing sensitive data. To tackle this, privacy preserving techniques could be employed. However, traditional privacy techniques, such as Homomorphic Encryption (HE) [15] and Secure Multiparty Computation (SMPC) [34], face operational challenges, while Differential Privacy (DP) introduces model inaccuracies [4,19].

Moreover, many healthcare institutions rely on cloud services for computational tasks [2]. With this in mind, our work proposes kernel-based learning on distributed medical data within a federated architecture using a semi-honest central server. This approach aligns with real-world scenarios where institutions aim to perform joint analysis for higher accuracy while maintaining data privacy and lacking on-premise computational resources. Specifically, we employ a one-shot federated learning approach, where a global model is computed in a single communication round [17,18].

This method ensures the central server cannot derive sensitive information from the data. We introduce OKRA (Orthonormal K-fRAmes), which uses randomized encoding to project input data into higher-dimensional spaces. This facilitates the training of kernel learning models as if the data were centralized, with lower computational and communication costs compared to traditional privacy techniques. Unlike the comparable state-of-the-art solutions, OKRA requires less computation time, especially for high-dimensional data.

We make three contributions:

1. We present the OKRA methodology to privately compute kernel functions on distributed data.
2. We report a privacy analysis confirming the ability of OKRA to maintain the privacy of both input data and the size of medical images.
3. We evaluate OKRA through extensive experimentation involving various combinations of datasets, kernels and Machine Learning techniques.

Paper structure: Sect. 2 introduces related work. Section 3 describes our methodology, followed by a privacy analysis in Sect. 4. Section 5 presents our experimental results, followed by conclusion in Sect. 6.

2   Related Work

2.1   One-Shot Federated Learning

Federated Learning (FL) [24] allows distributed nodes to train models on local data while preserving privacy. In traditional FL, nodes iteratively share local model parameters with a central server, which aggregates them to form a global model. However, this iterative communication introduces latency and security risks, such as data/model poisoning and membership inference attacks [25].

In contrast, one-shot federated learning approaches such as [18], restrict communication to a single round. Each participant sends its relevant parameters to the central server once, which subsequently undertakes the role of model training. For our methodology, the central server computes relevant kernels with a single round of communication.

2.2   Privacy Preserving Techniques

FL incorporates various techniques to safeguard the privacy of input data:

Cryptography-Based Approaches: These primarily constitute the usage of HE and SMPC. HE aims to protect privacy of aggregated models by encrypting the parameters of the local models. However, this is computationally intensive, often leading to slower processing speeds, which can be a bottleneck in real-time applications [12]. Meanwhile, SMPC splits the data into secret shares distributed among participants for shared computation. Despite being less computationally intensive compared to HE, it requires continuous communication, often resulting in high execution times [7,26].

Differential Privacy: FL methods using DP introduce noise to individual models to enhance privacy, making it challenging to deduce the original model or identify a specific data point's membership. However, the noise added is irremovable and often impacts model accuracy [1,4,19].

Randomized Encoding-Based Approaches: Contrary to more computationally demanding approaches, these methods present an efficient way of preserving data privacy, such as maintaining relationships between dot products [8,31]. Randomized encoding approaches [3,9,23,30] use a variety of techniques, including but not limited to geometric perturbations. Geometric perturbations introduce structured noise to the data. This noise can be factored out, thus differing from differential privacy methods in the context of preserving model accuracy. It should be noted that the security of randomized encoding-based approaches is dependent on the specific problem at hand since they are tailored to meet the unique demands and constraints of a privacy-preserving task.

2.3   Privacy-Preserving Kernel Learning Using Randomized Encoding

The integration of randomized encoding methods into kernel learning presents a promising avenue, particularly in their cost-effectiveness compared to traditional approaches. This subsection delves into two noteworthy contributions in this domain that achieve exact model predictions: ESCAPED [32] and FLAKE [20], each offering distinct methodologies and implications.

ESCAPED makes use of randomized encoding within a multi-party computational framework for kernel computations. Its core innovation lies in enabling secure, collaborative computing without compromising individual data privacy. However, this approach necessitates intricate communication channels among all participating data providers. This requirement becomes particularly cumbersome with the introduction of new data entities, posing significant logistical challenges in scalable deployments. On the other hand, FLAKE works towards optimizing communication overhead in kernel-based learning. It employs a one-shot federated learning architecture, requiring data providers to only communicate once with a central server. However, this methodology encounters its own set of challenges, particularly with computational efficiency and scalability. It especially demonstrates limitations in processing high-dimensional data, a common characteristic in complex datasets such as medical images.

In this work, we present an alternative to current randomized encoding methods for kernel computation in machine learning. We propose a one-shot algorithm, hence bypassing ESCAPED's communication overheads, while scaling better than FLAKE to accommodate higher dimensional datasets.

3   Methodology

In this section, we delve into the methodology of our approach, focusing on private kernel computation in a distributed environment. First, we detail the adversarial architecture we operate in. In our architecture, for simplicity, we consider participants Alice, Bob, and Charlie, along with a central server orchestrating the training of a machine learning model, as described in Fig. 1. However, this can be extended to any number of participants. We operate under the assumption of semi-honest participants and a non-colluding central server. A semi-honest party follows the protocol correctly but may attempt to infer additional information. This is in line with existing state-of-the-art [20,32].

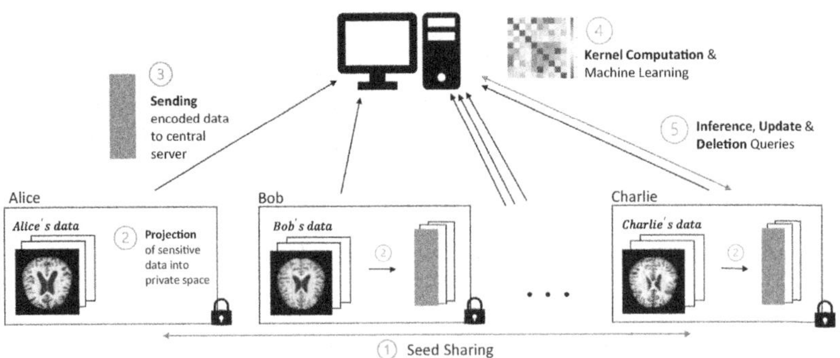

Fig. 1. Overview of OKRA.

Our method must satisfy four key requirements, with **privacy** being the first: Neither the central server, nor the corrupted participants can learn a non-corrupt participant's data, and the image sizes remain concealed from the server. Second, **correctness** is crucial, with the model's accuracy needing to align that of a centrally trained model. Third, **efficiency** is important, as communication costs and execution times should be practical even when input data scales up. Finally, **data adaptability** is essential, allowing the model to accommodate new data and participants with minimal overhead.

We will focus on four of the most widely used kernel functions: Linear, Gaussian, Polynomial, and Rational Quadratic kernels. We begin by briefly defining them.

3.1   Preliminaries

Kernel functions project the data into a Hilbert space using a feature map, and compute the similarity between the data points in this space. The trick here lies in the ability to compute this similarity without explicitly mapping the data - a consequence of Mercer's theorem, which provides necessary and sufficient conditions for a function from an input space to be a 'kernel', ensuring the mapped data resides in a Hilbert space. Below are four widely used kernels of interest: Linear Kernel (K_{lin}) [11], Gaussian Kernel (K_{RBF}) [6], Polynomial Kernel (K_{poly}) [27], and Rational Quadratic Kernel (K_{RQ}) [33] -

$$K_{lin}(x,y) = xy^T, \qquad K_{RBF}(x,y) = \gamma^2 \exp\left(-\frac{\|x-y\|^2}{2l^2}\right),$$

$$K_{poly}(x,y) = (1 + xy^T)^d, \qquad K_{RQ}(x,y) = \gamma^2 \left(1 + \frac{\|x-y\|^2}{2\alpha l^2}\right)^{-\alpha}.$$

Here, γ is the amplitude, l the length-scale, d the degree of the polynomial, and α the scale mixture. We now define orthonormal k-frames, that are central to our study.

Definition 1. *Orthonormal k-frame [10] A k-frame is an ordered set of k linearly independent vectors in a vector space of dimension n, with $k \leq n$. An ordered set of k linearly independent orthonormal vectors is called an* Orthonormal k-frame.

3.2   The OKRA Methodology

Data Preprocessing. Prior to delving into our methodology, we now detail the preprocessing steps for medical imaging datasets. Each dataset comprises n medical images, formatted into a four dimensional array of size $n \times h \times w \times c$, where h is the height, w the width, and c the number of color channels. We flatten each image into a 1D array, so that the dataset has the size $n \times (hwc := f)$. Participants Alice, Bob, and Charlie possess preprocessed data matrices A, B, and C of size $n_A \times f$, $n_B \times f$, and $n_C \times f$ respectively. While n_A, n_B, and

$n_\mathcal{C}$ may differ due to variable data availability among participants, we require all participants contribute more than one image. The participants share a seed using any robust Public Key Infrastructure (PKI) for secure communication. We assume a trusted third party for fair public key distribution, in line with standard privacy-preserving federated learning practices.

Seed Generation. A robust Public Key Infrastructure (PKI) ensures every participant possesses the public signing keys of all others. At the onset of the protocol, a participant is arbitrarily designated as the leader, responsible for the generation of a random seed. A secure seed is shared with participants using public-key encryption and verified with digital signatures. We assume a trusted third party for fair public key distribution, in line with standard privacy-preserving federated learning practices [5,36].

Randomized Encoding. The participants Alice ($\mathcal{A}$), Bob ($\mathcal{B}$) and Charlie ($\mathcal{C}$) with their shared seeds generate their own sets of j orthonormal k-frames $({}_\mathcal{A}O_1, \ldots, {}_\mathcal{A}O_j)$, $({}_\mathcal{B}O_1, \ldots, {}_\mathcal{B}O_j)$, and $({}_\mathcal{C}O_1, \ldots, {}_\mathcal{C}O_j)$ respectively, using the seed such that

$$_\mathcal{P}O_i = [_\mathcal{P}v_i^{1\,T} \ldots {}_\mathcal{P}v_i^{k_i\,T}],$$

for $\mathcal{P} \in \{\mathcal{A}, \mathcal{B}, \mathcal{C}\}$. Further, for any $p \leq j$ and $\mathcal{P}, \mathcal{Q} \in \{\mathcal{A}, \mathcal{B}, \mathcal{C}\}$, it holds that

$$\left(_\mathcal{P}O_p^\dagger\right)\left(_\mathcal{Q}O_p\right) = I,$$

where $\dagger$ denotes the conjugate transpose. Each $_\mathcal{P}v_i^l$ is an element of a corresponding vector space V_i defined over $\mathbb{C}$ such that $k_i \leq dim(V_i)$. Further, $V_1 \oplus \ldots \oplus V_j = V$, $dim(V) = f + k$, and $k_1 + \ldots + k_j = f$. Then we define $_\mathcal{P}\Gamma$ to be the direct sum

$$_\mathcal{P}\Gamma = {}_\mathcal{P}O_1 \oplus \cdots \oplus {}_\mathcal{P}O_j.$$

To enhance privacy, a random permutation matrix $\sigma \in S_f$ is generated using the shared seed by all three participants. Here S_f denotes the symmetric group of degree f. They then permute the rows of their matrices $_\mathcal{A}\Gamma, {}_\mathcal{B}\Gamma$, and $_\mathcal{C}\Gamma$ according to this permutation, and subsequently, encode their data to derive $A' = A(_\mathcal{A}\Gamma \circ \sigma)^\dagger \in \mathbb{C}^{n_\mathcal{A} \times (f+k)}$, $B' = B(_\mathcal{B}\Gamma \circ \sigma)^\dagger \in \mathbb{C}^{n_\mathcal{A} \times (f+k)}$, and $C' = C(_\mathcal{C}\Gamma \circ \sigma)^\dagger \in \mathbb{C}^{n_\mathcal{C} \times (f+k)}$.

Alice, Bob, and Charlie send A', B', and C' to the central server. Let A_i denote the ith row of matrix A - an image in our setting. We propose the following.

Theorem 1. *Given A', B' and C', the central server can correctly compute the Linear, Gaussian, the Polynomial and the Rational Quadratic Kernels for the distributed input data.*

Proof. Without loss of generality, let us consider the kernels to be computed between Alice and Bob. Let us denote an image from Alice as $a = A_i$, and Bob as $b = B_j$. Then representing the encoded images as $a' = A'_i$ and $b' = B'_j$, we propose that the functions to be computed by the central server are as follows.

$$K^\dagger_{lin}(a', b') = [(A'_i)(B'_j)^\dagger], \quad K^\dagger_{RBF}(a', b') = \exp\big(-\gamma d_{ij}(A', B')\big),$$

$$K^\dagger_{poly}(a', b') = \big(1 + [(A'_i)(B'_j)^\dagger]\big)^d, \quad K^\dagger_{RQ}(a', b') = \gamma^2 \left(1 + \frac{d_{ij}(A', B')}{2\alpha l^2}\right)^{-\alpha},$$

where

$$d_{ij}(A', B') := \big([(A'_i)(A'_i)^\dagger] + [(B'_j)(B'_j)^\dagger] - [(A'_i)(B'_j)^\dagger] - [(B'_j)(A'_i)^\dagger]\big).$$

To show that these are well defined kernels, we prove the correctness of the computed functions. Without loss of generality for $P, Q \in \{A, B\}$ and any p, q -

$$[(P'_p)(Q'_q)^\dagger] = \big[(P')\big(_{\mathcal{Q}}\Gamma\sigma(Q_q)^\dagger\big)\big]_p$$

$$= \left[P\sigma^\dagger \left(\bigoplus_{k=1}^{q} {}_{\mathcal{P}}O^\dagger_k {}_{\mathcal{Q}}O_k\right)\sigma Q_q\right]_p = [P_p Q_q].$$

Hence, it follows that

$$[(A'_i)(B'_j)^\dagger] = ab^T, \quad [(B'_j)(A'_i)^\dagger] = ba^T,$$

$$[(A'_i)(A'_i)^\dagger] = aa^T, \quad [(B'_j)(B'_j)^\dagger] = bb^T,$$

$$K^\dagger_{lin}(a', b') = K_{lin}(a, b), \quad K^\dagger_{RBF}(a', b') = K_{RBF}(a, b),$$

$$K^\dagger_{poly}(a', b') = K_{poly}(a, b), \quad K^\dagger_{RQ}(a', b') = K_{RQ}(a, b).$$

This shows the correctness of the computed functions, hence theoretically satisfying our **Correctness** requirement. Further, to show that these functions are well defined kernels, we note that the map $\Gamma : \mathbb{R}^f \to \mathbb{C}^{f+k}$ defined by $_{\mathcal{P}}\Gamma(v) = v'$ for $v \in \mathbb{R}^f$ and $P \in \{\mathcal{A}, \mathcal{B}, \mathcal{C}\}$ is a bijection, and since $K_{lin}, K_{RBF}, K_{poly}$ and K_{RQ} are well defined kernels, so are $K^\dagger_{lin} = K_{lin} \circ (\Gamma^{-1} \times \Gamma^{-1})$, $K^\dagger_{RBF} = K_{RBF} \circ (\Gamma^{-1} \times \Gamma^{-1})$, $K^\dagger_{poly} = K_{poly} \circ (\Gamma^{-1} \times \Gamma^{-1})$, and $K^\dagger_{RQ} = K_{RQ} \circ (\Gamma^{-1} \times \Gamma^{-1})$ on the range of $\Gamma \times \Gamma$.

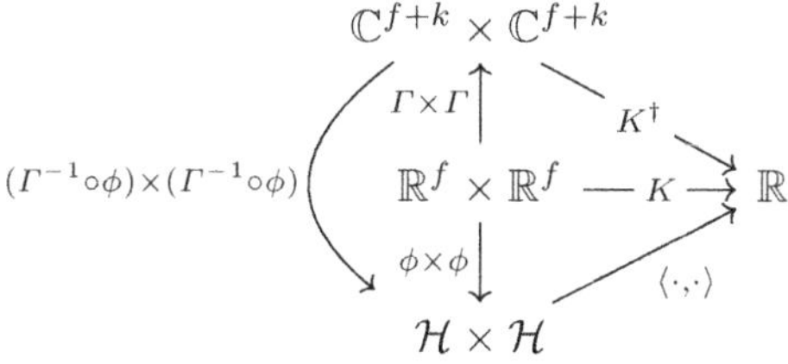

Hence we have the above commutative diagram.

3.3 Data Adaptability

In our methodology, Alice, Bob, and Charlie contribute datasets A, B, and C. Introducing additional data X from Charlie or another participant is seamless. The shared seed is used to generate X' for further kernel computations with the existing data. Only specific kernel portions are updated, saving computational resources. When data changes dynamically, OKRA creates new masks, denoted by Γ_t for training iteration t, ensuring privacy for each party's data alterations. If a participant withdraws data, the corresponding contribution is removed, complying with data protection standards such as the GDPR.

4 Privacy Analysis

As stated earlier, OKRA operates under the environment consisting of a proper subset of semi-honest participants, and/or a non-colluding semi-honest central server. For this purpose, we will show that it guarantees security in both settings.

It is essential to highlight that our privacy model excludes certain extreme adversarial conditions. We exclude data distributions where the number of features or the training data of one or more input parties can be guessed, and protocols with only one input party.

Theorem 2. *The proposed methodology is secure against a semi-honest adversary who corrupts the central server.*

Proof. A semi-honest central server is only the receiver of the encoded data from the input parties, and follows the protocol as intended. Without loss of generality, let there be two input parties Alice and Bob with input data $A \in \mathbb{R}^{n_A \times f}$ and $B \in \mathbb{R}^{n_B \times f}$, respectively, where $n_{\mathcal{P}}$ is the number of samples with the corresponding party $\mathcal{P} \in \{\mathcal{A}, \mathcal{B}\}$ and f is the number of features in each image. The semi-honest function party receives the projected data from Alice and Bob. These consist of $A' = A\gamma^{\dagger} \in \mathbb{C}^{n_A \times (f+k)}$ and $B' = B\gamma^{\dagger} \in \mathbb{C}^{n_B \times (f+k)}$ where $k > 0$. Then, it computes the kernel functions mentioned above in Theorem 1. The data to which the function party has access then includes

(a) A' and analogously, B',
(b) $AB^T = (BA^T)^T$, AA^T and analogously BB^T.

Regarding (a), it is evident that A' does not reveal the number of features of A. We now show that A' is not produced by a unique pair A and $_{\mathcal{A}}\Gamma \circ \sigma$.

Given a unitary matrix $U \in \mathbb{C}^{f \times f}$ with $f > 1$, for $\tilde{A} = AU$ and $_{\mathcal{A}}\tilde{\Gamma} = \Gamma \circ \sigma \circ U$, we have $A' = A(_{\mathcal{A}}\Gamma \circ \sigma)^{\dagger} = \tilde{A}_{\mathcal{A}}\tilde{\Gamma}^{\dagger}$. In the context of reconstruction attacks using analogous data, consider a sample dataset $\hat{A}$ that is drawn from a distribution similar to A. In the context of reconstructing A, the goal is to find a transformation matrix R such that $A'R = \hat{A}$ approximates A. However, since $\hat{A}$ and A do not exist within the same feature space, this leads to loss of information, rendering the approximation ineffective. Further, since Γ consists of orthogonal vectors, the inability to approximate A can also be attributed to the unbalanced orthogonal procrustes problem which is unsolved [16].

Regarding (b), we adopt the proof from [32]. The matrices that produce these symmetric matrices are not unique, since for any unitary matrix $U \in \mathbb{C}^{(f+k) \times (f+k)}$ where $f > 1$, labeling $\tilde{A} = A'U$ and $\tilde{B} = B'U$, we have

$$\tilde{A}\tilde{A}^T = AA^T, \quad \tilde{B}\tilde{B}^T = BB^T, \quad \tilde{A}\tilde{B}^T = AB^T.$$

Similar to the prior argument, these results can also be derived from data in a different feature space. As a result, the function party is limited to learning only the singular values and vectors of the matrices. This means it can derive U and S from the singular value decomposition $A = USV^T$ by eigen-decomposing AA^T. Yet, this information is inadequate to deduce A since V is unknown. Thus, the function party does not learn (i) the input data or (ii) the number of the features.

Theorem 3. *The proposed methodology is secure against a proper subset of semi-honest participants.*

Proof. Since the proposed methodology follows one-shot federated learning with no communication back from the central server, the uni-directional flow of data prohibits the corrupt participants from learning anything about the data of the non-corrupt participants.

Therefore, we've shown that the data of non-corrupt participants can not be learned by corrupt semi-honest participants, and that the central server doesn't learn the image sizes, satisfying our **Privacy** requirement.

5   Experiments

Experiments were conducted on an AMD 7713 at 2.0 GHz with one CPU core and 16 GB of RAM, simulating hospitals collaboratively performing kernel-learning with limited resources. Communication settings included a bandwidth of 1.25 MBps, an average latency of 0.1 s, and a packet loss of 2%. Unless stated otherwise, OKRA was tested with three participants. Each participant operated as an independent process using TCP connections. Data encoding and transmission to the central server were done via multiple threads. The server performed kernel functions on aggregated data for machine learning model training. Experiments were deemed successful if OKRA's accuracy matched non-private classification methods and surpassed the run-times of state-of-the-art randomized encoding algorithms, ESCAPED [32] and FLAKE [20].

5.1   Datasets

To demonstrate OKRA's applicability, we experimented with two clinical image datasets with different label types and distributions. The first dataset consists of preprocessed MRI images capturing various stages of Alzheimer's disease [22]. The second dataset features augmented images of white blood cells, representing four distinct cell types [29]. For run-time analysis, we used synthetic datasets, essential for benchmarking without real-world data irregularities. The generated data is balanced, with multi-class clusters drawn from a multivariate normal distribution (mean 0, variance 1 for each attribute). We generated 400 samples per participant with 1000 features.

5.2 Correctness of the Model

Before beginning the run-time experiments, we aimed to validate OKRA's correctness on kernel learning methods. We conducted classification using SVM and dimensionality reduction with Kernel-PCA. Each experiment was run 10 times, testing various combinations of datasets, methods, and kernels (Gaussian, Polynomial, and Linear). Seeds were used for deterministic behavior and reproducible results.

For the classification task, OKRA-SVM was tested against a Naive SVM. Both implementations shared identical hyperparameters $C \in \{2^{-4}, ..., 2^{10}\}$ (misclassification penalty) and $p \in \{1, ..., 5\}$ (degree), optimized using Grid Search. Both were trained with a 5-fold cross-validation. For this implementation, we used Sequential Minimal Optimization (libsvm) provided by scikit-learn [35]. We applied Macro Averaging when the datasets had an unbalanced distribution of classes. Correspondingly, we have evaluated the balanced datasets with Micro Averaging. Table 5.2 shows that OKRA-SVM and the naive classifier produce the same F1 and ROC AUC scores.

Table 1. Performance metrics for OKRA-PCA

Dataset	Kernel	OKRA	
		MSE	R^2
Alzheimer's Disease	Polynomial	0.0	1.0
Blood Cell	Gaussian	0.0	1.0

Dataset(Samples × Features)	Label Type	Kernel	Naive		OKRA	
			F1	ROC AUC	F1	ROC AUC
Alzheimer's Disease (6400 × 256 · 256)	Binary	Gaussian	0.91 ± 0.03	0.97 ± 0.01	0.91 ± 0.03	0.97 ± 0.01
	Multi-class	Gaussian	0.97 ± 0.04	1.00 ± 0.00	0.97 ± 0.04	1.00 ± 0.00
Blood Cell (12500 × 100 · 100 · 3)	Multi-class	Gaussian	0.88 ± 0.01	0.94 ± 0.01	0.88 ± 0.01	0.94 ± 0.01
Synthetic (200 − 51200 × 12800)	Multi-class	Linear	0.89 ± 0.03	0.95 ± 0.02	0.89 ± 0.03	0.95 ± 0.02

For the dimensionality reduction, we compared a naive Kernel-PCA and OKRA-PCA. A kernel is computed, followed by eigenvalue decomposition to sort eigenvalues and eigenvectors by magnitude. The top n eigenvectors are then chosen to transform the data into the new principal component space. To assess utility, the principal components of both Kernel-PCA and OKRA-PCA, utilizing the same kernel, were compared by computing the Mean Squared Error (MSE) and R-squared (R2).

This provides a quantitative comparison of the two PCA methods in terms of their ability to capture the data's variance and structure. Table 1 indicates that both OKRA-PCA and naive PCA are highly consistent and produce identical results, explaining the same patterns in the data. As expected, the encoding has no influence on the results. Hence we experimentally satisfy the **Correctness** requirement.

5.3   Performance Analysis

To ensure OKRA's computational overhead is manageable, especially with high-dimensional datasets like images, we compared its run-time with FLAKE and ESCAPED. We focused on two run-time types: *Encoding time*, the overhead of encoding data using OKRA for an individual participant, and *Total time*, which includes communication from when the server initiates connections to the completion of the kernel function computation. Our analysis examined the impact of varying the number of participants and image sizes. i

Scaling up the Number of Participants. Evaluating our method's adaptability across multiple participants is crucial. Figure 2 shows the cumulative time required for data encoding, transmission, and linear kernel function computation. Each participant encodes a dataset of 400 samples with 1000 features. OKRA performs exceptionally well: with 10 participants, the execution takes only 16 s on average.

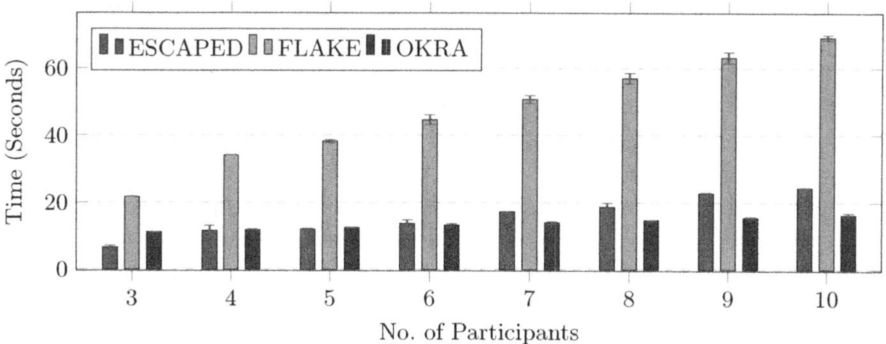

Fig. 2. Runtime comparison with increasing participants.

Scaling up the Image Size. Medical image data, essential for accurate diagnoses and treatment, is typically captured at high resolutions. However, for preprocessing, these images are often scaled down for memory and computational efficiency, with rescaled sizes ranging from 32×32 up to 256×256. To assess whether OKRA can encode preprocessed medical images without burdening the data holder, we experimented with increasing image sizes for one participant with 400 images (refer to Fig. 3). The results demonstrate that OKRA is applicable to medical imaging, whereas state-of-the-art methods are not. We, therefore, meet both requirements *Data Adaptability* and *Efficiency*.

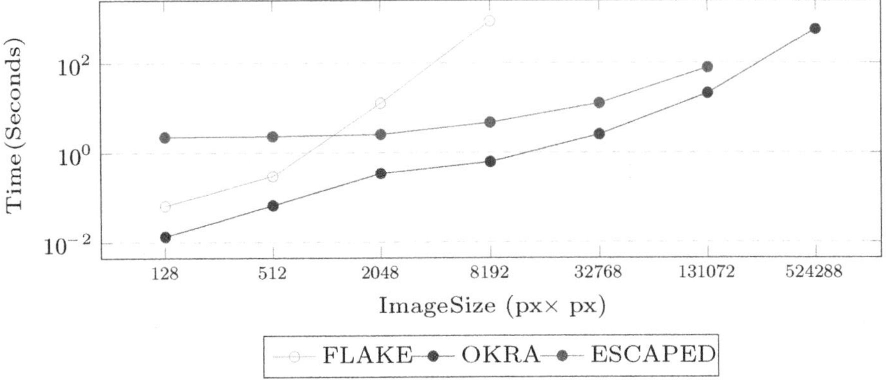

Fig. 3. Encoding times against image sizes.

5.4 Discussion

Randomized encoding methods, used for privacy-preserving kernel learning, offer the advantage of computational efficiency while maintaining model accuracy. In the field of randomized encoding methods, our proposed solution, OKRA, demonstrates significant advantages. It consistently outperforms both FLAKE and ESCAPED without compromising utility. Notably, OKRA minimizes the need for extended communication rounds and supports data adaptability, issues primarily encountered with ESCAPED. Additionally, OKRA achieves more efficient data encoding than FLAKE and can handle larger image sizes. With these strengths, OKRA emerges as a robust and efficient option within the domain of randomized encoding techniques for image data.

In subsequent work, we aim to explore the challenges associated with the colluding central server scenario. Further, we intend to address malicious participant behavior, where encoded data is altered to cause inaccuracies in the server's results. Additionally, we plan to explore the applicability of OKRA to other kernel methods, given its promising potential.

6 Conclusion

Privacy preservation for the analysis of distributed medical images remains a critical challenge. While randomized encoding offers a promising avenue, existing methodologies often grapple with the trade-off between accuracy and efficiency. Addressing this gap with regards to kernel functions applied to high-dimensional data, we introduced OKRA, a novel algorithm operating within a one-shot federated learning paradigm. OKRA not only privately computes exact kernel functions but also ensures that kernel-based machine learning algorithms are trained efficiently. Empirical evaluations further validate the superiority of OKRA, demonstrating its potential to achieve accuracy levels comparable to centralized machine learning models while minimizing computational burdens. As the landscape of machine learning continues to evolve, OKRA's robustness and efficiency position it as a pivotal tool in privacy-preserving endeavors.

Acknowledgments. This study is supported by the German Federal Ministry of Research and Education (BMBF), under project number 01ZZ2010.

References

1. Adnan, M., Kalra, S., Cresswell, J.C., Taylor, G.W., Tizhoosh, H.R.: Federated learning and differential privacy for medical image analysis. Sci. Rep. **12**(1), 1–10 (2022)
2. Al-Jaroodi, J., Mohamed, N., Abukhousa, E.: Health 4.0: on the way to realizing the healthcare of the future. IEEE Access **8**, 211189–211210 (2020)
3. Applebaum, B., Ishai, Y., Kushilevitz, E.: Computationally private randomizing polynomials and their applications. Comput. Complex. **15**(2), 115–162 (2006)
4. Bagdasaryan, E., Poursaeed, O., Shmatikov, V.: Differential privacy has disparate impact on model accuracy. In: Advances in Neural Information Processing Systems, vol. 32 (2019)
5. Bonawitz, K., et al.: Practical secure aggregation for privacy-preserving machine learning. In: proceedings of the 2017 ACM SIGSAC Conference on Computer and Communications Security, pp. 1175–1191 (2017)
6. Boser, B.E., Guyon, I.M., Vapnik, V.N.: A training algorithm for optimal margin classifiers. In: Proceedings of the Fifth Annual Workshop on Computational Learning Theory, pp. 144–152 (1992)
7. Chen, H., Ünal, A.B., Akgün, M., Pfeifer, N.: Privacy-preserving svm on outsourced genomic data via secure multi-party computation. In: Proceedings of the Sixth International Workshop on Security and Privacy Analytics, pp. 61–69 (2020)
8. Chen, K., Liu, L.: Geometric data perturbation for privacy preserving outsourced data mining. Knowl. Inf. Syst. **29**, 657–695 (2011)
9. Chen, K., Sun, G., Liu, L.: Towards attack-resilient geometric data perturbation. In: proceedings of the 2007 SIAM International Conference on Data Mining, pp. 78–89. SIAM (2007)
10. Christensen, O., et al.: An introduction to frames and Riesz bases, vol. 7. Springer (2003)
11. Cortes, C., Vapnik, V.: Support-vector networks. Mach. Learn. **20**, 273–297 (1995)
12. Costache, A., Smart, N.P.: Which ring based somewhat homomorphic encryption scheme is best? In: Sako, K. (ed.) CT-RSA 2016. LNCS, vol. 9610, pp. 325–340. Springer, Cham (2016). https://doi.org/10.1007/978-3-319-29485-8_19
13. Dou, B., et al.: Machine learning methods for small data challenges in molecular science. Chem. Rev. **123**(13), 8736–8780 (2023)

14. El Emam, K., Jonker, E., Arbuckle, L., Malin, B.: A systematic review of re-identification attacks on health data. PLoS ONE **6**(12), e28071 (2011)
15. Gentry, C.: Fully homomorphic encryption using ideal lattices. In: Proceedings of the Forty-First Annual ACM sySposium on Theory of Computing, pp. 169–178 (2009)
16. Gower, J.C., Dijksterhuis, G.B.: Procrustes problems, vol. 30. OUP Oxford (2004)
17. Griebel, L., et al.: A scoping review of cloud computing in healthcare. BMC Med. Inform. Decis. Mak. **15**(1), 1–16 (2015)
18. Guha, N., Talwalkar, A., Smith, V.: One-shot federated learning. arXiv preprint arXiv:1902.11175 (2019)
19. Hannemann, A., Friedl, B., Buchmann, E.: Differentially private multi-label learning is harder than you'd think. In: 2024 IEEE European Symposium on Security and Privacy Workshops (EuroS&PW), pp. 40–47. IEEE (2024)
20. Hannemann, A., Ünal, A.B., Swaminathan, A., Buchmann, E., Akgün, M.: A privacy-preserving framework for collaborative machine learning with kernel methods. In: 2023 5th IEEE International Conference on Trust, Privacy and Security in Intelligent Systems and Applications (TPS-ISA), pp. 82–90. IEEE (2023)
21. Hasan, B., Abdulazeez, A.M.: A review of principal component analysis algorithm for dimensionality reduction. J. Soft Comput. Data Min. **2**(1), 20–30 (2021)
22. LaMontagne, P.J., et al.: Oasis-3: longitudinal neuroimaging, clinical, and cognitive dataset for normal aging and alzheimer disease. MedRxiv, pp. 2019–12 (2019)
23. Lin, K.P., Chang, Y.W., Chen, M.S.: Secure support vector machines outsourcing with random linear transformation. Knowl. Inf. Syst. **44**, 147–176 (2015)
24. McMahan, B., Moore, E., Ramage, D., Hampson, S., y Arcas, B.A.: Communication-efficient learning of deep networks from decentralized data. In: Artificial Intelligence and Statistics, pp. 1273–1282. PMLR (2017)
25. Mothukuri, V., Parizi, R.M., Pouriyeh, S., Huang, Y., Dehghantanha, A., Srivastava, G.: A survey on security and privacy of federated learning. Futur. Gener. Comput. Syst. **115**, 619–640 (2021)
26. Mugunthan, V., Polychroniadou, A., Byrd, D., Balch, T.H.: Smpai: secure multi-party computation for federated learning. In: Proceedings of the NeurIPS 2019 Workshop on Robust AI in Financial Services (2019)
27. Schölkopf, B., Smola, A.J.: Learning with Kernels: Support Vector Machines, Regularization, Optimization, and Beyond. MIT Press (2002)
28. Seo, H., et al.: Machine learning techniques for biomedical image segmentation: an overview of technical aspects and introduction to state-of-art applications. Med. Phys. **47**(5), e148–e167 (2020)
29. Shenggan: Bccd dataset. https://github.com/Shenggan/BCCD_Dataset (2017), Accessed 27 July 2023
30. Swaminathan, A., Hannemann, A., Ünal, A.B., Pfeifer, N., Akgün, M.: Ppgwas: privacy preserving multi-site genome-wide association studies. arXiv preprint arXiv:2410.08122 (2024)
31. Ünal, A.B., Akgün, M., Pfeifer, N.: A framework with randomized encoding for a fast privacy preserving calculation of non-linear kernels for machine learning applications in precision medicine. In: Mu, Y., Deng, R.H., Huang, X. (eds.) CANS 2019. LNCS, vol. 11829, pp. 493–511. Springer, Cham (2019). https://doi.org/10.1007/978-3-030-31578-8_27
32. Ünal, A.B., Akgün, M., Pfeifer, N.: Escaped: Efficient secure and private dot product framework for kernel-based machine learning algorithms with applications in healthcare. In: Proceedings of the AAAI Conference on Artificial Intelligence, vol. 35, pp. 9988–9996 (2021)

33. Williams, C., Rasmussen, C.: Gaussian processes for regression. In: Advances in Neural Information Processing Systems, vol. 8 (1995)
34. Yao, A.C.: Protocols for secure computations. In: 23rd Annual Symposium on Foundations of Computer Science (sfcs 1982), pp. 160–164. IEEE (1982)
35. Zeng, Z.Q., Yu, H.B., Xu, H.R., Xie, Y.Q., Gao, J.: Fast training support vector machines using parallel sequential minimal optimization. In: 2008 3rd International Conference on Intelligent System and Knowledge Engineering, vol. 1, pp. 997–1001. IEEE (2008)
36. Zhang, C., Li, S., Xia, J., Wang, W., Yan, F., Liu, Y.: Batchcrypt: efficient homomorphic encryption for cross-silo federated learning. In: Proceedings of the 2020 USENIX Annual Technical Conference (USENIX ATC 2020) (2020)

Toward a Unified Graph-Based Representation of Medical Data for Precision Oncology Medicine

Davide Belluomo, Tiziana Calamoneri, Giacomo Paesani$^{(\boxtimes)}$, and Ivano Salvo

Computer Science Department, Sapienza University of Rome, Rome, Italy
{calamo,paesani,salvo}@di.uniroma1.it

Abstract. We present a new unified graph-based representation of medical data, combining genetic information and medical records of patients with medical knowledge via a unique knowledge graph. This approach allows us to infer meaningful information and explanations that would be unavailable by looking at each data set separately. The systematic use of different databases, managed throughout the built knowledge graph, gives new insights toward a better understanding of oncology medicine. Indeed, we reduce some useful medical tasks to well-known problems in theoretical computer science for which efficient algorithms exist.

Keywords: knowledge graph · precision oncology medicine · network medicine

1 Introduction

One of the recent and numerous applications of graph theory is *network medicine* [1] that aims to identify, prevent, and treat diseases: graph-based approaches have offered an effective tool to systematically explore the intrinsic complexity of diseases, leading to the identification of disease specificity, disease-associated genetic mutations and a new way to assign treatments to patients. A new approach in the framework of network medicine is the so-called *personalized* or *precision* medicine, that is, the systematic use of individual patient characteristics to determine which treatment option is most likely to result in a better average outcome for the patient.

In particular, a new trend in precision oncology aims to shape drug treatments based on the specific gene mutation profile of the particular patient (see, for example, [17]). In the last 20 years, medical practitioners tested this new approach and obtained an improvement regarding its effectiveness. However, precision medicine developments have not been in line with the expectations. A common belief among oncologists and researchers in bioinformatics is that the discrepancy between real and expected performance can be partially explained by looking at the role of gene mutations in cancer evolution: the ambitious long-term goal of our research is to contribute to filling the gap in medical knowledge

L. Cerulo et al. (Eds.): CIBB 2024, LNBI 15276, pp. 96–110, 2025.
https://doi.org/10.1007/978-3-031-89704-7_8

by examining and comparing genetic profiles of an extensive database of patients together with different treatment outcomes.

In computer science, graphs and networks are widely exploited data structures to store information as they can represent complex systems as sets of binary interactions or relations between various entities: in particular, it is possible to encode the information stored in different databases in a single graph.

Our research embraces this line of research: exploiting various databases at the same time, we represent all the available data regarding genetic information and medical records of a group of patients, together with medical knowledge in a unique *knowledge graph* and perform a guided analysis of some medical issues.

In particular, in Sect. 2, after giving some preliminary definitions, we formally describe how we construct our knowledge graph. Then, we show how some medical issues can be modeled as graph problems and solved through classical graph algorithms. Section 3 is devoted to showing the results of some preliminary experiments as a proof of concept. Finally, Sect. 4 concludes the paper.

2   Data and Methods

The main idea of this work is to collect together as much information as possible, coming from the structured data of different databases recording data from medical studies, and official documents of regulatory agencies, in order to study oncological diseases and, in particular, relationships among gene mutations, diseases, and treatment effectiveness, and try to infer vital information supporting the medical community.

In the following, we propose to use a graph (the knowledge graph H described in Subsect. 2.2) to represent all such information in a uniform way. This approach has several advantages: we can give support to study and reduce medical issues by means of well-studied graph problems that, in turn, have well-known solutions based on efficient graph algorithms. Moreover, we can exploit the flexibility of graphs as a data structure, and easily support a function for quickly updating the information stored in the graph H; this is especially useful when new medical studies are published or when a new drug is individuated, and it is useful to incorporate this information in the graph. It is worth noting that this approach is in contrast to the static graphs created after a training phase and using them as predictive models.

2.1   Preliminary Definitions

We choose to favor intuition over formalism so, in this subsection, we informally give some basic definitions concerning graphs that will be useful in the following. The reader interested in a more formal setting can refer, for example, to [7].

A *graph* $G = (V, E)$ is constituted by a finite set V of elements, known as *nodes*, and a collection of *edges* E, connecting pairs of nodes, representing some kind of binary relation. It is possible to label some nodes and/or edges to annotate them with additional information. The set of nodes that are connected through an edge to the same node v constitute the *neighborhood* of v, and a *path* of G is a sequence of edges that joins a sequence of nodes.

A graph $G' = (V', E')$ is a *subgraph* of $G = (V, E)$ if $V' \subseteq V$, $E' \subseteq E$, and every edge in E' has both endpoints in V'. The *subgraph induced by a node set* $U \subseteq V$ is the graph whose nodes are all the nodes in U and whose edges are all the edges present in G such that both endpoints are in U.

A graph $G = (V, E)$ is *k-partite* if its node set V can be partitioned into $k > 1$ subsets, and no edge connects two nodes from the same subset. A 2-partite graph is also known as *bipartite*.

2.2   The Knowledge Graph H

The graph H, which is the core of this work, is built as the union of three graphs: the graph G, storing *Genetic information* of a group of patients (whose edges are colored in Green); the graph R, storing information from patient *medical Records* (whose edges are colored in Red), and the graph M, storing some general information that we will call *Medical knowledge* (whose edges are colored in Magenta).

While graphs G, R, and M share some (set of) nodes (for example, the set of patients), their edge sets are pairwise disjoint.

The green graph is bipartite and is defined as $G = (Pa \cup Mu, E_G)$ where:

– *Pa* is a set of encrypted recorded *patients* in the database; a mapping $\rho : Pa \to \mathbb{N}$ labels every patient $p \in Pa$ with their *survival period*, that is the time interval that spans between the diagnosis of a specific disease and the time of the study, if the patient is still alive, or the time of the patient's death, otherwise. Patients are also labeled with a boolean mapping $\alpha : Pa \to \{T, F\}$ that represents whether the patient is alive or not at the time of the study: for every $p \in Pa$, $\alpha(p) = T$ if p is alive and F, otherwise.
– *Mu* is a set of *gene mutations*. Observe that a gene can have more than one mutation;
– the set of green dashed edges E_G contains an edge (p, m) if the patient p has the mutation m, and this edge is labeled with the variant allele frequency (VAF) associated with m for patient p;

The red graph is 3-partite and defined as $R = (Pa \cup Di \cup Dr, E_R)$ where:

– *Pa* is the set of patients as defined in graph G;
– *Di* is a set of *oncological diseases*, that affect patients in Pa;
– *Dr* is a set of *drugs* of interest, possibly labeled with a string β annotating adverse effects;
– the set of red solid edges E_R contains: • an edge (d, p) if the patient p is affected by the disease d; • an edge (p, f) if the patient p has been treated with the drug f; these edges are labeled with a pair of values (t, e) where $t \in \mathbb{N}$ is the number of treatments preceded it and $e \in \{p, u, r, n\}$ (standing for positive, unaltered, reduced, negative) represents the effectiveness of that

drug; clearly, more drugs could have the same value of t when a cocktail of drugs has been administrated;

The magenta graph is 3-partite and defined as $M = (Mu \cup Di \cup Dr, E_M)$ where:

- Mu, Di, and Dr are defined as in graphs G and R;
- the set of magenta dotted edges E_M contains: • an edge (d, m) if it is known that typically the disease d appears in presence of the gene mutation m; these edges can be labeled with a measure that quantifies the relevance of m with respect to d. • an edge (m, f) if it is known that the drug f has some effect on the mutation m.

The graph $H = (V, E)$ (see Fig. 1) is the union of its subgraphs G, R and M. Therefore the set of nodes is $V = Pa \cup Mu \cup Di \cup Dr$, and the set of edges is $E = E_G \cup E_R \cup E_M$, that is, the union of green, red, and magenta edges as defined above. Note that no edge of H has both endpoints in the same node subset; hence, H is 4-partite.

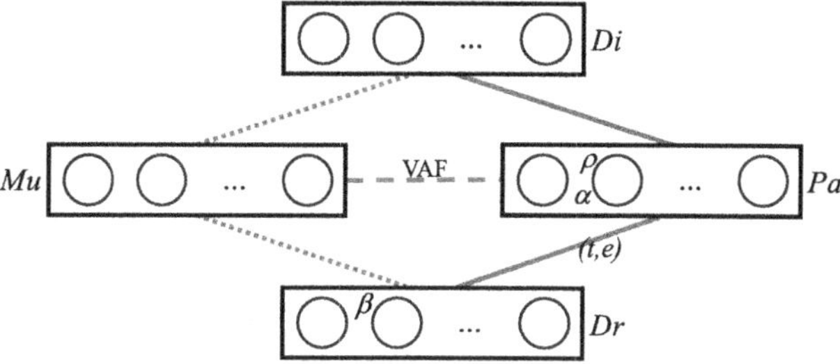

Fig. 1. A schematization of the knowledge graph H: green dashed lines represent edges of G (genetic information), red solid lines represent edges of R (information from medical records) and magenta dotted lines represent edges of M (medical knowledge). Di is the set of nodes representing the diseases, Pa the patients, Mu the genetic mutations, and Dr the drugs. (Color figure online)

2.3   Experiment Design

In our experiments, we constructed the green graph G by exploiting database $cBioPortal$ [3,4,8], from which we extract the genetic information of each patient. Gene mutations are derived from NGS (Next Generation Sequencing).

The information encoded in the red graph R can be obtained from the examination of a large number of medical records, which not only keeps track of the diseases affecting the patients, but also the treatment history, a quantitative estimation of the responses, and the survival period. In our experiments, we use cBioPortal again, to deduce this information.

At least in principle, the magenta graph M should store the vast and ever-evolving medical knowledge. In particular, in our experiments, we consider the *DisGeNET* database [22] and regulatory agencies databases (see, for example, *PharmGKB* [28,29]) for the information about the interaction between specific diseases and gene mutations, and for approved drugs targeting gene mutations, respectively.

Through knowledge graph H, we can answer many questions by exploiting graph algorithms. The underlying idea is that the involvement of multiple databases and subgraphs gives us robust and precise machinery to deduce meaningful information. Here, for each of the three listed examples, we describe the medical issue, we model it into a graph problem, propose an algorithmic solution, and highlight which part of the knowledge graph H is involved.

Comparing Medical Knowledge and Data Evidence. A natural issue is understanding whether the medical knowledge agrees with information that can be inferred from medical records. Using the information in the graph H, a possible question is the following: does data evidence from $G \cup R$ match the medical knowledge stored in M? An answer to this question can improve the understanding of the relationship between diseases and gene mutations.

On one hand, thanks to the red edges of H, fixing a disease d, we consider set $Pa(d)$ of all patients affected by d, corresponding to all the nodes in the neighborhood of d in Pa. We define $Mu_\cup(d)$ as the set of all gene mutations affecting at least one patient in $Pa(d)$ and $Mu_\cap(d)$ as the set of all mutations affecting every patient in $Pa(d)$ (through green edges).

On the other hand, we determine through magenta edges the set $Mu_{|d}$ of gene mutations known to be involved in disease d, which results from medical knowledge. In our experiments, we label magenta edges between a mutation and a disease with the so-called GDA score. This score, provided by the DisGeNET database, ranges between 0 and 1 and takes into account the number and type of sources (level of curation, model organisms), and the number of publications supporting the association between m and d. In this setting, the set $Mu_{|d}$ can be computed by considering the neighbors of d in the magenta graph having GDA score close to 1.

Provided that the sample of patients is sufficiently broad, we have that $Mu_{|d}$ should be contained in $Mu_\cup(d)$: every gene mutation known to be involved in d necessarily occurs in some patient with disease d. Next, we compare the two sets of mutations $Mu_{|d}$ and $Mu_\cap(d)$ and distinguish the following cases: if $Mu_{|d} = Mu_\cap(d)$ then the medical knowledge perfectly matches with the experimental evidence for disease d; otherwise, either the medical knowledge is incomplete for disease d because there are gene mutations that are present in every patient with disease d but are not anticipated by the current medical knowledge, or the evidence is inconsistent for disease d because some patients with disease d do not have a predicted gene mutation, or a combination of them. In such cases, deeper examinations are suggested.

For each $d \in Di$ computing $Mu_{|d}$, $Mu_{\cup}(d)$ and $Mu_{\cap}(d)$ can be efficiently done through standard graph search algorithms, such as breadth-first search.

Partitioning Patients Into Homogenous Groups. Medical evidence shows (*e.g.*, see [23]) that the percentage of patients that positively react to treatments is less than expected, although drugs are chosen based on the patient specific gene mutation profile. The general feeling among experts is that concentrating on driver gene mutations is not enough.

To face this general problem, here we propose some possible approaches, all based on the idea of recognizing groups of patients that, for some reason, can be considered as similar. Then, we can propose to medical doctors a deep analysis of the gene mutations of patients in the same groups, so that they can look for the presence of specific gene mutations that the drug treatment has not targeted: some of these gene mutations could inhibit the cure and be considered responsible for the treatment failure.

As a first approach we consider a *similarity function* between patient genetic profiles (see, for example, [16]); the idea is that if a patient has been successfully treated with a drug, patient with a similar genetic profile could be successfully administered with the same drug. More in detail, given a threshold k, exploiting green edges of H, we can determine the sets of patients such that their genetic profiles are within k from each other.

This approach is related to the so-called *agnostic paradigm*, in which patients with very similar genetic profiles are administered with the same drug, regardless of the tumor each patient has been diagnosed. Nevertheless, the agnostic paradigm, although biologically fascinating, did not produce significant effects apart from very few cases (*e.g.*, NTRK [18]).

As an example, note that in the previous section we implicitly considered groups of patients affected by the same disease. As a second aproach, we also propose to further partition patients with the same disease d according to their *survival period* (available from the magenta graph), clearly strongly related to the effectiveness of the administered drugs. The goal is to check if patients in the same group exhibit deeper similarities in the genetic profile with respect to the whole set of patients and provide evidence of some treatments' (in)effectiveness.

Optimized Drug Treatments. Here, we propose an algorithm joining the information obtained separately on the one hand from $G \cup M$ and on the other hand from R to suggest drug treatments optimizing the benefits and minimizing adverse effects for a specific patient.

For any gene mutation m, we can exploit magenta edges to deduce the set $Dr_{|m}$ of the drugs that have an effect on m. Hypothetically, administering to a patient p all the drugs in $\bigcup_{m_p} Dr_{|m_p}$ (where m_p is any gene mutation of p, selected through the green edges) would guarantee the best treatment for p.

Nevertheless, given the possible adverse effects of these drugs (possibly depending on their interactions), only few of them can be administered simultaneously to a patient, even at the cost of ignoring some gene mutations. Indeed, in practice, only very few mutations of a patient are being treated: current drugs are designed to deal with very specific gene mutations, known as *target*. Hence, given a patient p and a (small) subset Z of their gene mutations, the aim is to compute a drug subset W of minimum size that targets all gene mutations in Z.

This problem is related to the well-studied *minimum hitting set* problem, defined as follows. Let U be a finite set and $\mathcal{U} = \{U_1, U_2, \ldots\}$ a collection of subsets of U. A *hitting set* for $\mathcal{U}$ is a subset U' of U such that $U_i \cap U' \neq \emptyset$, for every i. The minimum hitting set problem consists of determining a hitting set of minimum size. Computing the minimum hitting set is known to be computationally hard [24].

Our problem can be modeled in terms of the minimum hitting set problem as follows: U coincides with the set of the drugs Dr and each U_i is a $Dr_{|m}$, for some $m \in Z$. Thus, solving the minimum hitting set problem on this instance gives a minimum-size drug treatment that targets all gene mutations in Z.

In the literature, some papers propose similar strategies. In particular, in [26], the authors solve the hitting set problem with a heuristic approach restricting to drug combinations of size at most three. Note that there are some papers that aim to solve the adequate drug treatment problem. The work of Johnson [15] highlights a polynomial-time heuristic approximation algorithm. Finally, in [19], it has been developed a statistical mechanics approach to attack this problem. We point out that these previous approaches are either non-deterministic or do not obtain exact solutions.

In contrast to the previous work, we propose a deterministic and exact algorithm to solve the adequate drug treatment problem, which can be generalized by taking into account the adverse effects and negative interactions of drugs, thus improving the precision and safety of drug treatments. Namely, we propose to modify graph H by adding to the drug nodes a weight related to their toxicity, and new edges between interacting drugs, weighted by the magnitude of the interaction. Now, the main idea consists of solving the problem by computing a hitting set that minimizes the sum of the weights of the nodes involved in the solution (corresponding to the global toxicity of the chosen drugs) and the weights of the new edges induced by the solution (corresponding to the interactions between such drugs). Since the set of target gene mutations is small in practice, the proposed algorithm is computationally tractable in practice.

3   Results

In this section, we show the results of some experiments. Due to the lack of some crucial information in the public databases and the difficulty of getting some part of it, we only partially address the objectives described in Subsect. 2.2; nevertheless, we try to keep the flavor of the underlying idea.

We focus on three different medical studies: Metastatic Non-Small Cell Lung Cancer [14] with 930 patients, MSK MetTropism [21] with 24755 patients, and MSK-IMPACT Clinical Sequencing Cohort [30] with 7091 patients. We chose these studies because they consider a sequencing technology guaranteeing a 500-gene panel for each patient. Nevertheless, for the sake of brevity, we report only the results concerning the first study, but the interested reader can find all the other results in the ArXiv version [2].

Analysis of Data in Public Databases. We observed that the databases we use as reference, CBioPortal and DisGeNET, are not coherent; indeed, while the former contains specific genetic mutations, the latter deals only with mutated genes. It follows that, in order to compare the extracted results, we have to downgrade the genetic mutations to simple mutated genes. In order to have an idea about how much information we are losing in this way, we compare the data extracted from CBioPortal, counting them in different ways. In Table 1, we show the following information for the study MSK MetTropism:

- the percentage of the 10 most frequent mutations with respect to the total number of mutations;
- the percentage of the 10 most frequent mutated genes, where all the mutations on the same gene are counted;
- the percentage of the 10 most frequent mutated genes, where multiple mutations on the same gene are counted as one.

Table 1. Results of an experiment performed on the 24755 patients of MSK Met-Tropism study: we show the 10 most frequent mutations (first columns), mutated genes with multiplicity (second columns), and mutated genes without multiplicity (third columns) together with the corresponding percentages.

gene mutations	%	genes (with mult.)	%	genes (w/o mult.)	%
KRAS_12_25398284_25398284	12.7	TP53	43.6	TP53	50.3
TERT_5_1295228_1295228	6.7	KRAS	21.7	KRAS	22.0
KRAS_12_25398285_25398285	4.8	PIK3CA	14.2	APC	14.7
PIK3CA_3_178936091_178936091	3.3	APC	9.8	PIK3CA	14.5
BRAF_7_140453136_140453136	3.2	TERT	9.6	TERT	11.5
PIK3CA_3_178952085_178952085	3.2	EGFR	4.8	ARID1A	10.2
TP53_17_7578406_7578406	2.7	BRAF	4.4	PTEN	7.2
PIK3CA_3_178936082_178936082	2.0	PTEN	3.4	KMT2D	7.1
TP53_17_7577120_7577120	1.7	ARID1A	3.1	EGFR	6.6
TP53_17_7577538_7577538	1.7	CDKN2A	2.8	BRAF	6.2

From this data, it is evident that not only it is completely different to consider percentages of mutated genes instead of specific mutations, but it is also different to take into account multiplicity instead of ignoring it. It follows that the setup used to extract each data from the knowledge graph H must be accurately detailed to medical doctors.

Comparing Medical Knowledge and Data Evidence: Lung Adenocarcinoma. We now consider only the patients characterized by the same disease $d \in Di$. Analogously to Table 1, Table 2 shows data w.r.t. the 3972 patients affected by one of the most frequent diseases included in the MSK MetTropism study, namely Lung Adenocarcinoma.

Table 2. Results of an experiment performed on the MSK MetTropism study on the 3972 patients affected by lung adenocarcinoma: we show the 10 most frequent mutations (first columns), mutated genes with multiplicity (second columns), and mutated genes without multiplicity (third columns) together with the corresponding percentages.

gene mutations	%	genes	%	genes	%
KRAS_12_25398285_25398285	14.7	TP53	49.2	TP53	48.5
KRAS_12_25398284_25398284	13.6	EGFR	34.9	KRAS	33.9
EGFR_7_55259515_55259515	9.1	KRAS	34	EGFR	29.4
EGFR_7_55242465_55242479	5.0	STK11	13.9	STK11	16.1
EGFR_7_55242466_55242480	2.9	KEAP1	11.1	KEAP1	14.1
EGFR_7_55249071_55249071	2.9	RBM10	7.7	RBM10	11.7
U2AF1_21_44524456_44524456	2.0	PIK3CA	5.5	PTPRD	8.8
PIK3CA_3_178936091_178936091	1.8	BRAF	4.7	SMARCA4	8.2
ERBB2_17_37880981_37880982	1.6	CDKN2A	4.2	ATM	7.8
KRAS_12_25380275_25380275	1.5	SMARCA4	3.9	NF1	7.5

Comparing Tables 1 and 2, one can observe sensible differences: for example, the gene mutation KRAS_12_25398285_25398285 appears in only 4.8% of all patients while in 14.7% of those affected by Lung Adenocarcinoma. Moreover, the gene mutation TERT_5_1295228_1295228 appears in 6.7% of the patients in Table 1 while is negligible in Table 2. An even more notable discrepancy can be observed in the gene EGFR: only 6.6% of the total population of patients has this gene mutated, while the percentage increases to 29.4 for the patients affected by Lung Adenocarcinoma. These considerations are not meant to infer any conclusion at the medical level but, especially if joined with similar studies, aim to suggest a direction for further research.

On the one hand, it is clear that no single gene mutation appears in all patients affected by lung adenocarcinoma. Therefore, $Mu_\cap(d)$ is trivially empty. On the other hand six of the genes appearing in Table 3, namely BRAF, KRAS, EGFR, STK11, ATM, and TP53, are also represented in Table 2 showing a level of agreement between medical knowledge on lung adenocarcinoma disease and the evidence collected on the patients. Anyway, some genes appearing in Table 3 with GDA score 1, namely ALK and ROS1, do not appear in Table 2 and hence in $Mu_\cup(d)$, and some genes appearing in Table 2 with frequency above 10% without multiplicity, namely KEAP1 and RBM10, do not appear in Table 3,

Table 3. Mutated genes with GDA score at least 0.8 in lung adenocarcinoma.

mutated gene	GDA Score	mutated gene	GDA Score	mutated gene	GDA Score
BRAF	1.0	FGFR2	0.85	MAP2K1	0.8
ALK	1.0	AKT1	0.85	CTNNB1	0.8
ROS1	1.0	MUC5AC	0.85	CDKN2A	0.8
KRAS	1.0	TYMS	0.85	RAF1	0.8
EGFR	1.0	CCND1	0.85	CHRNA3	0.8
ERBB2	0.95	STK11	0.8	FGFR3	0.8
PIK3CA	0.95	TERT	0.8	ATM	0.8
TP53	0.95	FGFR4	0.8	EGF	0.8
MYC	0.9	HRAS	0.8		

showing some inconsistencies between medical knowledge and data that we have considered.

Partitioning Patients Into Homogenous Groups: Survival Period. Estimating the effectiveness of drug treatment is a difficult task because it takes into account different parameters. One of these parameters is the survival period.

We partition the patient population into three sets: the first one $Pa_{\geq 36} = \{p \in Pa \mid \rho(p) \geq 36\}$ contains all the patients whose survival period is of at least 36 months, the second one $Pa_{\leq 6} = \{p \in Pa \mid \rho(p) \leq 6 \wedge \alpha(p) = F\}$ contains all the patients whose survival period is of at most 6 months and are marked as deceased, and the third set contains all the remaining patients.

We selected the 5295 patients of the MSK MetTropism study in $Pa_{\geq 36}$ and the 2768 patients in $Pa_{\leq 6}$ and summarized the results in Tables 4 and 5, respectively. Comparing Table 1 with Tables 4 and 5, one can observe that the distribution of the percentages of their mutations completely changes. As an example, TP53, KRAS, and TERT dramatically increase their percentages, whereas PIK3CA, EGFR, and STR11 decrease significantly their percentages. This behavior can be explained by the medical awareness that certain combinations of mutations indicate either a different response to treatments or a different evolution of the disease. A deep study of these results should be performed by medical doctors, who could individuate interesting combinations of gene mutations, both in the population of patients with a low survival period and in that one with a high survival period.

Partitioning Patients Into Homogenous Groups: Genetic Mutation Profile. Moreover, to understand whether there are some groups of patients that can be considered similar, we implemented two different similarity measures based on the genetic profiles. The *Hamming distance* [10] between two patients counts the number of gene mutations that affect only one of them. The *Jaccard distance* [13] is a variation of the Hamming distance where the value is

Table 4. Results of the experiment on the MSK MetTropism study on the 5295 patients with a survival period of at least 36 months: we show the 10 most frequent mutations (first columns), mutated genes with multiplicity (second columns), and mutated genes without multiplicity (third columns) together with the corresponding percentages.

gene mutations	%	genes	%	genes	%
KRAS_12_25398284_25398284	8.6	TP53	35.6	TP53	38.8
TERT_5_1295228_1295228	6.6	PIK3CA	16.9	PIK3CA	17
PIK3CA_3_178952085_178952085	4.7	KRAS	15.4	KRAS	15.7
KRAS_12_25398285_25398285	3.6	TERT	9.9	APC	13.7
PIK3CA_3_178936091_178936091	3.5	APC	9.7	TERT	11.5
BRAF_7_140453136_140453136	3.0	EGFR	6.6	ARID1A	9.3
PIK3CA_3_178936082_178936082	2.6	BRAF	4.8	EGFR	7.8
TP53_17_7578406_7578406	2.1	PTEN	4.7	PTEN	7.2
EGFR_7_55259515_55259515	1.8	ARID1A	3.7	FAT1	6.2
TP53_17_7577538_7577538	1.4	CTNNB1	3.0	PTPRT	6.1

normalized by the total number of gene mutations affecting the two considered patients, taking into account the inequalities due to the possibly imbalanced number of observed gene mutations or different gene sequencing (*e.g.*, different number of checked genes). Clearly, two patients having a similar genetic profile are also very close with respect to the considered measures.

The overall idea is that patients who are grouped together, whether they have either Hamming or Jaccard distance small, might experience the same disease, similar disease evolution, and comparable responses to drug treatments.

Regardless of the similarity measure used, our experiments show that most of the patients are isolated, that is, the groups of similar patients are very often singletons. As an example, we report the results obtained from the MSK Met-Tropism study when considering patients at Hamming distance at most 10 from each other. Out of 24755 patients, there are only 6 groups that are not singletons which include a total of 14 patients. This means, at least considering the data at our disposal, that it is very unlikely that any two patients are similar from the genetic profile point of view.

As expected, these results confirm that genetic similarities are not enough to explain different behaviors of the human body with respect to oncology medicine. Since the techniques we use to aggregate patients are not the most sophisticated nor the most appropriate for this specific task, in the following we propose more advanced clustering techniques.

Paritioning Patient in Homogenous Groups: Coexisting Mutations. We wonder whether there exist some combinations of gene mutations that appear simultaneously in significant portions of the patient population: we compute all the (maximal) k-coexisting-mutation sets, *i.e.*, sets of mutations simultaneously present in at least $k\%$ of patients, and return these sets of patients.

Table 5. Results of the experiment on the MSK MetTropism study on the 2768 patients that have a survival period of at most 6 months: we show the 10 most frequent mutations (first column), mutated genes with multiplicity (second column), and mutated genes without multiplicity (third column) together with the corresponding percentages.

gene mutations	%	genes (with mult.)	%	genes (w/o mult.)	%
KRAS_12_25398284_25398284	15.3	TP53	57.8	TP53	62.1
TERT_5_1295228_1295228	8.2	KRAS	27.9	KRAS	27.9
KRAS_12_25398285_25398285	7.1	TERT	12.1	TERT	13.3
BRAF_7_140453136_140453136	3.3	PIK3CA	11.4	PIK3CA	12.2
PIK3CA_3_178952085_178952085	2.8	APC	6.4	ARID1A	10.2
TP53_17_7578406_7578406	2.7	CDKN2A	6.0	APC	10.1
PIK3CA_3_178936091_178936091	2.6	BRAF	5.2	CDKN2A	9.5
PIK3CA_3_178936082_178936082	2.3	STK11	4.0	KEAP1	7.2
TP53_17_7577538_7577538	2.2	EGFR	3.7	STK11	6.9
TP53_17_7577094_7577094	2.0	SMAD4	3.3	RB1	6.8

From the evaluation of the data, we observe that k-coexisting-mutation sets are made of a single mutation, even for very small values of k. Considering the MSK MetTropism study again, there is only one k-coexisting-mutation set, with $k = 12$, which consists of the single mutation KRAS_12_25398284_25398284. It seems unrealistic to assume that every patient affected by KRAS_12_25398284_ 25398284 can be considered similar with respect to the affected disease, disease evolution, and response to drug treatments.

Some combinations of gene mutations are particularly relevant for medical doctors: contemporary mutation in genes EGFR and KRAS is one of them. So, we extracted all the patients with both these two genes mutated, whose 61.3% of them were alive at the time of the study. It is natural to wonder whether there is an explanation for the alive patients to survive. For each analyzed gene, we computed the fraction (expressed as a percentage) of the patients having that gene mutated in three different patient populations: all patients, the living, and the deceased ones. It turns out that the patients with certain further mutations (such as ARID1A, PIK3CA, AT1, or PTEN) are much more likely to survive, as shown in Table 6, indeed the percentage of living patients is more than 80% in the presence of these gene mutations. This kind of table could be of interest to better understand whether there are special combinations of gene mutations that significantly increase the survival probability. The results of this experiment for other combinations of genes, such as EGFR and T790R, do not provide the same interesting output: for example, in all three considered studies, no patient had these two genes mutated at the same time.

Table 6. Results of an experiment performed on the MSK MetTropism study on the 93 patients that have both EGFR and KRAS genes mutated: we show the 10 most frequent mutations (first column), the percentage of patients with that gene mutated (second column), and the percentages of living and deceased patients among those with that gene mutated (third and fourth columns).

gene mutations	% patients	% living patients	% deceased patients
KRAS	100	61.3	38.7
EGFR	100	61.3	38.7
TP53	52.7	51.0	49.0
APC	45.2	76.2	23.9
ARID1A	38.7	83.3	16.7
KMT2D	35.5	93.9	6.1
PIK3CA	35.5	90.9	9.1
FAT1	35.5	81.8	18.2
ATM	33.3	77.4	22.6
PTEN	31.2	89.7	10.3

4   Conclusions

In this paper, we designed a unified graph-based representation of medical data for precision oncology medicine and proposed three possible applications whose solutions exploit known results from theoretical computer science.

Our approach's novelty lies in how we store and deduce information. In particular:

- we develop a knowledge graph that exploits various databases to deduce fundamental information using graph-theoretic tools;
- we implement a deterministic framework to infer personalized medical information in contrast to past research strategies that have been using data aggregation [6,20,25], pattern recognition [5,11] and statistical performance [12,27];
- our knowledge graph model allows one for quick and efficient updates, whether there is a new node or some information has changed, in contrast to static models based on machine learning techniques (see for example [9]).

Acknowledgement. The authors would like to thank medical doctors Gennaro Daniele and Pasquale Lombardi for the exciting discussions on cancer handling from a medical point of view, Kilian Schulz for contributing to extracting data from databases during his honors program, Luciano Giacò and Federica Persiani for their helpful feedback on the databases to be used. This work was supported by *Sapienza* University of Rome, project title: *Graph models for precision oncology medicine*, grant number: RM122181612C08BB.

References

1. Barabási, A.L., Gulbahce, N., Loscalzo, J.: Network medicine: a network-based approach to human disease. Nat. Rev. Genet. **12**(1), 56–68 (2011). https://doi.org/10.1038/nrg2918
2. Belluomo, D., Calamoneri, T., Paesani, G., Salvo, I.: Toward a unified graph-based representation of medical data for precision oncology medicine. arXiv pp. 1–19 (2024). https://arxiv.org/abs/2410.14739
3. de Bruijn, I., et al.: Analysis and visualization of longitudinal genomic and clinical data from the aacr project genie biopharma collaborative in cbioportal. Can. Res. **83**(23), 3861–3867 (2023). https://doi.org/10.1158/0008-5472.CAN-23-0816
4. Cerami, E., Gao, J., Dogrusoz, U., Gross, B.E., Sumer, S.O., Aksoy, B.A., Jacobsen, A., Byrne, C.J., Heuer, M.L., Larsson, E., et al.: The cbio cancer genomics portal: an open platform for exploring multidimensional cancer genomics data. Cancer Discov. **2**(5), 401–404 (2012). https://doi.org/10.1158/2159-8290.CD-12-0095
5. Cheng, T., Zhan, X.: Pattern recognition for predictive, preventive, and personalized medicine in cancer. EPMA J. **8**(1), 51–60 (2017). https://doi.org/10.1007/s13167-017-0083-9
6. Cirillo, D., Valencia, A.: Big data analytics for personalized medicine. Curr. Opin. Biotechnol. **58**, 161–167 (2019). https://doi.org/10.1016/j.copbio.2019.03.004
7. Diestel, R.: Graph Theory, 4th Edition, Graduate texts in mathematics, vol. 173. Springer (2012)
8. Gao, J., et al.: Integrative analysis of complex cancer genomics and clinical profiles using the cbioportal. Sci. Signaling **6**(269), pl1–pl1 (2013). https://doi.org/10.1126/scisignal.2004088
9. Gong, F., Wang, M., Wang, H., Wang, S., Liu, M.: Smr: medical knowledge graph embedding for safe medicine recommendation. Big Data Res. **23**, 100–174 (2021). https://doi.org/10.1016/j.bdr.2020.100174
10. Hamming, R.W.: Error detecting and error correcting codes. Bell Syst. Tech. J. **29**(2), 147–160 (1950). https://doi.org/10.1002/j.1538-7305.1950.tb00463.x
11. Huang, Y., Zhao, Y., Capstick, A., Palermo, F., Haddadi, H., Barnaghi, P.: Analyzing entropy features in time-series data for pattern recognition in neurological conditions. Artif. Intell. Med. **150**, 102821 (2024). https://doi.org/10.1016/j.artmed.2024.102821
12. Indrayan, A.: Personalized statistical medicine. IJMR **1157**(1), 104–108 (2023). https://doi.org/10.4103/ijmr.ijmr_1510_22
13. Jaccard, P.: The distribution of the flora in the alpine zone. 1. New phytologist **11**(2), 37–50 (1912). https://doi.org/10.1111/j.1469-8137.1912.tb05611.x
14. Jee, J., et al.: Overall survival with circulating tumor dna-guided therapy in advanced non-small-cell lung cancer. Nat. Med. **28**(11), 2353–2363 (2022). https://doi.org/10.1038/s41591-022-02047-z
15. Johnson, D.S.: Approximation algorithms for combinatorial problems. Proc. STOCS **1973**, 38–49 (1973). https://doi.org/10.1145/800125.804034
16. Kosman, E., Jokela, J.: Dissimilarity of individual microsatellite profiles under different mutation models: empirical approach. Ecol. Evol. **9**(7), 4038–4054 (2019). https://doi.org/10.1002/ece3.5032
17. Langreth, R., Waldholz, M.: New era of personalized medicine: targeting drugs for each unique genetic profile. Cncologist **4**(5), 426–427 (1999). https://doi.org/10.1634/theoncologist.4-5-426

18. Marchetti, A., Ferro, B., Pasciuto, M.P., Zampacorta, C., Buttitta, F., D'Angelo, E.: NTRK gene fusions in solid tumors: agnostic relevance, prevalence and diagnostic strategies. Pathologica **114**, 199–216 (2022). https://doi.org/10.32074/1591-951X-787

19. Mézard, M., Tarzia, M.: Statistical mechanics of the hitting set problem. Phys. Rev. E **76**(4), 041124 (2007). https://doi.org/10.1103/PhysRevE.76.041124

20. Moscatelli, M., et al.: An infrastructure for precision medicine through analysis of big data. BMC Bioinf. **19**(Suppl 10) (2018). https://doi.org/10.1186/s12859-018-2300-5

21. Nguyen, B., et al.: Genomic characterization of metastatic patterns from prospective clinical sequencing of 25,000 patients. Cell **185**(3), 563–575 (2022). https://doi.org/10.1016/j.cell.2022.01.003

22. Piñero, J., et al.: The disgenet knowledge platform for disease genomics: 2019 update. Nucleic Acids Res. **48**(D1), D845–D855 (2020). https://doi.org/10.1093/nar/gkz1021

23. Plana, D., Palmer, A.C., Sorger, P.K.: Independent drug action in combination therapy: implications for precision oncology. Cancer Discov. **12**(3), 606–624 (2022). https://doi.org/10.1158/2159-8290.CD-21-0212

24. Shi, L., Cai, X.: An exact fast algorithm for minimum hitting set. Proc. IJCCSO 2010 **1**, 64–67 (2010). https://doi.org/10.1109/CSO.2010.240

25. Ullah, A., Azeem, M., Ashraf, H., Alaboudi, A.A., Humayun, M., Jhanjhi, N.: Secure healthcare data aggregation and transmission in IoT - a survey. IEEE Access (2021). https://doi.org/10.1109/ACCESS.2021.3052850

26. Vazquez, A.: Optimal drug combinations and minimal hitting sets. BMC Syst. Biol. **81**(3), 1–6 (2009). https://doi.org/10.1186/1752-0509-3-81

27. Venkatraman, D.L., Pulimamidi, D., Shukla, H.G., Hegde, S.R.: Tumor relevant protein functional interactions identified using bipartite graph analyses. Sci. Rep. **11**(1), 21530 (2021). https://doi.org/10.1038/s41598-021-00879-2

28. Whirl-Carrillo, M., et al.: An evidence-based framework for evaluating pharmacogenomics knowledge for personalized medicine. Clin. Pharmacol. Therapeutics **110**(3), 563–572 (2021). https://doi.org/10.1002/cpt.2350

29. Whirl-Carrillo, M., et al.: Pharmacogenomics knowledge for personalized medicine. Clin. Pharmacol. Ther. **92**(4), 414–7 (2012). https://doi.org/10.1038/clpt.2012.96

30. Zehir, A., et al.: Mutational landscape of metastatic cancer revealed from prospective clinical sequencing of 10,000 patients. Nat. Med. **23**(6), 703–713 (2017). https://doi.org/10.1038/nm.4333

FP-Elegans M1: Feature Pyramid Reservoir Connectome Transformers and Multi-backbone Feature Extractors for MEDMNIST2D-V2

Francesco Bardozzo[1]([✉]), Pierpaolo Fiore[1], Pietro Liò[2], and Roberto Tagliaferri[1]

[1] Neuronelab - DISAMIS, Università degli Studi di Salerno, Fisciano, Italy
fbardozzo@unisa.it
[2] Computer Laboratory, University of Cambridge, Cambridge, UK

Abstract. In medical imaging, several deep learning models, such as vision transformers (ViT), have shown improved capability in recognizing image patterns efficiently by enhancing model efficiency in parameter optimization and sample effectiveness. Our research introduces a fusion approach that leverages multi-backbone pre-trained models (ResNet, EfficientNet, VGG) as feature extractors and the *Caenorhabditis elegans'* pyramid connectome ViT in the tail. In most cases, the proposed model demonstrates superior performance on MEDMNIST2D V2.0 classification challenges. Moreover, our approach maintains comparable performance while utilizing fewer training parameters than conventional state-of-the-art models. This indicates a significant step toward more efficient deep-learning architectures in medical diagnostics.

Keywords: pyramid vision transformers · connectome transformers · medmnist2d · reservoir

1 Scientific Background

Recent advancements in medical diagnostics have demonstrated the effectiveness of applying Deep Learning (DL) techniques directly to raw or preprocessed images. Various models and strategies have been employed to address a wide range of tasks, including classification [5,13], reconstruction [1,13], and image enhancement [2]. Notably, transformers have recently achieved impressive results across these tasks. Such for example, one notable contribution relies on Swin UNet Transformers (Swin UNETR) for self-supervised pre-training in 3D medical image analysis [23]. Moreover, a comparative study on vision transformer encoders and few-shot learning for medical image classification highlights their effectiveness when combined with ProtoNets [18]. On the other hand, efficient convolutional neural network (CNN) showcase superior performance when dependent on transformers and multilayer perceptrons (MLPs) [27].

© The Author(s), under exclusive license to Springer Nature Switzerland AG 2025
L. Cerulo et al. (Eds.): CIBB 2024, LNBI 15276, pp. 111–120, 2025.
https://doi.org/10.1007/978-3-031-89704-7_9

However, within the scope of the MEDMNIST2D V2.0 competition [25], architectures employing various feature extractor backbones [9] have demonstrated their ability to outperform other CNN/Auto-configured models. The work of [12] on MEDMNIST2D V1.0 introduced the Feature Pyramid Vision Transformer (FPViT), merging the strengths of both multi-backbone residual networks and pyramid ViT for enhanced feature extraction maintaining MLP as classificators. Our recent work introduces Elegans-AI M1 and M2 models, which incorporate the structural principles of *C. elegans* connectomics with sophisticated transformer architectures, including both deep and reservoir transformers designed as artificial connectomes [3]. These models leverage transformers as Dense networks rewired according to connectome synapses, multi-head attention mechanisms, and advanced gating strategies, mirroring bio-plausible topological and functional complexities. Extending our previous works, in this paper, we introduce an advancement over the original Elegans-AI M1 in a reservoir configuration by integrating our connectome transformer with a pyramidal, multi-backbone feature extraction system. The reservoir configuration of the transformer connectome plays a unique role in the training by contributing only topologically to the learning process [3,17]. The proposed model has been tested on MEDMNIST2D v2.0 dataset [25]. Our results show that this model outperforms existing state-of-the-art models, including Pyramid and ResNet ViT, across benchmarks such as PathMNIST, OCTMNIST, etc.

2   Materials and Methods

Our models are trained, validated, and tested on the benchmark dataset MedMNIST2D V2.0 [25], ensuring they meet high standards of accuracy and reliability. This robust dataset allows us to rigorously evaluate our models, providing confidence in their performance and applicability in real-world medical imaging tasks. Consider an image $x \in \mathbb{R}^{H \times W \times Ch}$, where H and W are the spatial dimensions, and Ch is the number of channels which can be $Ch = 1$ or $Ch = 3$ depending on grey-scale or RGB configuration. The benchmark dataset is provided at resolution 28×28, 64×64 and 224×224, in our setup, we have $H = W = 64$ (for all the benchmarks excluded TissueMNIST which is imported at resolution 224×224). Adapting the methodology described by [12], we implement pyramidal pre-trained models to scale original inputs to $32 \times 32 \times Ch$, $64 \times 64 \times Ch$, and $128 \times 128 \times Ch$ using bilinear interpolation. Unlike the method proposed in [12], which employed the same feature extractor (ResNet18) across all scaled image sets, we utilized distinct backbones for each scale. In detail, for PathMNIST and OCTMNIST our model processes inputs at $32 \times 32 \times Ch$ through a trainable *ResNet50V2* model [10], inputs at $64 \times 64 \times Ch$ through a trainable *ResNet101V2*, and inputs at $128 \times 128 \times Ch$ through a *ResNet152V2* [10]. While for the other benchmarks, our model processes inputs at $32 \times 32 \times Ch$ through a trainable *Efficient-Net B0* [22] model, inputs at $64 \times 64 \times Ch$ through a trainable *VGG19* [20], and inputs at $128 \times 128 \times Ch$ through a *ResNet50* [10]. The total number of trainable parameters in the feature extraction part of the model amounts to ≈ 50

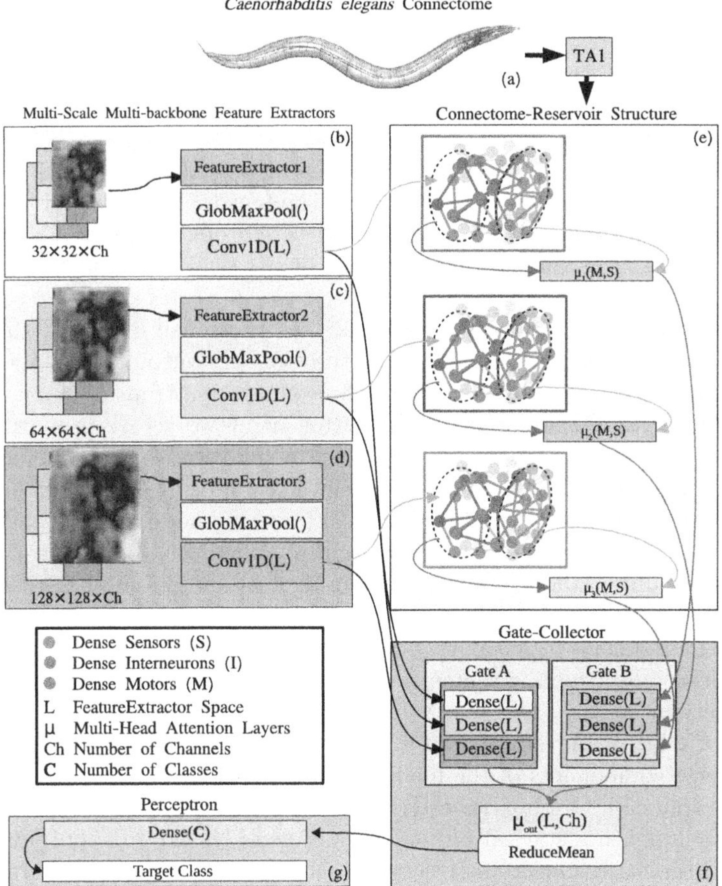

Fig. 1. Feature Pyramid Elegans M1 (FP-Elegans M1): Fig. 1 illustrates the comprehensive model architecture, starting from the original *C.elegans* structure in (**a**) to its artificial transformation in box (**e**) with Transformation Algorithm 1 (**TA1**) [3]. In Boxes (b-c-d) the Pyramid/Multi-scale configuration is shown. Input images, varying in the number of channels – with $Ch = 1$ OR 3, are fed at different resolutions into three different pretrained models. Outputs from these blocks are processed trough a 1D convolutional layer aligning the number of neurons. Then, these outputs follow two branches. In the first branch the outputs are directed to **Gate A**, where they are stacked. In the second branch, they are driven into three replicas of the Connectome-Reservoir structure in box (**e**). Within block (**e**), each reservoir applies trainable multi-head attention (μ_1, μ_2, μ_3) between the patches of input Dense connectome sensors and of output Dense motor neurons. The outputs of μ_1, μ_2 and μ_3 are stacked in **Gate B** of box (**f**). A cross multi-head attention μ_{out} is designed to work between the features extracted by the pretrained models (**Gate A**) and those extracted after the reservoir step (**Gate B**). Lastly, a single layer perceptron is employed for classification.

million. As illustrated in Fig. 1 - Box (**b, c, d**), to standardize output shapes from feature extractors across these scales, the output tensor are condensed into a $L = 16$ neuron layer using Global-Max pooling and Conv1D transformation. In Fig. 1, box (**a**) depicts a tensor network where each tensor represents a neuron in the connectome and each connection represents a synapse. This network is organized according to a sensor-interneuron-motor chemical-electrical synapse schema, as outlined in the legend of Fig. 1. The tensorial information flows from sensors to motors within this schema. The three replicas of the Connectome-Reservoir transformer, illustrated in Fig. 1 - box (**e**), are configured dynamically with a network of Dense layers containing $L = 16$ neurons and *Rectified Linear Units (ReLu)* as activation functions[1]. As previously detailed, the inputs to the sensor neurons are duplicates of the Conv1D outputs from the feature extractors. A corresponding replica of the connectome transformer network is allocated for each feature extractor. The tensor networks are primarily Dense networks that incorporate skip connections by-addition to represent synapses. Each replica complies with a type of reservoir computing by keeping all layers non-trainable. This results in a total of ≈ 95 thousand non-trainable parameters. Features flow from the sensor to motor tensors, mirroring movement through a flow graph, culminating in the aggregation and stacking of motor feature maps in the final layers. The top-scored feature maps are derived via cross multi-head attention mechanisms μ_1, μ_2 and μ_3. These mechanisms resemble cross-attention, where sensor tensors acting as queries (Q_S) engage with a separate set of tensor-motors serving as keys (K) and values (V_M). As depicted in Fig. 1 - box (**f**), the Gate-Collector plays a pivotal role in integrating and synthesizing features from diverse components of the model architecture. The feature extracted tensors in output (Fig. 1 - box (**b-c-d**)) are channelled into **Gate A**, where they are accumulated in a stacked format. Simultaneously, the same outputs are fed into the 3 replicas connectome reservoirs - as illustrated in box (**e**). In the tail of each connectome reservoir, trainable multi-head attention (μ_1, μ_2 and μ_3) is applied. The outputs from μ_1, μ_2 and μ_3 within the connectome reservoirs are then consolidated at **Gate B**. This gate serves as a critical junction where these processed features are prepared for another cross-attention mechanism (μ_{out}). This mechanism specifically examines and contrasts the features consolidated at **Gate A** from the feature extractors and those refined at **Gate B** from the connectome feed-forward pathways. The cross multi-head attention outputs are averaged and reduced to an L-dimensional vector with the tensorial reduce mean function by facilitating a dynamic interplay between these two gates. Ultimately, the outputs of refined and selected (Fig. 1 - box (**f**)) are forwarded to a single-layer perceptron (Fig. 1 - box (**g**)), which acts as the final decision-making layer in the classification pipeline. We initialized the models' weights using the *Glorot* initialization method [8]. This initialization strategy, combined with our choice of *AdamW* [14] enhances the stability and efficiency of the training process. As adopted in [12], we used a learning rate of 0.0001 and employed *sparse cate-*

---

[1] The transformation algorithm **TA1**, which converts biological connectomes into artificial reservoir ones, is detailed in our previous work [3].

gorical cross entropy as our loss function λ_{mc} for multi-class tasks. While the *binary cross entropy* λ_{bc} for binary classifications. In the multi-class problem, the *SoftMax* activation function in the final perceptron layer aids in outputting probability distributions for class predictions. The *Rectified Linear Unit (ReLU)* in other layers supports efficient learning without gradient issues. The batch size in our preliminary experiments is fixed at 64. During model training we collected the top last-best validation accuracy checkpoint over 100 epochs. The accuracy [19] results from the last best checkpoint on the test set are detailed in Table 3.

Table 1. The highest accuracy for models on the MEDMNIST2D-V2.0 test set is highlighted in bold.

Methods	PathMNIST	ChestMNIST	DermaMNIST	OCTMNIST	PneumoniaMNIST	RetinaMNIST
ResNet-18 (28) [25]	0.907	0.947	0.735	0.743	0.854	0.524
ResNet-18 (224) [25]	0.909	0.947	0.754	0.763	0.864	0.493
ResNet-50 (28) [25]	0.911	0.947	0.735	0.762	0.854	0.528
ResNet-50 (224) [25]	0.892	**0.948**	0.731	0.776	0.884	0.511
auto-sklearn11 [25]	0.716	0.779	0.719	0.601	0.855	0.515
AutoKeras [25]	0.834	0.937	0.749	0.763	0.878	0.503
Google AutoML Vision [25]	0.728	**0.948**	0.768	0.771	**0.946**	0.531
ViT [12]	0.785	0.947	0.745	0.679	0.885	0.565
Resnet+ViT [12]	0.915	0.947	0.748	0.807	0.897	0.548
FPViT [12]	0.918	**0.948**	0.766	0.813	0.896	**0.568**
FP-Elegans-M1 (ours)	**0.938**	0.947	**0.796**	**0.912**	0.906	0.460
Methods	BreastMNIST	BloodMNIST	TissueMNIST	OrganAMNIST	OrganCMNIST	OrganSMNIST
ResNet-18 (28) [25]	0.863	0.958	0.676	0.935	0.900	0.782
ResNet-18 (224) [25]	0.833	0.963	0.681	0.951	0.920	0.778
ResNet-50 (28) [25]	0.812	0.956	0.680	0.935	0.905	0.770
ResNet-50 (224) [25]	0.842	0.950	0.680	0.947	0.911	0.785
auto-sklearn [25]	0.803	0.878	0.532	0.762	0.829	0.672
AutoKeras [25]	0.831	0.961	**0.703**	0.905	0.879	0.813
Google AutoML Vision [25]	0.861	0.966	0.673	0.886	0.877	0.749
ViT [12]	0.865			0.830	0.835	0.657
Resnet+ViT [12]	0.841			0.929	0.900	0.783
FPViT [12]	0.891			0.935	0.903	0.785
FP-Elegans-M1 (ours)	**0.894**	**0.976**	0.640	**0.957**	**0.941**	**0.817**

3   Results and Conclusion

As shown in Table 1, our reservoir FP-Elegans M1 models in most cases (8 benchmarks over 12) have surpassed the state-of-the-art models such as ResNet-18, ResNet-50, auto-sklearn, AutoKeras, Google AutoML Vision, ViT, and ResNet+ViT, as well as FPViT. We observe poor performance on both RetinaMNIST and TissueMNIST. Further examination of the results from RetinaMNIST reveals that no model has yet achieved satisfactory performance. Conversely, the results from TissueMNIST indicate that performance issues on this dataset may be attributed to inherent uncertainty as proved in [26]. It is worth noting that, unlike other approaches, we do not employ any data augmentation techniques in our training processes. Our results are primarily achieved through the use of an optimized model that incorporates an extensive computational graph, of which only a minimal portion is trainable. In terms of parameter comparisons, ViT exhibit significant variation in scale, with ViT-B containing approximately 85 million parameters (small-scale), ViT-L reaching around 307 million parameters (large-scale), and ViT-22B, recently introduced in [6], comprising 22 billion parameters. In contrast, *Elegans* models leverage approximately 49 million of trainable parameters distributed across the EfficientNet, VGG19, and ResNet50 feature extractors, along with approximately 95,000 non-trainable parameters housed within the Connectome Reservoirs. Moreover, since connectomes function as reservoirs, they provide several insights from a neuroscientific point of view into the effectiveness of the model optimization, highlighting that differences in performance can often be attributed to the connectome topological characteristics - these features represent an optimized structure that has been shaped by years of evolutionary pressure [3]. Within model connectome replicas (Fig. 1 - block (e)), the non-linear pathways from sensors to motors, integrated into multi-head attention mechanisms and governed by complex synaptic interactions, significantly boost the diversity of extracted features. This diversification stems from each pathway processing information uniquely, due to its distinct synaptic connections. Consequently, the final output in motor neurons presents a wider range of feature representations, enhancing the model's comprehension of input data.

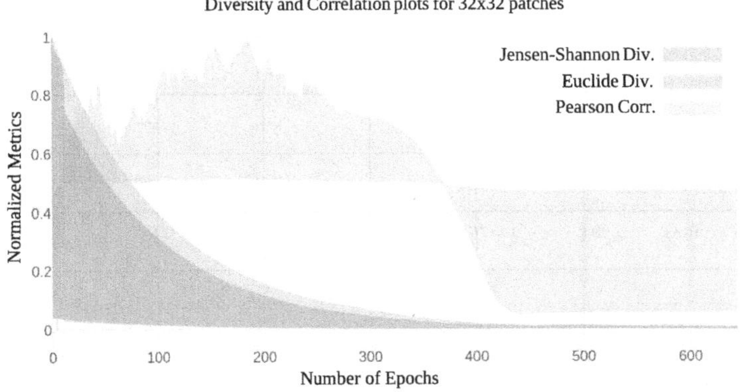

Fig. 2. Diversity and correlation metrics of the connectome replica's multi-head attention for 28x28xCh patches, averaged across epochs and models, for Feature Pyramid Elegans M1 models. The blue area represents the Jensen-Shannon Divergence, the red area tracks the Euclidean distance metric, and the green area indicates the Pearson correlation.

This enhanced feature diversity plays a pivotal role in improving the model's accuracy and generalizability across various datasets and tasks [4,11]. To further investigate its effects, we monitored the multi-head attention mechanism

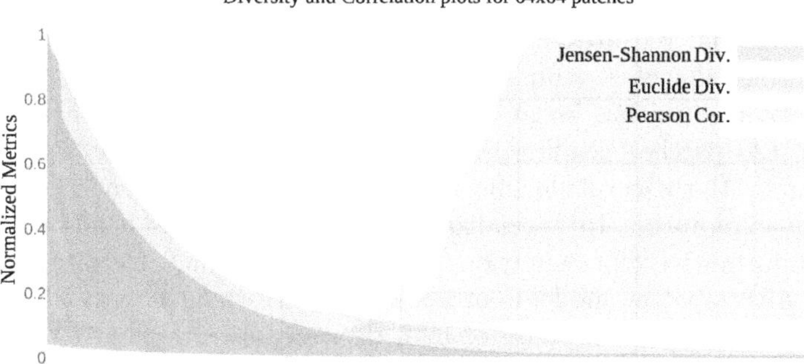

Fig. 3. Diversity and correlation metrics of the connectome replica's multi-head attention for 64x64xCh patches, averaged across epochs and models, for Feature Pyramid Elegans M1 models. The blue area represents the Jensen-Shannon Divergence, the red area tracks the Euclidean distance metric, and the green area indicates the Pearson correlation.

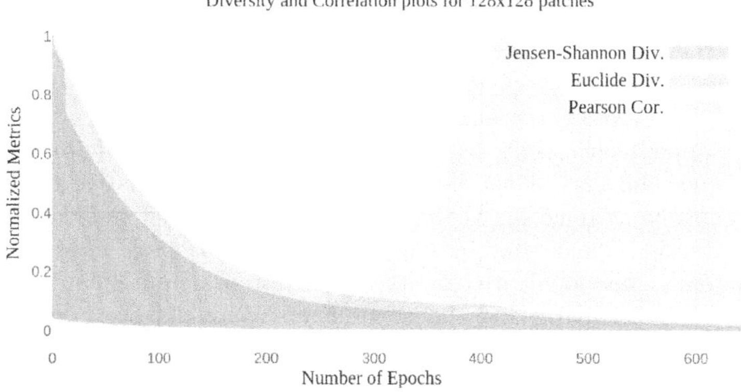

Fig. 4. Diversity and correlation metrics of the connectome replica's multi-head attention for 128x128xCh patches, averaged across epochs and models, for Feature Pyramid Elegans M1 models. The blue area represents the Jensen-Shannon Divergence, the red area tracks the Euclidean distance metric, and the green area indicates the Pearson correlation.

across the tail end of each connectome at three resolutions (32×32 - Fig. 2, 64×64 - Fig. 3, 128×128 - Fig. 4). These heads operate between sensory inputs and motor outputs within replicated reservoir-like connectome units. Our observations revealed a dynamic evolution in attention head diversity. Initially, the model exhibits high sample efficiency by leveraging diversity across heads to explore a broad spectrum of patterns. This diversity, measured using Jensen-Shannon Divergence and Euclidean distance, enables the model to effectively manage complex, non-linear representations within each connectome replica (see also blue area and red area in Figs. 2, 3 and 4 respectively). As training progresses and model accuracy improves, diversity naturally decreases, while Pearson correlation-indicating linear alignment-gradually increases in the 64×64 and 128×128 patches, leading to an anti-correlation with the 32x32 patches. On one hand, the general decline in diversity overall replica reflects a transition from exploration to specialization, where attention heads converge on common, informative representations in the larger patches. Despite the reduced non-linear diversity, the model maintains high performance without overfitting, suggesting that the initial phase of high diversity facilitates generalization. In contrast, the later phase ensures task optimization through focused and coordinated behaviour. On the other hand, the observed anti-correlation between the Pearson correlation coefficients for 64×64 and 128×128 patches relative to 32×32 patches guides the neural network toward convergence. This is achieved by emphasizing distinctions in smaller patches while maintaining a high degree of similarity for larger ones. In this paper, as in future research, we emphasize the importance of establishing post-hoc relational functions to better explain the complex non-linear behaviors emerging from these systems [7,24]. These approaches, centered on diversity analysis, are critical for advancing explainability and interpretability in both current and prospective applications of such models. In the Literature, several efforts are being made to enhance connectome-based models through experimentation with larger connectomes, including those of *Drosophila melanogaster* and other more complex organisms that are not yet fully mapped, such as those of *Homo sapiens*, *Macaca mulatta*, *Felis catus*, and *Mus musculus* [21]. Thus, in future findings, an alternative testing strategy could involve segmenting the connectome into distinct functional areas to examine whether, as indicated by some studies on *Drosophila melanogaster*, the visual system's functional area in its artificial representation might be suitable for visual tasks [16], and the olfactory area could be leveraged for classification tasks [15]. Finally, future developments could explore how the performance of these systems improves when configurations of artificial connectomes are not functioning as reservoirs but are fully trainable [3] and other baseline configurations.

Acknowledgments. This work was partly supported by the project 'Future Artificial Intelligence Research (FAIR),' Project Code PE00000013, Spoke 3 'Resilient AI,' under the National Recovery and Resilience Plan (PNRR), Mission 4 'Education and Research,' Component 2 'From Research to Business,' Investment 1.3, funded by the European Union - NEXTGENERATIONEU - RAISE - WP2 (CUP E63C22002150007). The authors wish to thank the CINECA Super Computing Application and Innovation department (SCAI) of Bologna (Italy) for granting MARCONI100.

Disclosure of Interests. The authors have no competing interests to declare that are relevant to the content of this article.

References

1. Azad, R., et al.: Advances in medical image analysis with vision transformers: a comprehensive review. Med. Image Anal. 103000 (2023)
2. Bardozzo, F., et al.: Enhanced tissue slide imaging in the complex domain via cross-explainable gan for fourier ptychographic microscopy. Comput. Biol. Med. **179**, 108861 (2024)
3. Bardozzo, F., Terlizzi, A., Simoncini, C., Lió, P., Tagliaferri, R.: Elegans-AI: how the connectome of a living organism could model artificial neural networks. Neurocomputing **584**, 127598 (2024)
4. Berg, A.: Applications of diversity and the self-attention mechanism in neural networks (2022)
5. Ciaparrone, G., et al.: Label-free cell classification in holographic flow cytometry through an unbiased learning strategy. Lab Chip **24**(4), 924–932 (2024)
6. Dehghani, M., et al.: Scaling vision transformers to 22 billion parameters. In: International Conference on Machine Learning, pp. 7480–7512. PMLR (2023)
7. Gilpin, L.H., Bau, D., Yuan, B.Z., Bajwa, A., Specter, M., Kagal, L.: Explaining explanations: An overview of interpretability of machine learning. In: 2018 IEEE 5th International Conference on Data Science and Advanced Analytics (DSAA), pp. 80–89. IEEE (2018)
8. Glorot, X., Bengio, Y.: Understanding the difficulty of training deep feedforward neural networks. In: Proceedings of the Thirteenth International Conference on Artificial Intelligence and Statistics, pp. 249–256. JMLR Workshop and Conference Proceedings (2010)
9. He, K., Zhang, X., Ren, S., Sun, J.: Deep residual learning for image recognition. In: Proceedings of the IEEE Conference on Computer Vision and Pattern Recognition, pp. 770–778 (2016)
10. He, K., Zhang, X., Ren, S., Sun, J.: Identity mappings in deep residual networks. In: Leibe, B., Matas, J., Sebe, N., Welling, M. (eds.) ECCV 2016. LNCS, vol. 9908, pp. 630–645. Springer, Cham (2016). https://doi.org/10.1007/978-3-319-46493-0_38
11. Li, J., Wang, X., Tu, Z., Lyu, M.R.: On the diversity of multi-head attention. Neurocomputing **454**, 14–24 (2021)
12. Liu, J., Li, Y., Cao, G., Liu, Y., Cao, W.: Feature pyramid vision transformer for medmnist classification decathlon. In: 2022 International Joint Conference on Neural Networks (IJCNN), pp. 1–8. IEEE (2022)
13. Liu, Z., Lv, Q., Yang, Z., Li, Y., Lee, C.H., Shen, L.: Recent progress in transformer-based medical image analysis. Comput. Biol. Med. 107268 (2023)
14. Loshchilov, I., Hutter, F.: Decoupled weight decay regularization. arXiv preprint arXiv:1711.05101 (2017)
15. Morra, J., Daley, M.: A fully-connected neural network derived from an electron microscopy map of olfactory neurons in drosophila melanogaster for odor classification. In: 2020 IEEE International Conference on Systems, Man, and Cybernetics (SMC), pp. 4504–4509. IEEE (2020)
16. Morra, J., Daley, M.: Using connectome features to constrain echo state networks. In: 2023 International Joint Conference on Neural Networks (IJCNN), pp. 1–8. IEEE (2023)
17. Morra, J., Flynn, A., Amann, A., Daley, M.: Multifunctionality in a connectome-based reservoir computer. In: 2023 IEEE International Conference on Systems, Man, and Cybernetics (SMC), pp. 4961–4966. IEEE (2023)
18. Nurgazin, M., Tu, N.A.: A comparative study of vision transformer encoders and few-shot learning for medical image classification. In: Proceedings of the IEEE/CVF International Conference on Computer Vision, pp. 2513–2521 (2023)

19. Pedregosa, F., et al.: Scikit-learn: Machine learning in python. J. Mach. Learn. Res. **12**, 2825–2830 (2011)
20. Simonyan, K., Zisserman, A.: Very deep convolutional networks for large-scale image recognition. arXiv preprint arXiv:1409.1556 (2014)
21. Suarez, L.E., et al.: A connectomics-based taxonomy of mammals. Elife **11**, e78635 (2022)
22. Tan, M., Le, Q.: EfficientNet: Rethinking model scaling for convolutional neural networks. In: Chaudhuri, K., Salakhutdinov, R. (eds.) Proceedings of the 36th International Conference on Machine Learning. Proceedings of Machine Learning Research, vol. 97, pp. 6105–6114. PMLR, 09–15 June 2019. https://proceedings.mlr.press/v97/tan19a.html
23. Tang, Y., et al.: Self-supervised pre-training of swin transformers for 3d medical image analysis. In: Proceedings of the IEEE/CVF Conference on Computer Vision and Pattern Recognition, pp. 20730–20740 (2022)
24. Wattenberg, M., Viégas, F.B.: Relational composition in neural networks: a survey and call to action. arXiv preprint arXiv:2407.14662 (2024)
25. Yang, J., et al.: Medmnist v2-a large-scale lightweight benchmark for 2d and 3d biomedical image classification. Sci. Data **10**(1), 41 (2023)
26. Yang, K., Li, D., Hu, M., Zhai, G., Yang, X., Zhang, X.P.: Uncertainty-aware sampling for long-tailed semi-supervised learning. arXiv preprint arXiv:2401.04435 (2024)
27. Yue, W., Liu, S., Li, Y.: Eff-pcnet: an efficient pure cnn network for medical image classification. Appl. Sci. **13**(16), 9226 (2023)

Natural Language Processing (NLP) and Large Language Models (LLM) for Unstructured Data in Health Informatics

Driver Gene Detection via Causal Inference on Single Cell Embeddings

Chengbo Fu[1] and Lu Cheng[1,2]

[1] Department of Computer Science, School of Science, Aalto University, Espoo, Finland

[2] Institute of Biomedicine, University of Eastern Finland, Kuopio, Finland
lu.cheng.ac@gmail.com

Abstract. Driver genes are pivotal in different biological processes. Current methods generally identify driver genes by associative analysis. Leveraging on the development of current large language models (LLM) in single cell genomics, we proposed a **C**ausal inference based approach to **I**dentify **D**river genes (called CID) from scRNA-seq data. Through experiments on three different datasets, we show that CID can (1) identify biologically meaningful driver genes that have not been captured by current associative-analysis based methods, and (2) accurately predict the direction of expression changes in downstream target genes if a driver gene is knocked out. This study presents a resource-efficient in silico framework for identifying key regulatory genes from learned cell embeddings and simulated perturbations, thereby streamlining follow-up experiments and deepening our understanding of how gene-level interventions influence cellular phenotypes and disease.

Keywords: driver gene · causal inference · scBERT · LLM · scRNA-seq

1 Introduction

In cancer biology, a driver gene is defined as a gene that significantly affects the expression of downstream genes, altering cellular behavior and contributing to the malignant phenotype. Therefore, driver genes play critical roles in cancer development. However, identifying driver genes is not an easy task.

Traditional methods identify potential driver genes in cancer are mainly based on DNA-seq or RNA-seq data. The DNA-seq based approaches, such as MutSigCV [9], OncodriveFM [5], and OncodriveCLUST [15], are based on the idea that cancer cells try to gain survival advantage over normal cells by mutating the driver genes, so the mutation frequencies of driver genes in the cancer tissue are higher than that in the normal tissue. The RNA-seq based approaches, such as CONEXIC [1] and TieDIE [13], are based on the idea that potential driver genes are differentially expressed in the cancer tissue, so differential gene expression (DE) analysis could be used for potential driver gene detection.

L. Cerulo et al. (Eds.): CIBB 2024, LNBI 15276, pp. 123–133, 2025.
https://doi.org/10.1007/978-3-031-89704-7_10

Differentially expressed genes may not be real driver genes because (1) they may be downstream genes of the driver gene and (2) the driver gene may not necessarily be highly differentially expressed as a small change of driver gene is enough to cause the malignant phenotype, e.g. transcription factors. Therefore, real driver gene identification relies on further wet lab experiments, which are laborious and expensive to perform. To validate a driver gene, researchers usually knock out (KO) or overexpress a potential driver gene in cancer cell lines to see if the malignant phenotype disappears or is enhanced. Over the years, a manually curated database NCG (Network of Cancer Genes) [14] has collected cancer driver genes reported in the literature, which could be used as a reference for potential driver genes.

As we are entering the era of single cell genomics, it is natural to identify potential driver genes using DE analysis on single cell RNA-seq (scRNA-seq) data. The advantage of scRNA-seq is that we could classify cells into different cell types to decipher the tissue heterogeneity and perform differential analysis between cell types. The differentially expressed driver genes are called marker genes. Since we are now comparing the cell types and scRNA-seq data may not be generated from cancer tissues, the concept of cancer driver gene is not defined. However, we want to borrow the term "**driver gene**" and use it to refer to genes that affect the expression of lots of downstream genes and determine cell type differentiation. Note that the driver genes are not equivalent to marker genes that are generated by DE analysis.

There exists several limitations if we directly use marker genes derived from scRNA-seq data as driver genes. First, marker genes are derived by DE analysis, which is associative rather than causal. As a result, the marker genes may be the downstream genes of driver genes, while driver genes might not appear on the top of the DE gene list. Another limitation is that validation experiments of the potential driver genes are laborious and expensive. It will significantly reduce the costs if we could perform in silico validation of the potential driver genes.

Current developments of large language models (LLMs) on scRNA-seq data, such as scBERT [18] and scGPT [3], provide a promising direction to overcome these challenges. Traditionally, LLMs are neural network models-originally designed for natural language processing (NLP)-that learn rich contextual representations of words and sentences from vast textual corpora. In the case of single-cell genomics, these models have been adapted to treat genes and cells as "tokens" and "sentences," respectively. By training on a diverse range of scRNA-seq datasets, LLMs capture intrinsic patterns and correlations across numerous genes and cell types, ultimately embedding these biological elements into a common, high-dimensional space.

Within this embedding space, techniques for perturbation and masking–originally employed in natural language processing (NLP) to investigate relationships between characters or words (e.g., [12,17]) have been adapted to explore the interactions and influences among genes. By strategically manipulating the input data, such as by masking specific genes, researchers can probe the relationships and influences between genes. By comparing how gene embeddings shift in response to these manipulations, it becomes possible to infer directed regulatory impacts and identify candidate driver genes. This adaptation of LLM-based approaches, therefore, enables a novel, in silico causal inference framework, allowing researchers to extract meaningful biological insights directly from large-scale single-cell data.

Current LLMs serve as the foundation models rather than directly solve the driver gene identification problem. scBERT uses attention scores to find marker genes, which is an associative analysis. The attention score depicts the correlation between two genes. Causal analysis, on the contrary, considers the directed relation between two genes. The impact of gene A to gene B is not the same as vice versa. However, we have not been able to find such a causal analysis based approach from the literature that addresses the driver gene identification problem.

To fill the gap, we propose a Causal inference approach to Identify Driver genes (CID) by leveraging the power of LLM scBERT. Let us illustrate the idea using an example where we want to study the impact of gene A to gene B. First, we will mask gene B from an input cell and predict its embedding e_1. Next we mask both gene A and B from the input cell and again predict gene B's embedding e_2. If gene A is an upstream gene, i.e. driver gene, then the second embedding e_2 will be far away from the first embedding e_1 as gene A has a huge impact on gene B; otherwise the two embeddings will be very close as gene A has little impact to gene B. The operation of masking gene A and B together mimics the $do(X)$ operation in causal inference, i.e. the intervention.

We demonstrate CID's new insights on three datasets. We first illustrate that CID can identify new driver genes with biological interpretations in delta pancreas islet cells, which are not picked by scBERT. Then, we show CID accurately predicts the impact of driver genes towards their target genes in a perturb-seq data consisting of 19 knockout experiments. Finally, we compare CID with a recent driver gene identification method CSDGI on a scRNA-seq data of breast cancer, where CID has identified more driver genes ($n = 12$) that have annotations in the driver gene database NCG 7.1 [14].

The consequence of this study is a framework that not only reduces the reliance on resource-intensive wet-lab experiments but also provides a roadmap to prioritize candidate driver genes for further functional studies. By streamlining the identification of these pivotal genes, CID can accelerate the discovery of novel therapeutic targets and enhance our fundamental understanding of gene regulatory networks underlying complex cellular states.

2 Data and Methods

Pancreas islet cells scRNA-seq data (Muraro dataset) was downloaded from Gene Expression Omnibus (GEO) GSE85241 [10]. Perturb-seq datasets (Dixit dataset) was downloadded from (GEO) GSE90063 [4]. Breast cancer data was downloaded from (GEO) GSE75688 [2].

All downloaded data were provided in the form of a gene expression matrix. We preprocessed the gene expression matrices following the same steps as scBERT. The expression of each cell is mapped to a 16,906-dimensional vector, with each dimension representing a unique human gene. Log-normalization is performed on the data using a size factor of 10,000. Cells with fewer than 200 expressed genes are filtered out. By default, the pre-trained weights of scBERT are used in CID analysis.

All DE analysis were performed using Scanpy (v1.9.8) [16] with the `sc.tl.rank_genes_groups` function (`wilcoxon` as the test method).

In pancreas islet cells analysis, we pooled all the marker genes identified by scBERT both as the potential driver genes and target genes in CID analysis. In case of calculating the impact of a driver gene to itself, we directly set it to 0. For each cell type, we take the top 10 driver genes and compare it with scBERT.

The perturb-seq data (Dixit dataset) contains 19 driver gene knock-out (KO) and wild-type (WT) pairs. After DE analysis, we filter genes that have at least 20 reads in both KO and WT samples for downstream analysis. Up-regulated genes are selected by $pvalue < 0.05$ and $log2fc > 1$. Down-regulated genes are $pvalue < 0.05$ and $log2fc < -1$. No-change genes are $|log2fc| < 0.5$. From each category, 10 target genes are uniformly sampled for CID prediction. The `nn.softmax` function in `predict.py` of scBERT was used to convert embeddings into discrete gene expression levels.

In breast cancer data analysis, we feed the gene expression matrix to the scBERT script `pretrain.py` with default parameters to re-train scBERT. Cells are divided into cancer and normal cells based on the `information.csv` file.

3   Results

CID performs causal inference based on the framework of scBERT. We denote the input scRNA-seq gene expression matrix by $X = (\boldsymbol{x}_1, \boldsymbol{x}_2, \ldots, \boldsymbol{x}_N)$, where $\boldsymbol{x}_i = (x_{i1}, x_{i2}, \ldots, x_{iG})$ denotes the expression profile of the i-th cell and $x_{ij} \in \{0, 1, 2, 3, 4, \#\}$ denotes the expression level of j-th gene in the i-th cell. Note that scBERT categorizes the continuous gene expression levels into 5 discrete levels and $\#$ is reserved for the masking operation.

The scBERT network, denoted by f_{scBERT}, transforms the gene expression matrix X into a latent space $\boldsymbol{Z} = (\boldsymbol{z}_1, \boldsymbol{z}_2, \ldots, \boldsymbol{z}_N)$:

$$f_{\text{scBERT}}(\boldsymbol{x}_i) = \boldsymbol{z}_i = (\boldsymbol{y}_{i1}, \boldsymbol{y}_{i2}, \ldots, \boldsymbol{y}_{ig}, \ldots, \boldsymbol{y}_{iG}) \tag{1}$$

where $\boldsymbol{y}_{ig} = (y_{ig1}, y_{ig2}, \ldots, y_{igD})$ is a D-dimensional embedding for the g-th gene in the i-th cell and $\boldsymbol{z}_i$ is a concatenated vector of $\boldsymbol{y}_{ig}$ for $g = 1, 2, \ldots, G$.

To quantify the impact of gene j to gene k, we introduce the mask notations $\boldsymbol{x}_i^{-\{k\}}$ and $\boldsymbol{x}_i^{-\{j,k\}}$, which refers to masking the gene expression level of gene sets $\{k\}$ and $\{j, k\}$, respectively. $\boldsymbol{x}_i^{-\{k\}}$ and $\boldsymbol{x}_i^{-\{j,k\}}$ are defined as

$$\boldsymbol{x}_i^{-\{k\}} = (x_{i1}, x_{i2}, \ldots, x_{i,k-1}, \#, x_{i,k+1}, \ldots, x_{iG}) \tag{2}$$

$$\boldsymbol{x}_i^{-\{j,k\}} = (x_{i1}, x_{i2}, \ldots, x_{i,j-1}, \#, x_{i,j+1}, \ldots, x_{i,k-1}, \#, x_{i,k+1}, \ldots, x_{iG}) \tag{3}$$

The predicted embeddings of scBERT based on $\boldsymbol{x}_i^{-\{k\}}$ and $\boldsymbol{x}_i^{-\{j,k\}}$ are given by

$$f_{\text{scBERT}}(\boldsymbol{x}_i^{-\{k\}}) = \boldsymbol{z}_i^{-\{k\}} = (\boldsymbol{p}_{i1}, \boldsymbol{p}_{i2}, \ldots, \boldsymbol{p}_{ik}, \ldots, \boldsymbol{p}_{iG}) \tag{4}$$

$$f_{\text{scBERT}}(\boldsymbol{x}_i^{-\{j,k\}}) = \boldsymbol{z}_i^{-\{j,k\}} = (\boldsymbol{q}_{i1}, \boldsymbol{q}_{i2}, \ldots, \boldsymbol{q}_{ik}, \ldots, \boldsymbol{q}_{iG}) \tag{5}$$

Therefore, the impact score of gene j to gene k in cell i is given by the Euclidean distance $\|\boldsymbol{p}_{ik} - \boldsymbol{q}_{ik}\|$ between the two predicted embeddings of gene k. The average impact of gene j to gene k over all N cells is given by

$$D_{j \to k}(X) = \frac{1}{N} \sum_{n=1}^{N} D_{j \to k}(\boldsymbol{x}_n) = \frac{1}{N} \sum_{n=1}^{N} \|\boldsymbol{p}_{nk} - \boldsymbol{q}_{nk}\| \tag{6}$$

In real data analysis, due to the high computational load, we have to select a set of potential driver gene $\boldsymbol{s} = (s_1, s_2, \ldots, s_r, \ldots, s_R)$ and a set of target genes $\boldsymbol{t} = (t_1, t_2, \ldots, t_m, \ldots, t_M)$, where $s_r, t_m \in \{1, 2, \ldots, G\}$. Usually the potential driver genes and target genes are chosen as DE genes with a relaxed threshold. We will compute the following marginal impact for each driver gene s_r over all target genes $\boldsymbol{t}$ given by

$$D_{s_r \to t}(X) = \sum_{m=1}^{M} D_{s_r \to t_m}(X), \tag{7}$$

where $D_{s_r \to t_m}(X)$ is calculated by Eq. 6. The potential driver genes are then ranked by their marginal impact (Eq. 7) and used as the final result of CID. The higher the impact, the more likely the gene is a driver gene. Figure 1 shows the general workflow of CID. For each target gene, we derive its impact to all target genes, from which we calculate its marginal impact. We rank the potential driver genes by their marginal impacts and select the top ones as the final candidate for wet lab validation.

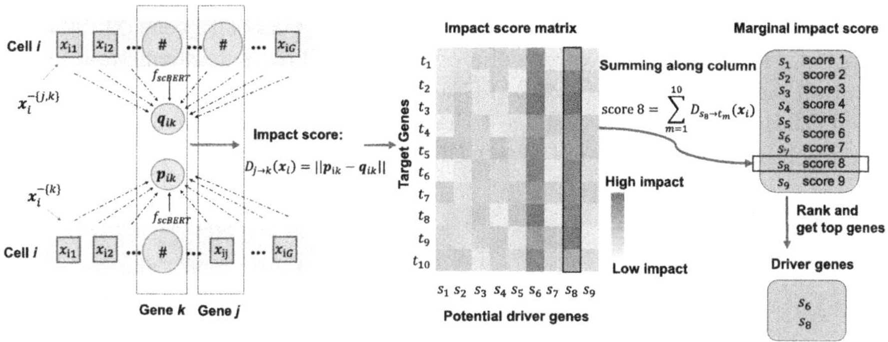

Fig. 1. CID workflow.

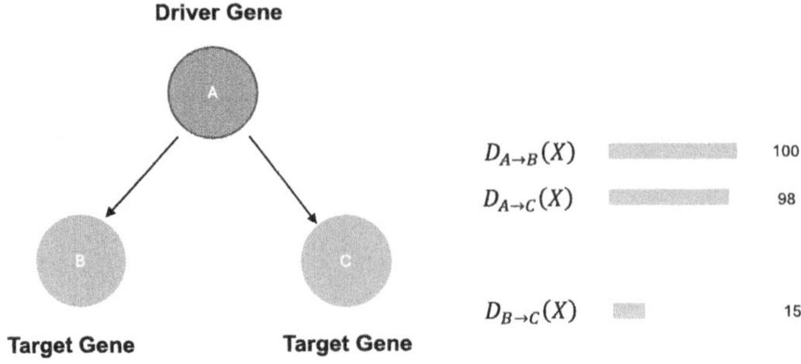

Fig. 2. Toy example of driver gene.

As illustrated in Fig. 2, the driver gene A have large impact scores to downstream genes B and C, while the impact score of B to C is relatively small as B is not a parent of C.

3.1   CID vs ScBERT on Pancreas Islet Cells Data

We applied CID to a scRNA-seq dataset of pancreas islet cells that had been analysed by scBERT, where gene expression matrix and cell type annotation were provided. Following the analysis of scBERT, we focused on the driver gene detection on 4 cell types: alpha cells, beta cells, delta cells and gamma cells. The gene expression matrices of the four cell types were fed into CID independently to identify cell-type specific driver genes. Figure 3 shows the Venn diagram between the driver genes of CID and the marker genes given by scBERT based on attention scores, where genes in red color are marker genes given by CellMarker 2.0 database [6]. It is obvious that scBERT finds more marker genes than CID, which suggests the attention score approach in scBERT is an associative analysis.

CID identified multiple potential driver genes that were not marker genes and not found by scBERT in different cell types. We performed a survey of the literature of these genes and found that HADH and SCD5 in delta cells (blue fonts in Fig. 3) were reported as driver genes in previous publications [8,11]. HADH is reported to cause monogenic diabetic cell disorders via delta cells [8]. SCD5 is reported [11] to be highly expressed in delta cells and regulate downstream transcription factors such as SOX9, MYC and HES1, which are important for cell differentiation. These reports suggest that CID identifies biologically meaningful driver genes, which are missed by scBERT.

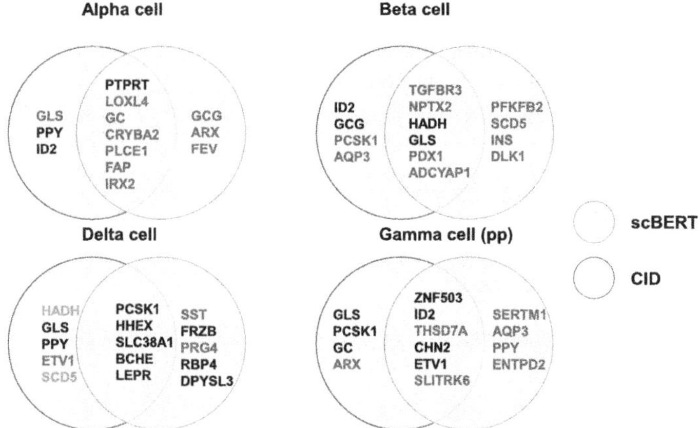

Fig. 3. CID vs scBERT predicted driver genes. Red and blue fonts indicate database (CellMarker 2.0) annotated marker genes and literature-supported driver genes. (Color figure online)

3.2 Validation of CID Prediction on Perturb-Seq Data

We validate the up-regulation and down-regulation of target genes based on CID using perturb-seq data. The perturb-seq data is generated from 19 experiments, in each of which a driver gene is knocked out using CRISPR-Cas9 in K562 cells. scRNA-seq data is generated for the cells before (WT) and after (KO) the driver gene knock-out.

Based on the gene expression matrix of WT K562 cells, i.e. before the driver gene knockout, we use CID to predict the embedding e_1 of masking the target gene, as well as the embedding e_2 of masking both the target gene and the driver gene. We then use the gene expression classifier of scBERT to convert $e1$ and $e2$ into gene expression level x_1 and x_2. If x_1 is less than x_2, then the driver gene knockout up-regulates the target gene. Similar analogy applied to down-regulation if $x_1 > x_2$. In case $x_1 = x_2$, the driver gene has no impact on the target gene.

We obtained the up/down-regulation ground truth by comparing the scRNA-seq data of KO versus WT. DE analysis were performed for each driver gene knockout experiment, from which we classified the target genes into 3 categories: up-regulated, no-change, down-regulated. From each category, we randomly sampled 10 genes to test the performance of CID. Figure 4 shows the predicted regulation directions versus the ground truth on the sampled target genes across all 19 experiments and the corresponding log2fc values. It can be seen that the log2 fold change (log2fc) of target genes in the ground truth nicely agrees with CID's predictions. To further illustrate the difference among the three categories, we plotted boxplots and performed a Kruskal-Wallis test, which revealed highly significant differences ($P = 1.3124 \times 10^{-30}$, $P = 1.9004 \times 10^{-108}$, $P = 1.9311 \times 10^{-97}$) among the three groups. These results suggest that CID can accurately predict the regulation directions and the differences are statistically significant, thereby reinforcing the reliability of CID's predictions without conducting actual knockout experiments.

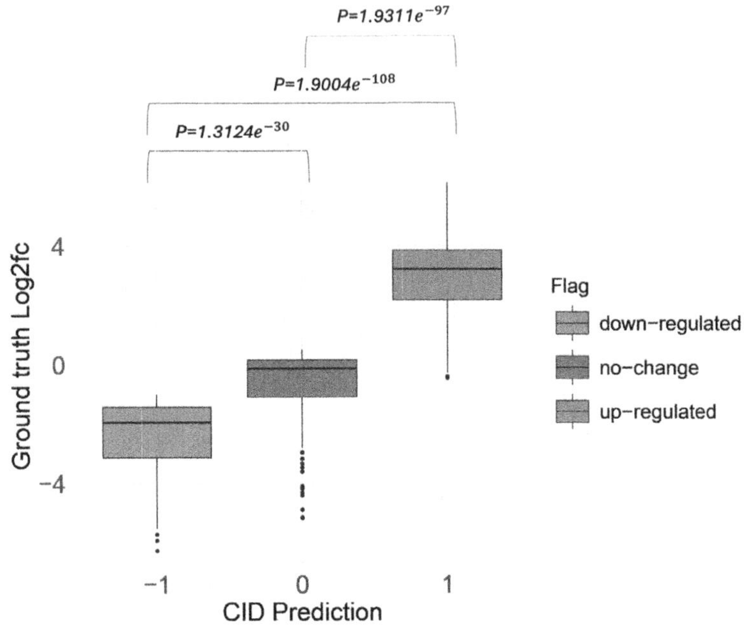

Fig. 4. CID predicted gene expression change vs ground truth log2fc.

3.3   CID vs CSDGI on Breast Cancer Data

CSDGI is a recent method that aims to find driver genes from scRNA-seq data based on ResNet. CSDGI has identified 70 driver genes in a breast cancer scRNA-seq data [7]. Here we re-analysed the scRNA-seq data using CID. First we re-trained scBERT on this data, where pre-trained weights (default setting) were used to initialize scBERT. Top 100 DE genes between caner cells and normal cells in the dataset were chosen both as the potential driver genes and target genes. Next we evaluated the marginal impact of the potential driver genes to the target genes using CID, with the impact of a gene to itself set to 0. Given the ranked driver genes by CID, we calculated the number of real driver genes by taking the top $K \in \{1, 2, \ldots, 70\}$ driver genes. If a candidate driver gene had an annotation in the NCG 7.1 database [14], we treated it as a real driver gene. Note that CSDGI only provided 70 driver genes, so we maximally take 70 potential driver genes.

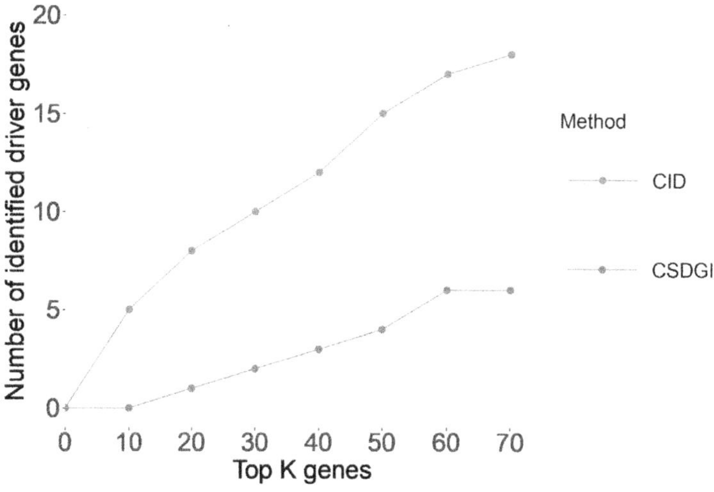

Fig. 5. CID vs CSDGI in driver gene discovery.

Figure 5 shows the number of real driver genes identified by CID and CSDGI in the top K potential driver genes. It can be seen that CID identifies more annotated driver genes than CSDGI.

4 Conclusion

Owing to the development of LLMs, researchers could use the generated cell and gene embeddings for various causal inference based tasks. In this paper, we demonstrate CID's superior performances in driver gene identification in three different settings. CID's results generally offer better biological interpretations.

Acknowledgments. We acknowledge the computational resources provided by the Aalto Science-IT project. This work was supported by the Research Council of Finland [grant number 335858, 358086].

Author contributions. F.C. conceptualized and conducted the analyses. L.C. supervised the project.

Data Availability Statement. The code is freely available at https://github.com/Dionysos-o/CID.git

Disclosure of Interests. The authors have no competing interests to declare that are relevant to the content of this article.

Appendix

Below is a list of abbreviations used in this manuscript.

Abbreviation	Explanation
LLM	Large Language Model
NLP	Natural Language Processing
scRNA-seq	Single-cell RNA sequencing
DNA-seq	DNA sequencing
RNA-seq	RNA sequencing
KO	Knockout
WT	Wild-Type
DE	Differentially Expressed
NCG	Network of Cancer Genes
scBERT	Single-cell Bidirectional Encoder Representations from Transformers
scGPT	Single-cell Generative Pre-trained Transformer
CID	Causal inference approach to Identify Driver genes
CSDGI	Causal Single-cell Driver Gene Identification
log2fc	log2 Fold Change

References

1. Akavia, U.D., et al.: An integrated approach to uncover drivers of cancer. Cell **143**(6), 1005–1017 (2010)
2. Chung, W., et al.: Single-cell RNA-seq enables comprehensive tumour and immune cell profiling in primary breast cancer. Nat. Commun. **8**(1), 15081 (2017)
3. Cui, H., et al.: scGPT: toward building a foundation model for single-cell multiomics using generative AI. Nature Methods **21**, 1–11 (2024)
4. Dixit, A., et al.: Perturb-seq: dissecting molecular circuits with scalable single-cell RNA profiling of pooled genetic screens. Cell **167**(7), 1853–1866 (2016)
5. Gonzalez-Perez, A., Lopez-Bigas, N.: Functional impact bias reveals cancer drivers. Nucleic Acids Res. **40**(21), e169–e169 (2012)
6. Hu, C., et al.: CellMarker 2.0: an updated database of manually curated cell markers in human/mouse and web tools based on scRNA-seq data. Nucl. Acids Res. **51**(D1), D870–D876 (2023)
7. Huang, M., et al.: Unravelling cancer subtype-specific driver genes in single-cell transcriptomics data with CSDGI. PLoS Comput. Biol. **19**(12), e1011450 (2023)
8. Lawlor, N., et al.: Single-cell transcriptomes identify human islet cell signatures and reveal cell-type-specific expression changes in type 2 diabetes. Genome Res. **27**(2), 208–222 (2017)
9. Lawrence, M.S., et al.: Mutational heterogeneity in cancer and the search for new cancer-associated genes. Nature **499**(7457), 214–218 (2013)
10. Muraro, M.J., et al.: A single-cell transcriptome atlas of the human pancreas. Cell Syst. **3**(4), 385–394 (2016)
11. Oshima, M., et al.: Stearoyl COA desaturase is a gatekeeper that protects human beta cells against lipotoxicity and maintains their identity. Diabetologia **63**, 395–409 (2020)
12. Paganelli, M., Del Buono, F., Baraldi, A., Guerra, F., et al.: Analyzing how BERT performs entity matching. Proc. VLDB Endow. **15**(8), 1726–1738 (2022)

13. Paull, E.O., et al.: Discovering causal pathways linking genomic events to transcriptional states using tied diffusion through interacting events (TieDIE). Bioinformatics **29**(21), 2757–2764 (2013)
14. Repana, D., et al.: The network of cancer genes (NCG): a comprehensive catalogue of known and candidate cancer genes from cancer sequencing screens. Genome Biol. **20**, 1–12 (2019)
15. Tamborero, D., et al.: Oncodriveclust: exploiting the positional clustering of somatic mutations to identify cancer genes. Bioinformatics **29**(18), 2238–2244 (2013)
16. Wolf, F.A., et al.: Scanpy: large-scale single-cell gene expression data analysis. Genome Biol. **19**, 1–5 (2018)
17. Wu, Z., Chen, Y., Kao, B., Liu, Q.: Perturbed masking: parameter-free probing for analyzing and interpreting BERT. arXiv preprint arXiv:2004.14786 (2020)
18. Yang, F., et al.: scBERT as a large-scale pretrained deep language model for cell type annotation of single-cell RNA-seq data. Nat. Mach. Intell. **4**(10), 852–866 (2022)

Assessing and Comparing Free Large Language Models' Responses to a Clinical Case: Accuracy, Safety, and Reliability

Elena Sblendorio[1,2] , Alessio Lo Cascio[2,3] , Daniele Napolitano[2,4] ,
Francesco Germini[2,5] , Vincenzo Dentamaro[6(✉)] , Michela Piredda[7] ,
and Giancarlo Cicolini[8]

[1] Azienda Ospedaliero-Universitaria Consorziale Policlinico di Bari, Piazza Giulio
Cesare, 11, 70124 Bari, Italy
`elena.sblendorio@students.uniroma2.eu`
[2] Department of Biomedicine and Prevention, University of Rome "Tor Vergata",
Rome, Italy
`alessio.locascio@students.uniroma2.eu,`
`daniele.napolitano@policlinicogemelli.it,`
`francesco.germini@students.uniroma2.eu`
[3] La Maddalena Cancer Center, Via San Lorenzo 312, 90146 Palermo, Italy
[4] CEMAD Digestive Disease Center, Fondazione Policlinico Universitario A. Gemelli
IRCCS, Università Cattolica del Sacro Cuore, 00168 Rome, Italy
[5] Direttore di Distretto Sociosanitario, ASL Bari, Bari, Italy
[6] Department of Computer Science, University of Bari "Aldo Moro", Bari, Italy
`vincenzo.dentamaro@uniba.it`
[7] Department of Medicine and Surgery Research Unit Nursing Science, Università
Campus Bio-Medico di Roma, Via Alvaro del Portillo, 21, 00128 Rome, Italy
`m.piredda@unicampus.it`
[8] Department of Innovative Technologies in Medicine and Dentistry, G.d'Annunzio
University of Chieti, Pescara, Italy
`g.cicolini@unich.it`

Abstract. This research examines how five free of charge Large Language Models (LLMs)-Zephyr, Mistral 7B, LLAMA 2 7B, ChatGPT 3.5 and Copilot Precise-perform when faced with a nursing clinical scenario involving a neuropsychiatric emergency. Their responses were evaluated based on established guidelines by a Delphi consensus using a 5-point Likert scale to rate safety, accuracy, reliability and the potential for improvement. The findings underscore the greatest importance of safety and accuracy metrics. LLAMA 2 7B exhibits balanced but poor performance, scoring 3 out of 5 in Safety, Accuracy, and References, and 4 out of 5 in providing Improvement suggestions. ChatGPT 3.5 demonstrates adequate performance in Safety, Accuracy, and References, each with a score of 4 out of 5, indicating its proficiency in generating accurate, reliable content and ensuring patient safety, though there is room for improvement in enhancement suggestions (3 out of 5). Copilot Precise shows a unique profile, with balanced scores of 3 out of 5 in Safety, Accuracy, and Improvements, and a perfect score of 5 out of 5 only in

L. Cerulo et al. (Eds.): CIBB 2024, LNBI 15276, pp. 134–149, 2025.
https://doi.org/10.1007/978-3-031-89704-7_11

References, highlighting its high accuracy in generating references. Reliability was reported in terms of both reference precision criteria and consistency over time computed through automated assessment. These preliminary results underscore the importance of developing language models that focus on ensuring safety and precision, in clinical decision-making scenarios. Further studies should aim to improve the accuracy and dependability of these models by examining a range of situations and incorporating real-time feedback mechanisms from experts. This will enhance their usefulness in clinical environments.

Keywords: large language models · clinical decision-making · accuracy · safety · reliability

1   Introduction

Artificial Intelligence (AI) has made strides in the field of computer science and technology evolving into a complex and dynamic area. Its focus is on developing systems and algorithms that can perform tasks traditionally handled by human intelligence. The primary goal is to equip machines with abilities to learn, reason, solve problems, and adapt to their surroundings. This realm includes specialized areas like machine learning, computer vision, natural language processing, and robotics.

Machine learning and large language models (LLMs) play a key role in AI advancement. Machine learning enables systems to improve their performance by learning from data without explicit programming for specific tasks [1]. Moreover, LLMs are sophisticated AI models adept at comprehending and generating human language. Trained on textual datasets, these models grasp common word usage patterns to facilitate more natural interactions between humans and machines [2].

The integration of AI into healthcare has transformed procedures, treatment methods, and patient care practices. Through its capability to swiftly analyze great amounts of data with efficiency, AI opens new avenues for improving the effectiveness and efficiency of healthcare services [3]. The World Health Organization recently updated its guidelines regarding the governance aspects of LLMs, acknowledging their growing use in healthcare [4].

In the healthcare field, LLMs play a role in various applications. One notable use is in natural language processing for managing information. These models can interpret large amounts of text found in clinical records, medical reports, and scientific publications. This aids in organizing information, extracting data, and creating decision support systems for healthcare professionals. Another significant application of LLMs is seen in the development of chatbots and virtual assistants. These tools can offer responses to common patient queries, assist in appointment scheduling, and provide guidance on disease management. This enhances the accessibility and efficiency of healthcare services.

LLMs are also instrumental in monitoring social media platforms and online forums. They can track these channels to identify signs of diseases or outbreaks, assisting in epidemiological surveillance and resource allocation. Furthermore, LLMs can generate materials tailored to patients' needs, covering topics such as disease management, post-operative care instructions, and health advice.

In the field of translation, LLMs have proven to be extremely useful. These sophisticated models aid in translating documents and patient information automatically, facilitating effective communication between healthcare providers and patients who speak different languages [5]. Additionally, LLMs in research can analyze vast clinical and research datasets to uncover patterns, relationships, and new findings that drive forward medical research and identify potential treatments [6].

Within care management systems, LLMs are pivotal in offering treatment suggestions, monitoring patient adherence to therapies, and enhancing overall care effectiveness [7]. They also elevate the quality and personalization of interactions by making them more meaningful and tailored [8]. Furthermore, in decision-making and emergency response scenarios, LLMs can provide valuable assistance by empowering healthcare professionals to make well-informed decisions [9].

Although leveraging LLMs, such as GPT-3, for healthcare decision support brings advantages, it is crucial to address ethical, legal, and security considerations. In order to respect patient privacy laws, ensuring data protection measures, confirming the accuracy of results, offering clear and easy-to-understand recommendations, continuously monitoring model performance, encouraging ongoing training for healthcare professionals, and acknowledging the limitations of these models are all vital factors according to several scholars [10,11].

The primary goal of this study is to assess the effectiveness of tools in aiding clinical decision-making and advocating for the careful, thoughtful, and integrated utilization of large language models (LLMs) in healthcare. The paper is organized in the following way: Sect. 2 sketches the state of the art, Sect. 3 describes materials and methods, Sect. 4 shows preliminary results and discussion, while conclusions are presented in Sect. 5.

2   State of the Art

The integration of large language models (LLMs) into healthcare, specifically in the context of clinical decision-making, has generated significant scholarly attention. The evolution of LLMs such as GPT-3 and BERT, trained on extensive internet data, has shown potential to revolutionize healthcare by providing sophisticated capabilities in generating human-like text, offering real-time recommendations, and supporting clinical decisions with plain-text explanations for diagnoses [12]. Researchers like Wang et al. [13] have demonstrated that tailored prompting strategies can enhance LLM's performance in clinical tasks, thereby underscoring the need for specialized training and evaluation frameworks to gauge and refine their effectiveness.

A substantial body of research has delved into various facets of LLM application in healthcare. Studies by Umerenkov et al. [14] and Lee et al. [15] focus on the model's ability to articulate and clarify diagnostic decisions, which has been shown to improve physicians' diagnostic consensus. Conversely, Gottlieb and Silvis [16] discuss the imperative of ensuring the accuracy and reliability of these models before their integration into clinical practice, pointing to the risk of errors that could affect patient outcomes.

Beyond general clinical applications, the use of LLMs has also been explored in more nuanced areas like patient-physician communication and personalized therapy. Velupillai et al. [17] suggest that while LLMs can significantly contribute to clinical NLP systems, there is a critical need for more rigorous and context-specific evaluations to advance the field. This is echoed in research by Perlis et al. [18], who emphasize the potential of LLMs in clinical research, pushing for robust methodologies to ensure their utility and efficacy.

The implications of these studies are profound, suggesting not only the potential enhancements LLMs can bring to clinical decision-making but also the challenges and responsibilities inherent in their deployment. It is evident that while LLMs hold promise, their integration into healthcare settings demands careful consideration, rigorous testing, and an ongoing commitment to improving their accuracy and utility in clinical environments. This collective body of work lays a solid foundation for future explorations and technological advancements, guiding the next steps in the responsible and effective use of AI in medicine.

The study by Turpen et al. [19] demonstrates the use of natural language processing (NLP) to analyze patient experience comments, revealing significant differences in language that can impact clinical outcomes. Such findings underscore the necessity for LLMs to be sensitive to the nuances of patient language to enhance the quality of care delivered. The research by Hayakawa et al. [20] examines how language choice in bilingual settings affects medical judgments and health decisions, providing crucial insights into the linguistic and cultural factors that LLMs must account for in global healthcare settings This research points to the importance of developing LLMs that are adaptable to the multilingual and multicultural contexts in which they are deployed.

Furthermore, the use of LLMs for clinical decision support is not without ethical implications. The work by Hochberg et al. [21] on modeling and annotating clinical decision-making styles highlights the potential of LLMs to influence decision-making processes in healthcare. Such capabilities necessitate a careful approach to ensure that these models support clinicians without inadvertently introducing biases or errors that could compromise patient care.

The integration of LLMs into healthcare holds substantial promise for enhancing clinical decision-making, patient communication, and research methodologies. However, as the scholarly work suggests, this technology must be implemented with a rigorous understanding of its limitations and a strong commitment to ethical principles. The ongoing research and development in this field are crucial for realizing the full potential of LLMs in healthcare, ensuring they serve as valuable aids to clinicians and patients alike, enhancing outcomes and the efficiency of care delivery.

In this work it has been proposed the evaluation of 5 LLMs on a clinical prompt with respect to specific guidelines on a 5-point Likert scale given by Delphi experts.

3   Material and Methods

In this work, clinicians have proposed a clinical case to five different state-of-the-art large language models and a Delphi consensus team measured the answers provided by the five large language models with respect to the guidelines in a five-point Likert scale.

The proposed clinical case is the following:

A neurology patient experienced a manic crisis characterized by agitation and aggression. The on-duty physician ordered the emergency administration of diazepam (10 mg) diluted in saline solution to be given intravenously (IV) over 10–15 min. The physician preferred to prescribe the medication via IV because the patient already had a peripheral IV catheter in place and to protect the nurses from risks of accidental needle sticks when administering the drug intramuscularly, given the patient's sudden movements and aggressive state [22].

Prior to the diazepam, the patient was conscious but disoriented, had an oxygen saturation of 94% on supplemental oxygen, heart rate of 100 bpm, and was normotensive.

Following the intervention, the crisis resolved. However, the patient developed respiratory depression with oxygen desaturation to 88%, hypotension at 70/45 mmHg, bradycardia at 59 bpm, and reduced Glasgow coma scale score of 7 indicating decreased consciousness.

The antidote that could be considered is flumazenil given as an IV bolus of 0.2 mg over 30 s, repeatable up to a total dose of 3 mg if needed. It must be administered through a dedicated lumen to avoid incompatibilities with other medications. Flumazenil is diluted in sodium chloride or 5% dextrose solution and infused slowly at a rate not exceeding 1 mg/min due to risk of adverse reactions like re-sedation, seizures, and withdrawal symptoms [23].

While flumazenil can effectively reverse the effects of benzodiazepines, especially in cases of long-term benzodiazepine use it can also precipitate withdrawal and other adverse effects [24]. These can include: benzodiazepine withdrawal risk syndrome, confusion, nausea, seizures, re-sedation, and, in rare cases, cardiac arrhythmias [25]. It is crucial to monitor the patient continuously for these potential effects [26].

Nursing care should focus on ensuring the patient's airway remains open and protected, monitoring vital signs rigorously, and preparing to manage potential complications, such as seizures or cardiac disturbances. Regular reassessment of hemodynamic parameters, including peripheral oxygen saturation with a pulse oximeter and ECG, as well as the Glasgow Coma Scale (GCS), is fundamental to determine the patient's status and responsiveness to the antidote [27].

A standardized protocol for such situations along with staff education can improve patient safety [28]. Continuous monitoring systems for patients receiving high-risk medications like benzodiazepines should be implemented [29].

Flumazenil is an antidote that should be uniformly included in the "drugs" drawer of the emergency cart worldwide, ensuring it is regularly stocked and maintained [22,28].

Psychological support and counseling services should be offered to nursing staff to help them manage the stress and emotional burden associated with high-risk patient care [30,31].

It is advocated that free master's programs in nursing in neuropsychiatric disorders should be made available to nurses [32].

The work environment directly affects the psycho-cognitive state of the patient through the setting and indirectly through its impact on the nurse, which in turn influences the nurse's performance and, consequently, patient outcomes [33–35].

Encourage the incorporation of evidence-based, non-pharmacological interventions to minimize the dependence on benzodiazepines and optimize patient outcomes through a multidisciplinary approach that prioritizes patient safety and well-being [31,36–38].

The two submitted prompts were:

1. Act as an expert neurology nurse. A neurology patient experienced a manic crisis characterized by agitation and aggression. The on-duty physician ordered the emergency administration of a full vial of diazepam (10 mg), diluted to 20 ml with saline solution, administered slowly intravenously (IV). This route was chosen because the patient already had a peripheral IV catheter in place, which also minimized the risk of accidental needle sticks for nurses given the patient's sudden movements and aggressive state. Following the intervention, the crisis was resolved. Prior to the IV administration of diazepam the patient was conscious and disoriented, with an oxygen saturation of 94% while receiving low-flow oxygen therapy (2 L/min). Additionally, before the crisis, the patient was normotensive, with a heart rate (HR) of 100 bpm. Post-administration, the patient exhibited respiratory depression (oxygen saturation decreased to 88%), a drop in blood pressure to 70/45 mmHg, and a reduced HR of 59 bpm. The Glasgow Coma Scale (GCS) score was calculated at 7, indicating a significant decrease in the level of consciousness.

What antidote could be prescribed by the physician for diazepam? Through what route should it be administered? What are the guidelines for its reconstitution and in which solution should it be diluted? At what rate should it be administered in mg per minute? What adverse reactions might be expected from the antidote?

Conduct necessary search using scientific sources, focusing on high-evidence sources from peer-reviewed journals and official healthcare guidelines.

Draft an initial response, ensuring all arguments are justified and based on the highest level of evidence rating. Review the draft for any inaccuracies or unnecessary information and revise accordingly.

Prioritize nursing best practices and patient safety in your rationale. Specify any parameter cutoffs considered for procedural operational guidance to align with the State of the Art.

Provide several reliable references from the last 5 years in APA format with the associated links.

At the end create a chapter inserting all the references in APA format.

2. After reviewing the critical issues inherent in the clinical case previously presented, provide recommendations for enhancing patient care, improving team performance, and optimizing hospital organizational processes.

The guidelines used as ground truth are present in references [22–38]. The Assessment Methodology framework adopted for the evaluation of the selected free LLMs was described in [39]. In particular, in this work, we focused on the domain 1 (patient safety based on State of the Art alignment), the first and second item of domain 2, i.e. accuracy and reference precision criteria, the domain 4, namely consistency, that is expressed by a unique item, i. e the Automated Assessment of Temporal Variability of Responses, computed by the selected technique MPNet V2 Metric, and domain 7, that is focused on the evaluation of LLMs' ability to Drive Evolution in Healthcare providing suggestions for improvements for patient safety, healthcare team wellness, and for the hospital organization. The Delphi commission was composed by a multidisciplinary team, including Nursing scholars, PhD nursing students engaged in both academic research and clinical practice, as well as AI experts. The evaluation of each response was performed with respect to the following Likert Scale. A 5-point Likert scale is a popular rating system used to measure attitudes or behaviors across a spectrum of agreement or disagreement. Each point on the scale typically corresponds to a respondent's degree of agreement with a statement. The standard breakdown of each point on a 5-point Likert scale is as follows:

1. **Strongly Disagree**: This response indicates that the respondent completely disagrees with the statement or finds it absolutely inaccurate.
2. **Disagree**: This response signifies that the respondent generally disagrees with the statement, although not as vehemently as with "Strongly Disagree."
3. **Neutral**: This choice reflects a middle ground where the respondent neither agrees nor disagrees with the statement, indicating indifference or uncertainty.
4. **Agree**: This response indicates that the respondent generally agrees with the statement but with some reservations.
5. **Strongly Agree**: This response shows that the respondent fully agrees with the statement and finds it very accurate.

This scale allows researchers to quantify qualitative data and analyze trends or patterns in attitudes and opinions efficiently.

The generated prompts are submitted to the following free of charge LLMs:

- **Zephyr**: Zephyr is a user-friendly model accessible via Hugging Face Spaces, designed for general-purpose chatbot applications. It employs advanced NLP techniques to offer responsive and context-aware dialogues, enhancing user interaction with AI [39].
- **Mistral**: Mistral 7B, available on Replicate via an API, is a 7-billion-parameter language model optimized for dialogue and general language tasks. It employs innovative attention mechanisms like grouped-query and sliding window attention for efficient handling of long sequences, providing faster inference and high-quality outputs [40].
- **LLAMA2**: Llama 2 is a suite of models ranging from 7 billion to 70 billion parameters, fine-tuned specifically for chat applications. Known for its robustness across various benchmarks, Llama 2-Chat demonstrates superior performance in dialogue tasks, focusing on safety and helpfulness in user interactions [41].
- **ChatGPT 3.5**: Developed by OpenAI, ChatGPT 3.5 is a variant of the GPT (Generative Pre-trained Transformer) model tailored for conversational AI. It excels in generating coherent and contextually relevant text based on user prompts, making it effective for a range of applications from customer service to entertainment [42].
- **Microsoft Copilot (Precise)**: Microsoft's Copilot, integrated into products like GitHub and Dynamics CRM, leverages AI to enhance productivity by providing contextual assistance, code suggestions, and document automation. It uses real-time data processing to recommend actions and insights, helping users navigate complex tasks efficiently.

The LLMs will be compared to assess the following criteria:

1. **Security**: The LLM is able to ensure security in decision support;
2. **Accuracy**: The LLM demonstrates high accuracy and the ability to focus on responses;
3. **Reliability** in terms of reference precision criteria and consistency over time.

4. **Drive Improvements**: The LLM is able to provide useful suggestions for improvements for the patient, the healthcare team and the Hospital Organization, driving the evolution of'Healthcare.

The tasks conducted for testing the initial feasibility of the LLMs' generated answers were the following:

1. Comparison with sector guidelines and the best available scientific evidence regarding the clinical case to evaluate their accuracy, safety, and coherence: Each generated response covers multiple topics-clinical and organizational-so the responses were divided by topic, and the comparison was made between each segment and the guideline related to the topic at hand. The comparison was conducted twice, using two different methodologies:
   - Manual Comparison: The research team conducting this study personally read all the responses generated by the LLMs and evaluated them against the reference guidelines, considering their professional experience.
   - Automated Comparison Using "MPNet V2": To evaluate the alignment of LLMs respect to Delphi response, the MPNetV2 technique is utilized [44]. This technique employs a transformer-based language model that leverages Masked and Permuted Pre-training (MPP), combining autoregressive modeling and an attention mechanism to capture contextual information, thus comparing two blocks of text reporting their similarity score. The guideline has been compared with the T0 answer. Regarding the temporal distance between the same questions resubmitted to various LLMs, they were submitted sequentially without waiting for determined time intervals. Waiting hours or days to resubmit the same question to an LLM does not change the intrinsic nature of the response model. Conversely, care was taken to close the previous session before sequentially resubmitting the same question, and care was taken to preserve the same sampling parameters.
2. Comparison with other responses generated by the same LLM to evaluate temporal variability: The prompt submission to each LLM was repeated four times to assess the level of consistency of responses. Prompts were submitted four times at different intervals, and the responses were compared using the "Inference API Metric", which compared "Source Sentences" (response generated at T0) with "Sentences to compare to" (responses generated at T1, T2, T3, T4), generating a "Sentences Similarity" value to evaluate the percentage variability of responses generated by the same LLM at different times. This yielded three "Sentences Similarity" values, treated as a statistical sample of responses generated by the LLM.
3. Evaluation of the scientific references of each LLM: Each reference provided was individually verified through Google Scholar, which is freely accessible and can identify academic literature texts.

4 Results and Discussion

Figure 1 illustrates the average performance scores of five different Large Language Models (LLMs)-Chat GPT 3.5, Zephyr Chat, LLAMA2, Microsoft Copilot, and Mistral-across the evaluation criteria of Accuracy, Safety, and reliability. Reliability was intended in terms of consistency over time and reference precision criteria. Zephyr and Mistral 7B display identical performance profiles, scoring 4 out of 5 in Safety and Improvements, 3 out of 5 in Accuracy, and 2 out of 5 in References. This uniformity suggests that these models require significant enhancement in accuracy and reference generation. LLAMA 2 7B exhibits a balanced but poor performance, with scores of 3 out of 5 in Safety, Accuracy, and References, and 4 out of 5 in Improvements.

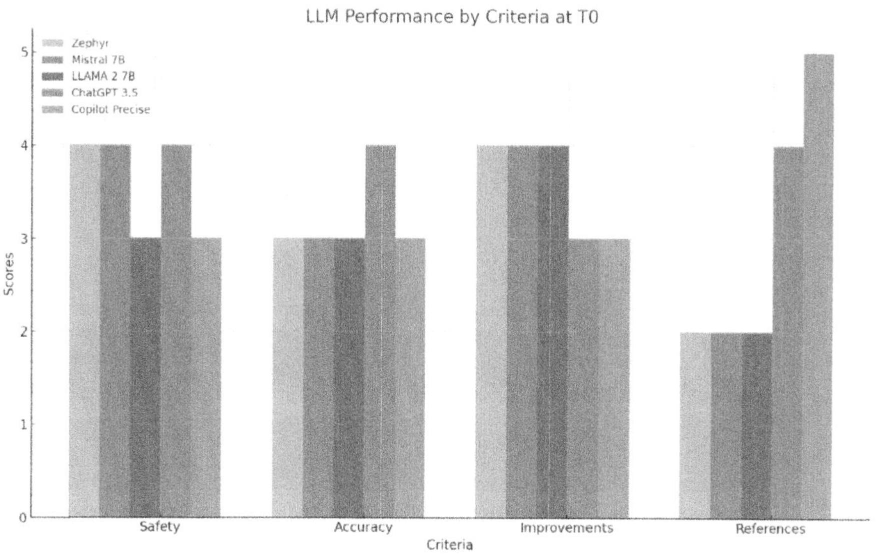

Fig. 1. Bar chart showing performance scores of LLMs across Safety, Accuracy, and Improvements and References criteria.

LLAMA 2 7B exhibits a balanced but moderate performance, with scores of 3 out of 5 in Safety, Accuracy, and References, and 4 out of 5 in Improvements. This indicates that LLAMA 2 7B is reliable in providing improvement suggestions but shows only moderate proficiency in maintaining safety, accuracy, and generating references.

ChatGPT 3.5 demonstrates strong performance in Safety, Accuracy, and References, each with a score of 4 out of 5, while scoring slightly lower in Improvements (3 out of 5). This suggests that ChatGPT 3.5 excels in generating accurate and reliable content and ensuring safety but shows some room for improvement in providing enhancement suggestions.

Copilot Precise shows a unique profile with balanced scores of 3 out of 5 in Safety, Accuracy, and Improvements, but achieves a perfect score of 5 out of 5 in References. This indicates that Copilot Precise is particularly adept at generating accurate references, although it could benefit from improvements in other evaluated areas.

Additionally, in the provided prompt, critical patient demographic details such as age, height, and weight were intentionally omitted, presenting potential biases in the responses. Nevertheless, none of the LLMs addressed the ambiguity inherent in the prompt, and all assumed that the patient was an adult. This oversight highlights a significant limitation in the handling of contextual cues by current LLMs, potentially affecting the appropriateness and specificity of their clinical recommendations. This incident underscores the necessity for enhancing the sensitivity of LLMs to missing but clinically pivotal information in healthcare settings, ensuring that their responses remain accurate and tailored to the individual characteristics of each patient.

The plot in Fig. 2 provides a comparative analysis of the similarity scores (computed using Inference API Metric) for responses generated by five different LLMs at the initial time point (T0), with respect to the official guidelines. The similarity scores range from 0 to 1, where higher scores indicate a greater alignment with the official guidelines.

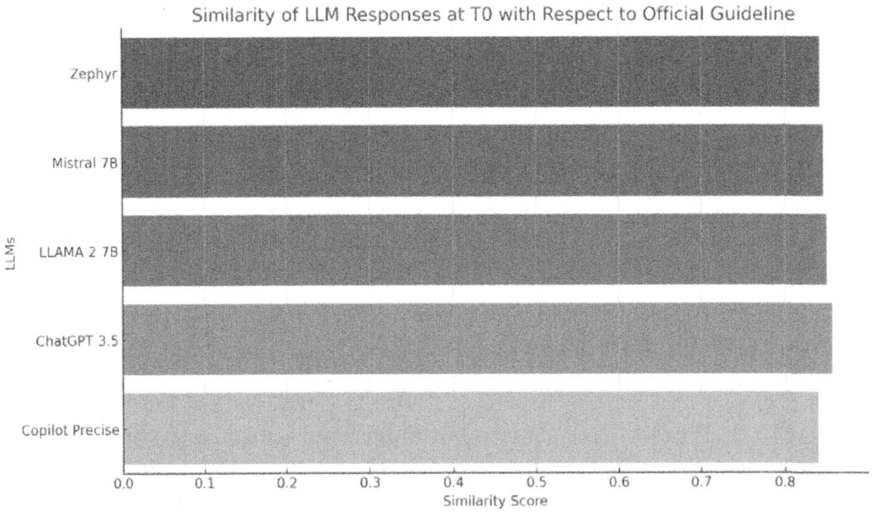

Fig. 2. Similarity scores of LLM responses at T0 compared to the official guideline.

As a novelty, we also use the selected automated method mpnet v2 to compare and compute the similarity of the responses provided by the LLMs to the clinical case respect to the Delphi experts' guidelines. The plot in Fig. 2 provides a comparative analysis of the similarity scores (computed using Inference API Metric) for responses generated by five different Large Language Models (LLMs) at the initial time point (T0), with respect to the official guidelines. The similarity scores range from 0 to 1, where higher scores indicate a greater alignment with the official guidelines. The LLMs evaluated are Zephyr, Mistral 7B, LLAMA 2 7B, ChatGPT 3.5, and Copilot Precise. The acceptable threshold for similarity is set at 0.85, indicating that scores below this value may not meet the desired level of alignment with the guideline. ChatGPT 3.5 achieves the highest similarity score of 0.8570, surpassing the acceptable threshold and indicating an appropriate adherence to the official guidelines. LLAMA 2 7B follows closely with a similarity score of 0.8500, meeting the threshold and reflecting reliable performance. Mistral 7B also demonstrates a similarity score of 0.8460, slightly below the acceptable threshold, suggesting that its responses are generally consistent with the expected standards. Zephyr, with a similarity score of 0.8420, falls marginally below the threshold. Although it demonstrates reasonable alignment, there is a slight deviation from the desired level of consistency. Copilot Precise records the lowest score of 0.8400, just below the threshold, indicating that its responses may require some refinement to adequate align with the guideline. Overall, ChatGPT 3.5, LLAMA 2 7B, and Mistral 7B exceed or meet the threshold, indicating high reliability. In contrast, Zephyr and Copilot Precise, while close, fall slightly short of the acceptable similarity score, suggesting areas for improvement to enhance their guideline adherence.

Figure 3 quantitatively illustrates the temporal variability of responses generated by five different LLMs: Chat GPT 3.5, Zephyr, LLAMA2 7B, Microsoft Copilot, and Mistral 7B. Higher variability scores indicate greater fluctuations in the responses over time, reflecting less consistency.

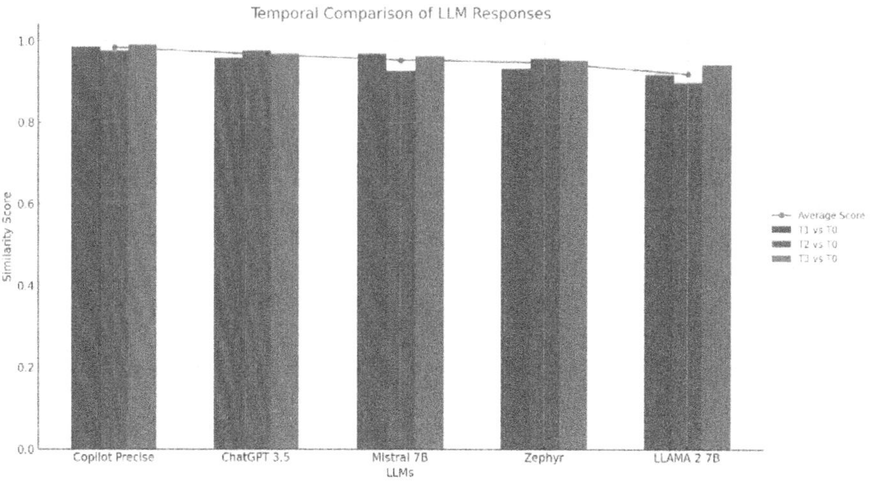

Fig. 3. Temporal response variability of five LLMs, indicating their consistency over time.

Figure 3 quantitatively illustrates the temporal variability of responses generated by five different Large Language Models (LLMs): Chat GPT 3.5, Zephyr,

LLAMA2 7B, Microsoft Copilot, and Mistral 7B. Higher variability scores indicate greater fluctuations in the responses over time, reflecting less consistency. Copilot Precise achieves the highest average similarity score, with individual scores of 0.9850 for T1 vs T0, 0.9760 for T2 vs T0, and 0.9910 for T3 vs T0. This demonstrates exceptional consistency and stability in its responses over time, indicating high reliability. ChatGPT 3.5 follows closely, exhibiting high similarity scores of 0.9580, 0.9760, and 0.9690 for T1, T2, and T3 comparisons, respectively, indicating strong alignment and consistent performance across different time points. Mistral 7B shows strong temporal consistency with similarity scores of 0.9690, 0.9270, and 0.9610 for T1, T2, and T3 comparisons. Despite a slight decrease at T2, Mistral 7B maintains a high overall similarity, suggesting reliable performance. Zephyr demonstrates reasonably high similarity scores of 0.9310, 0.9560, and 0.9500 for T1, T2, and T3 comparisons, reflecting a reliable level of consistency, although slightly lower than the top performers. LLAMA 2 7B exhibits the lowest average similarity scores among the evaluated LLMs, with scores of 0.9170, 0.8970, and 0.9410 for T1, T2, and T3 comparisons. This indicates more variability in its responses over time, suggesting areas for improvement to enhance stability and reliability. The addition of the average similarity scores in red provides a clear visual comparison of the overall performance of each LLM, highlighting the importance of temporal consistency in evaluating LLM performance. Copilot Precise and ChatGPT 3.5 emerge as the most reliable models, while LLAMA 2 7B shows the highest fluctuations, indicating potential areas for enhancement.

5   Conclusion

This study assessed five free of charge advanced Large Language Models (LLMs)-Zephyr, Mistral 7B, LLAMA 2 7B, ChatGPT 3.5, and Copilot Precise-by evaluating their responses to a clinical case against established guidelines using a Delphi consensus and a five-point Likert scale. The models were tested on their ability to ensure safety, accuracy, provide real references, and drive improvements. Preliminary results indicate that ChatGPT 3.5 and Copilot Precise showed respectively acceptable and poor adherence to guidelines and consistent performance over time. Mistral 7B, Zephyr and LLAMA 2 7B showed potential but required further enhancements, particularly in providing accurate references and maintaining consistency. However, it is imperative to underscore the necessity for exhaustive and detailed evaluations of LLM responses to all potential critical issues within each clinical domain. Furthermore, rigorous verification of their specific fine-tuning, in accordance with established best practices, is essential before these systems could be practically implemented in clinical settings. These findings underscore the importance of selecting LLMs that not only adhere to clinical guidelines but also maintain consistency over time. Future research should focus on refining these models to improve their accuracy and stability. Additionally, exploring more diverse clinical scenarios and incorporating real-time feedback mechanisms could further enhance their applicability in clinical decision-support systems, ultimately improving patient safety and care outcomes.

Conflict of Interests. The authors declare no conflicts of interest.

References

1. Goodfellow, I., Bengio, Y., Courville, A.: Deep Learning. MIT Press, Cambridge (2016)
2. Brown, T., et al.: Language models are few-shot learners. Adv. Neural. Inf. Process. Syst. **33**, 1877–1901 (2020)
3. Topol, E.: Deep Medicine: How Artificial Intelligence Can Make Healthcare Human Again. Basic Books, New York (2019)
4. World Health Organization: Ethics and Governance of Artificial Intelligence for Health: WHO Guidance. World Health Organization (2022)
5. Patel, S., Thakar, S., Esposito, A.: Leveraging AI for medical translation: challenges and opportunities. Health Inform. J. **28**(3), 14604582221112364 (2022)
6. Miotto, R., Wang, F., Wang, S., Jiang, X., Dudley, J.T.: Deep learning for healthcare: review, opportunities and challenges. Brief. Bioinform. **24**(2), bbx044 (2023)
7. Wang, Y., Kung, L., Byrd, T.A.: Big data analytics: understanding its capabilities and potential benefits for healthcare organizations. Technol. Forecast. Soc. Change **146**, 344–352 (2023)
8. Kocaballi, A.B., Laranjo, L., Coiera, E.: Measuring the impact of conversational interfaces on health communication: a systematic review. J. Med. Internet Res. **24**(6), e37908 (2022)
9. Steinberg, E., Lee, K.S., Saleh, M.N.: The role of AI in clinical decision-making: challenges and future directions. J. Med. Syst. **47**(2), 12 (2023)
10. Floridi, L., et al.: AI4People-an ethical framework for a good AI society: opportunities, risks, principles, and recommendations. Minds Mach. **28**(4), 689–707 (2022)
11. Leslie, D., Holmes, D., Hitrova, C., Floridi, L.: Ethics of AI in health care: a mapping review. BMC Med. Ethics **23**(1), 1–17 (2022)
12. Arora, A., Arora, A.: The promise of large language models in health care. Lancet **401**(10377), 641 (2023)
13. Wang, Y., Zhao, Y., Petzold, L.: Are large language models ready for healthcare? A comparative study on clinical language understanding. In: Machine Learning Healthcare Conference, pp. 804–823 (2023)
14. Umerenkov, D., Zubkova, G., Nesterov, A.: Deciphering diagnoses: how large language models explanations influence clinical decision making. arXiv preprint arXiv:2310.01708 (2023)
15. Lee, P.C., Sharma, S.K., Motaganahalli, S., Huang, A.: Evaluating the clinical decision-making ability of large language models using MKSAP-19 cardiology questions. JACC Adv. **2**(9), 100658 (2023)
16. Gottlieb, S., Silvis, L.: How to safely integrate large language models into health care. JAMA Health Forum **4**(9), e233909–e233909 (2023)
17. Velupillai, S., et al.: Using clinical natural language processing for health outcomes research: overview and actionable suggestions for future advances. J. Biomed. Inform. **88**, 11–19 (2018)
18. Perlis, R.H., Fihn, S.D.: Evaluating the application of large language models in clinical research contexts. JAMA Netw. Open **6**(10), e2335924–e2335924 (2023)

19. Turpen, T., Matthews, L., Guney, C.: Beneath the surface of talking about physicians: a statistical model of language for patient experience comments. Patient Exp. J. **6**(2), 51–58 (2019)
20. Hayakawa, S., Pan, Y., Marian, V.: Language changes medical judgments and beliefs. Int. J. Biling. **26**(1), 104–121 (2022)
21. Hochberg, L., et al.: Towards automatic annotation of clinical decision-making style. In: Proceedings of the LAW VIII-The 8th Linguistic Annotation Workshop, pp. 129–138 (2014)
22. U.S. Food and Drug Administration: Valium (diazepam) medication guide (2020). https://www.accessdata.fda.gov/drugsatfda_docs/label/2020/013263s094lbl.pdf. Accessed 14 May 2024
23. Hikma Pharmaceuticals USA: Flumazenil Injection, USP. MedLibrary.org. https://medlibrary.org/lib/rx/meds/flumazenil-19/. Accessed 15 May 2024
24. Razavizadeh, A.S., Zamani, N., Ziaeefar, P., Ebrahimi, S., Hassanian-Moghaddam, H.: Protective effect of flumazenil infusion in severe acute benzodiazepine toxicity: a pilot randomized trial. Eur. J. Clin. Pharmacol. **77**, 547–554 (2021)
25. MacDonald, T., Gallo, A., Basso-Hulse, G., Bennett, K., Hulse, G.K.: A double-blind randomised crossover trial of low-dose flumazenil for benzodiazepine withdrawal: a proof of concept. Drug Alcohol Depend. **236**, 109501 (2022)
26. Gallo, A.T., Hulse, G.: Pharmacological uses of flumazenil in benzodiazepine use disorders: a systematic review of limited data. J. Psychopharmacol. **35**(3), 211–220 (2021)
27. Richmond, J.S., et al.: Verbal de-escalation of the agitated patient: consensus statement of the American association for emergency psychiatry project BETA de-escalation workgroup. West. J. Emerg. Med. **13**(1), 17–25 (2012)
28. Dart, R.C., et al.: Expert consensus guidelines for stocking of antidotes in hospitals that provide emergency care. Ann. Emerg. Med. **71**(3), 314–325 (2018)
29. Hussein, M., Pavlova, M., Ghalwash, M., Groot, W.: The impact of hospital accreditation on the quality of healthcare: a systematic literature review. BMC Health Serv. Res. **21**, 1–12 (2021)
30. Pohl, S., Battistelli, A., Djediat, A., Andela, M.: Emotional support at work: a key component for nurses' work engagement, their quality of care and their organizational citizenship behaviour. Int. J. Afr. Nurs. Sci. **16**, 100424 (2022)
31. Chiappinotto, S., Palese, A., Longhini, J., et al.: Le videochiamate tra pazienti e familiari: Una revisione narrativa. Assist. Inferm. Ric. **41**(3), 120–128 (2022)
32. LaSala, C.A., Connors, P.M., Pedro, J.T., Phipps, M.: The role of the clinical nurse specialist in promoting evidence-based practice and effecting positive patient outcomes. J. Contin. Educ. Nurs. **38**(6), 262–270 (2007)
33. Copanitsanou, P., Fotos, N., Brokalaki, H.: Effects of work environment on patient and nurse outcomes. Br. J. Nurs. **26**(3), 172–176 (2017)
34. Zaghini, F.: The influence of work context and organizational well-being on psychophysical health of healthcare providers. Med. Lav. **111**(4), 306 (2020)
35. Palese, A., et al.: A path analysis on the direct and indirect effects of the unit environment on eating dependence among cognitively impaired nursing home residents. BMC Health Serv. Res. **19**, 1–14 (2019)
36. Sblendorio, E., et al.: Assessment of stress levels using technological tools: a review and prospective analysis of heart rate variability and sleep quality parameters. Neurodegener Dis **4**, 5 (2023)
37. Verrusio, W., Moscucci, F., Cacciafesta, M., Gueli, N.: Mozart effect and its clinical applications: a review. Br. J. Med. Med. Res. **8**(8), 639–650 (2015)

38. Gualandi, R., Masella, C., Piredda, M., Ercoli, M., Tartaglini, D.: What does the patient have to say? Valuing the patient experience to improve the patient journey. BMC Health Serv. Res. **21**, 1–12 (2021)
39. Tunstall, L., et al.: Zephyr: direct distillation of LM alignment. arXiv preprint arXiv:2310.16944 (2023)
40. Jiang, A.Q., et al.: Mistral 7B. arXiv preprint arXiv:2310.06825 (2023)
41. Touvron, H., et al.: Llama 2: open foundation and fine-tuned chat models. arXiv preprint arXiv:2307.09288 (2023)
42. Abdullah, M., Madain, A., Jararweh, Y.: ChatGPT: fundamentals, applications and social impacts. In: Proceedings of the SNAMS, pp. 1–6 (2022)
43. Ormerod, M., Martínez del Rincón, J., Devereux, B.: Predicting semantic similarity between clinical sentence pairs using transformer models: evaluation and representational analysis. JMIR Med. Inform. **9**(5) (2021)

Three-Stage Data Science Methodology to Explore Genetic Heterogeneity of Diseases

Silvia Cascianelli[(✉)][iD], Cristina Iudica, and Marco Masseroli[iD]

Dipartimento di Elettronica, Informazione e Bioingegneria, Politecnico di Milano, Milano, Italy
{silvia.cascianelli,marco.masseroli}@polimi.it,
cristina.iudica@mail.polimi.it

Abstract. Genetic heterogeneity poses a significant challenge in understanding complex diseases, as variations in the genetic makeup of individuals can lead to diverse disease manifestations and treatment responses. Here, we propose a three-stage data science methodology designed to systematically explore and analyze the genetic heterogeneity of a given disease, particularly focusing on critical patient subgroups. The proposed approach consists of a feature engineering phase, where various feature space options are devised and compared, a supervised learning framework for accurately classifying the patient subgroup of interest, and a final stage devoted to feature prioritization to identify gene variants with predictive and, potentially, therapeutic value. To this final aim, our methodology includes feature importance analysis and further exploration of clinically relevant and actionable genes involved in mutational events contributing to patient differentiation. As an application use case, we apply this methodology to investigate the mutational landscape of the critical subgroup of Triple-Negative Breast Cancer patients, demonstrating its validity in uncovering significant gene variants with possible therapeutic implications. This three-stage methodology offers a robust approach for advancing research into disease genetic heterogeneity and contributing to improving personalized treatment for patients.

Keywords: Feature engineering · Supervised models · Feature importance · Gene variant prioritization · Actionability

1 Introduction

The rapid advancement of Next-Generation Sequencing (NGS) technologies has dramatically increased the availability of mutational data, providing new opportunities to explore the genetic complexity and heterogeneity of diseases through Machine Learning (ML) techniques. By leveraging these techniques and the growing size of genomic datasets, predictive models can be developed to identify

L. Cerulo et al. (Eds.): CIBB 2024, LNBI 15276, pp. 150–164, 2025.
https://doi.org/10.1007/978-3-031-89704-7_12

genes associated with clinical outcomes or drug sensitivity, including therapeutic target (actionable) genes, potentially relevant in contributing to improving patient personalized treatments and management.

However, large-scale analyses, particularly from Whole-Exome Sequencing (WES) studies, pose significant challenges by generating extensive and sparse lists of mutated genes for each profiled patient. Using such data as input for ML models leads to highly numerous and sparse features, even if a gene-level feature space is adopted. Thus, subsets of relevant genes, based on prior knowledge or known disease drivers, are often selected to reduce feature space sizes, the curse of dimensionality and the overfitting risks. However, this limits the discovery of meaningful roles for novel, previously overlooked mutational events. Alternatively, the selection can be based on the statistical significance of gene mutations. Yet, many analytical methods estimating the statistical significance of mutated genes in a population yield numerous false positives, which can jeopardize the identification of truly relevant mutational events. Lastly, gene-level mutational information, despite being the most commonly extracted from mutation data, does not consider further details like the types of mutations or their distribution along the genes.

In this article, we propose a novel three-stage methodology that combines feature engineering, supervised classification, and feature importance analysis to effectively explore the genetic heterogeneity of critical patient subgroups. Unlike traditional approaches, our method leverages the full information content of mutational data to differentiate the mutational landscape of patient subgroups of complex clinical handling. The primary goal of our supervised analysis is not to classify patients, but to identify clinically relevant gene variants that could be potential therapeutic targets, relying on classification results and predictive features for well-known stratifications.

We validated this approach on Triple-Negative Breast Cancer (TNBC) patients, a particularly challenging subgroup known for poor responses to standard treatments. The collected results and findings emphasise the value of our three-stage Data Science methodology to identify:

- the most important variants for diseased patient stratification
- clinically relevant information about the presence of noteworthy mutational hotspots (i.e., genomic regions exhibiting a significantly higher frequency of mutations compared to surrounding areas)
- actionable genes, which could be potential therapeutic targets.

Particularly, training supervised models on relevant mutational-based features emerged as a paramount strategy for determining the optimal mutational feature space among those under examination. This enables the subsequent identification of the most influential variants and actionable gene candidates, offering new hints to advance treatment options for a critical patient subgroup.

2 Proposed Three-Stage Data Science Methodology

We conceived and implemented a three-stage methodology able to convert the search for clinically relevant or actionable gene variants into a Data Science process within a mutational-based patient stratification scenario. The proposed methodology includes:

- a comparative evaluation of multiple feature engineering options, obtained by considering significantly mutated genes, variant differentiation, and analysis of mutational hotspots;
- a supervised learning framework to evaluate the performance of different classifiers in recognizing critical patient subgroups of interest based on the alternative mutational feature spaces previously built;
- a feature importance analysis to prioritize the most crucial mutational events for the patient stratification of interest, and further search for gene variants with potential clinical relevance and therapeutic actionability.

Each of these three stages, respectively named 'Mutational feature engineering', 'Supervised stratification', and 'Variant prioritization and search for potential actionability', is detailed in the following subsections.

2.1 Mutational Feature Engineering

To obtain distinct mutational-based feature spaces, offering increasing levels of information, we use three feature engineering strategies. These leverage statistical tests to recognise significantly mutated genes and then possibly extract variant types or mutational hotspots. Our goal is indeed to compare feature spaces considering and combining a pre-selection of significantly mutated genes within the critical patient subgroup of interest, the variant type based on the translation effect of each mutation, and the mutational hotspots detected within the products of the significantly mutated genes. Genes can indeed contain mutational hotspots when solitary amino acids or stretches of amino acids in their gene products show elevated mutation frequency.

For significantly mutated gene identification, we apply the MutSig2CV [1] algorithm, which analyses mutational data in the Mutation Annotation Format (MAF) and identifies genes with mutation rates higher than the background mutation rate expected by random chance within a given patient population. Indeed, MutSig2CV uses three statistical tests to evaluate the gene mutation rates, the distributions of gene mutations and their presence in evolutionarily conserved regions. Then, it aggregates the so-obtained p-values into a joint p-value to assess mutation significance across the population and computes a q-value using the Benjamini-Hochberg correction [2] to control for false positives. Given the importance of exploring a wide mutational landscape for the critical patient subgroups, which are typically a limited portion of the entire disease cohort, in our methodology we opt to exclude only genes deemed as clearly non-significant in the subgroup of interest, retaining those with a q-value smaller than 1 based on MutSig2CV analysis. This prevents an overly restrictive initial selection of genes, ensuring a comprehensive investigation of mutational variants and hotspots potentially contributing to patient stratification.

In addition, our feature engineering strategy can interestingly consider the variant type, reported as *Variant_Classification* in MAF files, to build a feature space offering information about variant differentiation. This attribute indicates the translational effect of the variant allele on the gene product (i.e., missense, nonsense, etc.), and is employed to differentiate gene variants based on their consequence on the gene products. In fact, taking into account the functional impact and severity of a mutational event can be precious in better understanding disease progression mechanisms and aggressiveness in a critical patient subpopulation.

Finally, our feature engineering strategy can use hotspot detection to further increase the information content for the last feature space under examination. Indeed, hotspot localization within the significantly mutated genes of a critical patient population can be crucial for understanding the functional consequences of mutations on gene products and elucidating the molecular mechanisms driving, among others, disease onset and treatment resistance or sensitivity. To detect significant mutational hotspots within the subset of mutated genes previously identified by MutSig2CV, we employ the Mutation Cluster Smith-Waterman (MutClustSW) algorithm [3]. MutClustSW is designed to identify and estimate the correct size of mutational hotspots in the amino acid sequence of a mutated gene product (i.e., in a polypeptide). MutClustSW transforms the polypeptide into a one-dimensional vector representing mutation frequency scores and uses a scoring system similar to the one in the Smith-Waterman algorithm but tailored for a single sequence. Each position on the polypeptide is assigned a score that increases as the mutation frequency at that position rises, while non-mutated positions are given negative scores. MutClustSW addresses this as a maximum subarray problem, identifying segments with positive scores. It iteratively merges nearby clusters in a single hotspot based on a defined threshold: this process continues until no further positive-scoring segments are identified. Ultimately, from the algorithm we extract the location of the statistically significant hotspots, which are selected based on a p-value smaller than 0.05, specifying their starting and ending positions on the polypeptide.

All the previously obtained information enables the conversion of mutational data into alternative mutational-based feature spaces, all starting from the identified significantly mutated genes. Specifically, we consider the following options:

- A gene-level feature space specifying for each mutational feature only the *Hugo gene symbol* of the involved significantly mutated gene.
- A feature space specifying *Hugo gene symbol and Variant classification* for each mutational feature from all the involved significantly mutated genes (e.g., *TP53_Nonsense* or *TP53_Missense*).
- A feature space specifying *Hugo gene symbol, Variant classification, and (if any) hotspot location*, in the corresponding protein, for each mutational feature from all the involved significantly mutated genes (e.g., *TP53*

Nonsense noClust, if not in a hotspot, or *BRAF_Missense_600_600*, if in a unitary length hotspot).

The first one is a filtered feature space, adopting commonly used gene-level summarization of mutational events and focusing only on the significantly mutated genes identified in the subpopulation of critical patients. In contrast, the second feature space innovatively treats each distinct variant type within these significantly mutated genes as a separate feature, thereby capturing granular mutational patterns often overlooked in state-of-the-art gene-level analyses. Finally, the last space goes even further by incorporating additional features that represent specific mutational hotspots detected within the amino acid products of any considered variant type for all the significantly mutated genes. These second and third feature spaces significantly enhance the specificity and depth of information compared to the current state-of-the-art approaches, which typically rely solely on using gene-level mutational data. By capturing variant-level and hotspot-level information, our approach provides a more detailed and actionable view of the mutational landscape. While this growing information enriches the investigation, it also progressively increases the size and sparsity of the corresponding feature spaces used for the subsequent Machine Learning-based classification models. Therefore, to tackle these limitations, the initial selection of significantly mutated genes in the critical patient subpopulation plays a key role in the feasibility and reliability of the downstream classification analysis, limiting the genes under exam without being overly stringent.

2.2 Supervised Stratification

In the second stage of our methodology, we perform patient stratification through a supervised learning framework to learn to recognize the critical patient subgroup of interest. Indeed, the patient stratification of interest is converted into a binary classification task, taking as input an occurrence matrix at a time, corresponding to each feature space under evaluation and used to train alternative classification models. These occurrence matrices include patients along the rows with their mutational features along the columns and are filled in by counting and normalizing the number of mutational events occurring on each patient, as formally indicated in Fig. 1. Notably, certain patients may be excluded during the construction of each occurrence matrix, if they do not have any of the mutational features selected.

Several Machine Learning-based classifiers, including different Logistic Regression, Support Vector Machine (SVM) and Tree-based models, are assessed to compare feature space contributions regardless of the performance of a specific classification model. Indeed, searching for the classifier and feature space whose combination yields the best performance is not the ultimate goal of our proposed methodology: it serves to guide the following feature importance analysis for variant prioritization.

2.3   Variant Prioritization and Search for Potential Actionability

The last stage of our methodology is devoted to variant prioritization and search for potentially actionable gene variants. Variant prioritization involves feature importance analysis based on SHapley Additive exPlanations (SHAP) method [4], which is used to analyse the best classification solution, i.e., its mutational features and classification model. SHAP calculates the contribution of each feature to the classification output by examining the effect of including or excluding that feature across all possible subsets of features. Features are ranked based on their contribution to the patient stratification, to be then further investigated. Indeed, the search for clinically relevant and potentially actionable gene variants can extract gene candidates to help improve patient stratification and find alternative therapeutic options for difficult-to-treat patient subgroups. This search for actionable gene candidates relies primarily on two valuable open-access resources: the Therapeutic Target Database (TTD) [5] and the Precision Oncology Knowledge Base (OncoKB) [6,7]. The TTD provides comprehensive information on molecular targets of therapies, corresponding drugs/ligands and already associated alterations and diseases, all sourced from state-of-the-art literature and therapies. OncoKB is a curated collection of detailed information from clinical trials, scientific research, and Food and Drug Administration (FDA) approved therapies on the effects and treatment implications of specific cancer mutations; these latter ones are categorized based on their potential as therapeutic targets, their prognostic significance, and whether they are biomarkers for drug resistance or sensitivity.

	Mut_feature_1	Mut_feature_2	...	Mut_feature_k	...	Mut_feature_M
Patient_1	$count_{1,1}/\Sigma_j\, count_{1,j}$	$count_{1,2}/\Sigma_j\, count_{1,j}$	...	$count_{1,k}/\Sigma_j\, count_{1,j}$	...	$count_{1,M}/\Sigma_j\, count_{1,j}$
Patient_2	$count_{2,1}/\Sigma_j\, count_{2,j}$	$count_{2,2}/\Sigma_j\, count_{2,j}$	...	$count_{2,k}/\Sigma_j\, count_{2,j}$	...	$count_{2,M}/\Sigma_j\, count_{2,j}$
...	...	...	...	...	...	...
Patient_i	$count_{i,1}/\Sigma_j\, count_{i,j}$	$count_{i,2}/\Sigma_j\, count_{i,j}$	...	$count_{i,k}/\Sigma_j\, count_{i,j}$	...	$count_{i,M}/\Sigma_j\, count_{i,j}$
...	...	...	...	...	...	...
Patient_N	$count_{N,1}/\Sigma_j\, count_{N,j}$	$count_{N,2}/\Sigma_j\, count_{N,j}$	...	$count_{N,k}/\Sigma_j\, count_{N,j}$	...	$count_{N,M}/\Sigma_j\, count_{N,j}$

Fig. 1. Occurrence matrix for each feature space under exam, considering a total of N patients and M mutational features.

3   Application Use Case

Our application use case is focused on analysing the genetic heterogeneity of the critical Triple-Negative Breast Cancer patient subgroup. Despite the patient partition being easily obtainable based on transcriptomic data due to the TNBC lack of expression for estrogen receptor (ER), progesterone receptor (PR), and human epidermal growth factor receptor-2 (HER2), the impact on this critical subgroup of many occurring mutations is largely unknown. Similarly, the unexplored potential drug actionability of some of the most involved gene variants could be precious for alternative treatment options. In fact, TNBC is a particularly aggressive form of breast cancer, and the affected patients are typically unresponsive to endocrine and HER2-targeted therapies, leading to a poorer overall prognosis. In this relevant application use case, our three-stage methodology can investigate how to reconstruct this binary patient stratification based on mutational data and identify the most interesting mutational features and corresponding mutated genes in the TNBC subgroup, towards the discovery of new therapeutic options.

3.1   Data and Their Pre-processing

Somatic mutation datasets of breast cancer patients from Whole-Exome Sequencing (WES) experiments and aligned to the hg19 reference genome were used as input data. Specifically, we integrated datasets in Mutation Annotation Format (MAF files) from different sources [8–15]. We investigated mutation rate distribution across all the patients, by computing mutation rates as the total sum of mutation lengths for each patient divided by the total sum of gene lengths for all the genes under exam. Overall mutation rate analysis was used to exclude hyper- and hypo-mutated patients, who could alter subsequent steps of our investigation.

Table 1 reports the summary information for the integrated dataset, which accounts for more than 1.5 thousand patients. Yet, it also clearly highlights the elevated imbalance between TNBC (19.4%) and non-TNBC (80.6%) patients, which represents an additional challenge for Machine Learning classification. Imbalanced data can indeed lead to overfitting classifiers with biased predictions towards the majority class, failing to recognize the minority class as well as to generalize over unseen data. Notably, to address this strong class imbalance challenge, in our binary classification task we employed random undersampling of the majority class (non-TNBC) to achieve a balance with the minority class (TNBC), before the training (75%) and test (25%) sets splitting. Despite each undersampling being randomized, the approach was meant to ensure that each Machine learning model under exam had the greatest extent possible of the same training and testing patients to facilitate comparisons. Machine Learning models included non-regularized Logistic Regression, Lasso Logistic Regression, Ridge Logistic Regression, SVM with linear, polynomial or radial kernels, Random Forest and Extreme Gradient Boosting (XGBoost) models. A 5-fold cross-validation approach was then used in conjunction with grid search for hyperparameter tuning and training of all the classifiers under exam by optimizing their accuracy metric. Particularly, these classifiers were trained and tested using all the available samples and corresponding class labels and considering each of the three mutational-based feature space options, one at a time. This process ensured that each feature space option was examined independently and exhaustively, with a thorough evaluation of different classifiers when varying input representations based on the resolution level adopted for the mutational features.

Table 1. Summary information for the integrated dataset.

Integrated dataset	Mutated patients	Somatic mutations	Mutated genes
Total patients	1,554	182,830	19,985
non-TNBC	1,252 (80.6%)	131,498 (71.9%)	19,219
TNBC	302 (19.4%)	51,332 (29.1%)	15,570

TNBC: Triple-Negative Breast Cancer.

3.2 Results and Discussion

In this section, we present and discuss the findings obtained by applying our proposed Data Science approach to the aforementioned application use case.

Mutational-Based Feature Spaces and Supervised Stratification. Overall, all models based on the most informative feature space of *Hugo gene Symbol, Variant classification and (if any) hotspot location* reached good performances, especially considering the generalization in testing, where they resulted superior compared to the same models tuned and trained using the other two feature space options. The cross-validation results reported in Table 2 highlight other advantages of leveraging this richer representation of mutational data for predictive modeling. The models trained on the most informative (henceforth called preferred) feature space mainly overcome those trained on a simpler gene-level space (Hugo gene Symbol encoding), which is commonly adopted at the current state-of-the-art and was therefore used as a benchmark. Indeed, on the preferred feature space, we can appreciate very often improved performance across multiple classifiers, except for some SVM models. Particularly, XGBoost and Ridge Logistic Regression emerge as top classifiers across several metrics. Despite Ridge Logistic Regression achieving higher recall in recognizing the TNBC class, it struggled more with the non-TNBC class. Conversely, the XGBoost provided the most balanced performance across both classes, with strong values for all metrics, making it the most reliable and robust candidate for subsequent investigations.

By examining the independent testing results reported in Table 3 for the models trained on the preferred feature space, we can notice that the accuracy

Table 2. 5-fold cross-validation results. *Hugo gene Symbols, Variant classification and location (if any) of the hotspot* as PREFERRED feature space option (146 features from 23 mutated genes) compared with GENE-LEVEL (Hugo gene Symbol) feature space as benchmark. Training set (75%): 408 patients; 1-fold: 82 patients. The most promising classifier/s are in bold

Feature space: Classifier	Accuracy	Recall		Precision		F1-score	
		Sensitivity	Specificity	1	0	1	0
GENE-LEVEL: Full Logistic Regression	0.69 (± 0.034)	0.78 (± 0.012)	0.60 (± 0.08)	0.66 (± 0.029)	0.72 (± 0.09)	0.71 (± 0.051)	0.66 (± 0.031)
PREFERRED: Full Logistic Regression	0.69 (± 0.094)	0.79 (± 0.033)	0.59 (± 0.17)	0.67 (± 0.086)	0.72 (± 0.1)	0.72 (± 0.061)	0.64 (± 0.15)
GENE-LEVEL: Lasso Logistic Regression	0.70 (± 0.031)	0.80 (± 0.1)	0.60 (± 0.071)	0.67 (± 0.028)	0.77 (± 0.08)	**0.73 (± 0.043)**	0.67 (± 0.031)
PREFERRED: Lasso Logistic Regression	0.71 (± 0.071)	0.79 (± 0.062)	0.62 (± 0.18)	0.69 (± 0.079)	0.74 (± 0.04)	**0.73 (± 0.038)**	0.66 (± 0.14)
GENE-LEVEL: Ridge Logistic Regression	0.70 (± 0.031)	0.79 (± 0.13)	0.60 (± 0.082)	0.67 (± 0.024)	0.77 (± 0.095)	0.72 (± 0.053)	0.68 (± 0.022)
PREFERRED: Ridge Logistic Regression	**0.72 (± 0.078)**	0.80 (± 0.035)	0.63 (± 0.15)	0.69 (± 0.08)	0.75 (± 0.066)	**0.74 (± 0.053)**	0.68 (± 0.12)
GENE-LEVEL: Linear SVM	0.69 (± 0.043)	0.81 (± 0.1)	0.58 (± 0.05)	0.66 (± 0.026)	0.74 (± 0.091)	**0.73 (± 0.051)**	0.66 (± 0.038)
PREFERRED: Linear SVM	0.69 (± 0.069)	0.80 (± 0.072)	0.57 (± 0.14)	0.66 (± 0.064)	0.74 (± 0.067)	0.72 (± 0.05)	0.64 (± 0.12)
GENE-LEVEL: Polynomial SVM	0.71 (± 0.045)	0.82 (± 0.097)	0.60 (± 0.027)	0.67 (± 0.026)	0.78 (± 0.083)	**0.74 (± 0.055)**	0.68 (± 0.034)
PREFERRED: Polynomial SVM	0.69 (± 0.056)	**0.85 (± 0.087)**	0.53 (± 0.13)	0.65 (± 0.053)	**0.79 (± 0.074)**	0.74 (± 0.044)	0.68 (± 0.098)
GENE-LEVEL: Radial SVM	0.71 (± 0.041)	0.64 (± 0.12)	**0.77 (± 0.073)**	0.74 (± 0.036)	0.69 (± 0.061)	0.68 (± 0.066)	**0.72 (± 0.033)**
PREFERRED: Radial SVM	0.67 (± 0.068)	**0.94 (± 0.02)**	0.40 (± 0.013)	0.61 (± 0.056)	**0.86 (± 0.071)**	0.74 (± 0.041)	0.53 (± 0.14)
GENE-LEVEL: Random Forest	0.69 (± 0.042)	0.73 (± 0.10)	0.66 (± 0.059)	0.68 (± 0.031)	0.71 (± 0.071)	0.70 (± 0.057)	0.68 (± 0.056)
PREFERRED: Random Forest	0.70 (± 0.075)	0.79 (± 0.079)	0.61 (± 0.15)	0.68 (± 0.073)	0.74 (± 0.072)	0.72 (± 0.057)	0.66 (± 0.12)
GENE-LEVEL: XGBoost	0.71 (± 0.048)	0.81 (± 0.078)	0.61 (± 0.076)	0.68 (± 0.044)	0.77 (± 0.076)	**0.74 (± 0.046)**	0.68 (± 0.056)
PREFERRED: XGBoost	**0.72 (± 0.036)**	0.77 (± 0.045)	**0.68 (± 0.034)**	**0.70 (± 0.035)**	0.75 (± 0.041)	**0.73 (± 0.038)**	0.71 (± 0.035)

Table 3. Test results. *Hugo gene Symbols, Variant classification and location (if any) of the hotspot* as preferred feature space option (146 features from 23 mutated genes). Training set (75%): 408 patients; Test set (25%): 136 patients. The most promising classifier/s are in bold.

Classifier	Accuracy	Recall		Precision		F1-score		MCC	nMCC
		Sensitivity	Specificity	1	0	1	0		
Full Logistic Regression	0.72	0.76	0.68	0.70	0.74	0.73	0.71	0.44	0.72
Lasso Logistic Regression	**0.74**	**0.76**	**0.71**	**0.72**	**0.75**	**0.74**	**0.73**	**0.47**	**0.74**
Ridge Logistic Regression	**0.74**	**0.76**	**0.71**	**0.72**	**0.75**	**0.74**	**0.73**	**0.47**	**0.74**
Linear SVM	0.74	0.79	0.69	0.72	0.77	0.76	0.73	0.49	0.75
Polynomial SVM	0.74	0.79	0.68	0.71	0.77	0.75	0.72	0.47	0.74
Radial SVM	0.65	0.85	0.44	0.60	0.75	0.71	0.56	0.32	0.66
Random Forest	0.71	0.72	0.69	0.70	0.71	0.71	0.70	0.41	0.71
XGBoost	**0.73**	**0.74**	**0.72**	**0.72**	**0.73**	**0.73**	**0.73**	**0.46**	**0.73**

MCC: Matthews Correlation Coefficient;
nMCC: normalized MCC

is fairly consistent across the vast majority of models. This consistency is also confirmed by the Matthews Correlation Coefficient (MCC), the only metric that optimizes all four basic positive and negative rates simultaneously, providing a reliable balanced performance measure [16]. In addition, we reported also the nMCC, a normalized version of the MCC scaled in the [0; 1] range, commonly adopted to have the same value range and interpretation as the other well-established performance metrics. Overall the classification performances are not as high as those seen in many other machine learning-based classification tasks, but they remain particularly valuable given our investigation scenario and overall aim. In fact, the initial set of significantly mutated genes is derived solely from the MutSig2CV analysis of a critical patient subgroup of interest (the TNBC population), which accounted for only less than one-third of the total somatic mutations, as indicated in Table 1. Although the perspective for learning the supervised task was intentionally limited to the TNBC perspective, the tree-based XGBoost and the logistic regression models demonstrated their capability to handle such a challenging classification problem successfully.

Particularly, also in testing, besides the top-performing regularized logistic regressors, XGBoost remains one of the promising classifiers, confirming its robustness. This model indeed consistently achieved strong and stable performance metrics for both classes during the independent testing, as done in cross-validation experiments, where logistic regression models exhibited less robustness with higher variability between the two classes.

Given the convincing and consistent results for the majority of the assessed models, we confirmed the most informative *Hugo gene Symbol, Variant classification, and (if any) hotspot location* feature space as the preferred option and considered its mutational features for further investigation. This feature space comprised a total of 146 mutational features, representing variants from 23 different genes and including 41 identified hotspots. Subsequent feature importance analysis was conducted employing these mutational features together with the XGBoost predictor previously trained on them, to perform the final key step of variant prioritization.

Feature Prioritization and Promising Actionable Genes. The feature importance analysis step for variant prioritization was essential for identifying the most relevant mutational features and understanding their contribution to the critical patient stratification of interest. Through this step, we were able to inspect all the features in the selected feature space, and highlight crucial genetic variants that may drive future studies for improving TNBC patient clinical handling. The SHAP feature importance analysis revealed that 25 mutational features significantly contributed to patient stratification, for a total of 10 distinct involved genes. As we can observe in Fig. 2, which indicates the Shapley value of each influential feature, there is a majority of missense mutations (17), either sparse along the involved genes or localized in precise hotspots, as we can clearly notice for TP53 and PIK3CA genes.

TP53 and PIK3CA genes confirmed their great involvement in cancer disease, with 12 and 5 distinct variant types and localizations, respectively. Besides TP53 and PIK3CA mutations being well-established drivers of breast cancer, previous studies have also linked co-occurring TP53 and PIK3CA mutations to shorter disease-free survival, aligning with the aggressive nature of TNBC [17,18]. Furthermore, we retrieved other interesting mutational events concerning cancer-related genes. In fact, BRCA1 mutations are well-known for their role in hereditary breast and ovarian cancers [19]. PTEN is a tumour suppressor gene that regulates the crucial PI3K/AKT signaling pathway; thus, its loss of function mutation can result in uncontrolled cell growth [20]. Similarly, RB1 and

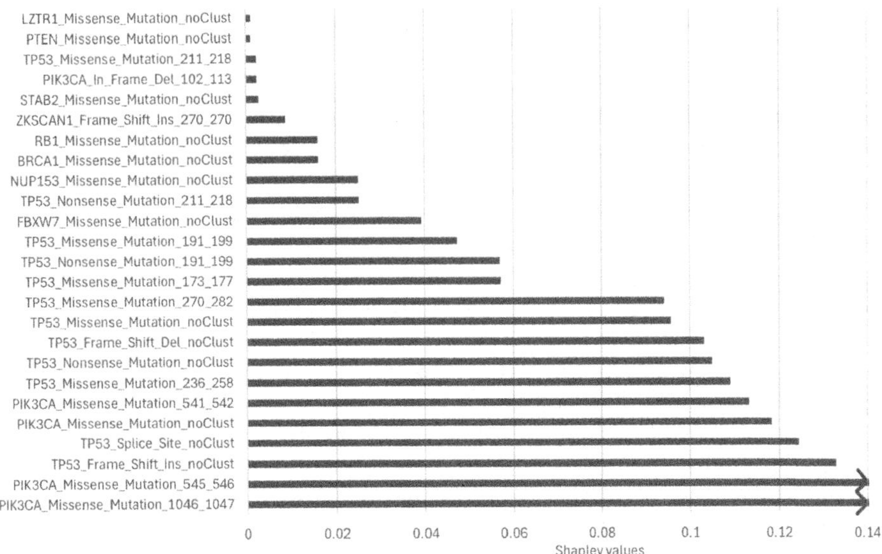

Fig. 2. Mutational features of major impact to patient stratification from the SHAP analysis of the fine-grained feature space of *Hugo gene Symbol, Variant classification, and (if any) hotspot location.*

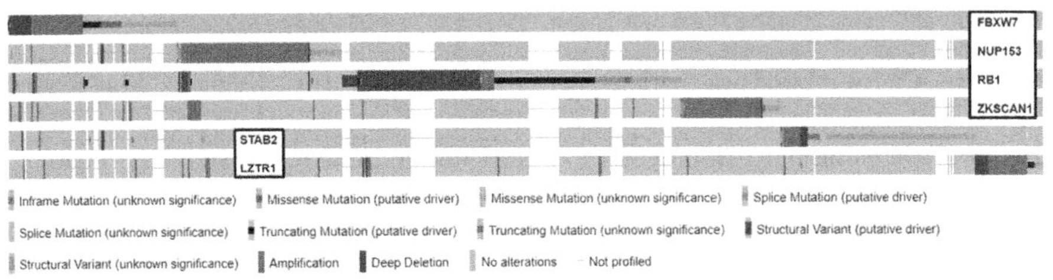

Fig. 3. All the alterations involving FBXW7, NUP153, RB1, ZKSCAN1, STAB2 and LZTR1 genes across 5,648 mutational profiles of invasive breast cancer patients from non-redundant studies available on CBioPortal [25, 26].

FBXW7 are crucial tumour suppressors involved in cell cycle regulation, and their loss of function is strongly associated with carcinogenesis, tumor metastasis and poorer outcomes in cancer, including more aggressive breast cancer subtypes [21–23]. Also, albeit less commonly discussed, NUP153 mutations and upregulation have been noted in some cancer studies (e.g., in hepatocellular carcinoma [24]) since they affect gene regulation, contributing to tumor onset and worse disease progression.

In addition, for all the emerged mutational features that do not involve the most well-established breast cancer-related genes, we further investigated the alteration rates and types of the corresponding genes (FBXW7, NUP153, RB1, ZKSCAN1, STAB2, LZTR1) across 5,648 mutational profiles of invasive breast cancer patients, obtained by WES or targeted sequencing, from the wide set of non-redundant studies available in cBioPortal [25,26]. Overall, these genes resulted in being altered in 12% of the assessed patients, with non-negligible evidence of missense mutations of still unknown significance, as we can see in Fig. 3. Most of them also show amplifications, except for RB1, which is mainly characterized by loss of function due to deep deletion and putative driver truncating mutations. Lastly, after this variant prioritization stage and the analysis of the emerged variants, we assessed the therapeutic potential (a.k.a. actionability) of the most relevant mutated genes using the TTD and OncoKB resources. Intriguingly, in TTD we identified two genes affected by non-clustered Missense mutations, FBXW7 and STAB2, which are both literature-reported potential therapeutic targets and should be further studied.

In OncoKB, FBXW7 is reported as a level 3A target (i.e., *Compelling clinical evidence supports the biomarker as being predictive of response to a drug in this indication but neither biomarker and drug are standard of care*) for a combination of Lunresertib + Camonsertib drugs in Ovarian and Endometrial Cancer [27]. Also, from further exploration in The Human Gene Database GeneCards) [28], FBXW7 emerged as strongly associated with colorectal cancer despite its mutations having already been detected in breast cancer cell lines. Specifically, FBXW7, as a tumor suppressor, promotes the degradation of various oncoproteins, including c-Myc, Notch, and cyclin E. Thus, FBXW7 mutations and loss of function can contribute to tumorigenesis and demonstrated clinical significance in several cancers, including breast cancer [23,29]. Furthermore, targeting or modulating FBXW7 activity holds promise for overcoming drug resistance in cancer treatment, particularly through strategies that enhance FBXW7-mediated degradation of anti-apoptotic proteins and sensitize cancer cells to therapy [22,30]. Recent studies also indicate STAB2 transcriptional co-factor as a biomarker and potential therapeutic target in various cancers, particularly associating its expression and epigenetic activity with tumour progression and metastasis [31]. In renal cell carcinoma, STAB2 emerged as an oncogenic chromatin organizer and targeting STAB2 showed effectiveness in suppressing tumor growth [32], suggesting its potential as a therapeutic target also for other cancer types. Yet, at the time of writing STAB2 has not been clearly linked to breast cancer, despite being already associated with another gynecological cancer, the Malignant Ovarian Brenner Tumor.

All these promising findings highlight the need for further in-depth investigation into these genes as potential therapeutic targets, exemplifying the effectiveness of our three-stage methodology in guiding future research to possibly improve the clinical management of a critical patient subgroup.

4   Conclusions

Our proposed Data Science computational framework has been designed to address the entire investigation workflow for exploring genetic heterogeneity in diseases, from mutational feature engineering to variant prioritization and assessment of actionability through supervised stratification strategies.

When applied to the case study of Triple-Negative Breast Cancer (TNBC), the methodology successfully identified influential genetic alterations, shedding light on new insights for patient stratification and subgroup characterization. Comparing the traditional gene-level encoding, commonly employed at the state-of-the-art, with the more comprehensive feature encoding preferred in this study demonstrates that the use of this latter not only enriches the information available for identifying potentially actionable gene variants but also improves the performance of the most promising classifiers. However, this increased resolution of information comes at the cost of a larger and sparser feature space, which represents the primary limitation to be addressed in our approach.

The exploratory analysis of TNBC highlighted particularly two actionable gene candidates (FBXW7 and STAB2), offering potential hints for advanced studies and novel targeted therapies that may improve treatment options for high-risk TNBC patients. Importantly, the fine-grained resolution brought by our methodology enabled us to pinpoint also specific details about many other relevant mutational events occurring in TNBC, which could be further explored, even from a pan-cancer perspective.

To enhance the robustness of the provided findings, further refinements of our methodology may include the joint evaluation of the significantly mutated genes and hotspots by leveraging a wider spectrum of statistical approaches, as well as the development of more sophisticated classification strategies, possibly also combining different feature engineering options and resolution levels. These improvements are expected to increase the sensitivity and reliability in detecting key genetic variations, not only improving classification performance but, mostly, the noteworthiness of the obtained findings. Moreover, incorporating innovative computational approaches that leverage advancements in AI-driven techniques for multi-omics data integration [33, 34] and cutting-edge feature selection strategies [35–38] could further enhance our methodology and the robustness of the obtained results.

Therefore, future applications of this three-stage methodology to different complex diseases and patient stratifications of interest hold significant potential. They could indeed elucidate novel insights into heterogeneous genetic traits of a given disease, paving the way for new research directions toward more effective treatments and improved clinical outcomes for patients affected by aggressive and difficult-to-treat disease subtypes.

Conflict of Interests. The authors declare to have no conflicts of interest.

References

1. Lawrence, M., et al.: Discovery and saturation analysis of cancer genes across 21 tumour types. Nature **505**(7484), 495–501 (2014)
2. Benjamini, Y., Hochberg, Y.: Controlling the false discovery rate: a practical and powerful approach to multiple testing. J. Roy. Stat. Soc.: Ser. B (Methodol.) **57**(1), 289–300 (1995)

3. Rhee, J.K., et al.: Identification of local clusters of mutation hotspots in cancer-related genes and their biological relevance. IEEE/ACM Trans. Comput. Biol. Bioinf. **16**(5), 1656–1662 (2018)
4. Lundberg, S.M., Lee, S.I.: A unified approach to interpreting model predictions. In: Advances in Neural Information Processing Systems, vol. 30 (2017)
5. Chen, Y., et al.: TTD: therapeutic Target Database describing target druggability information. Nucleic Acids Res. **52**, D1465–D1477 (2023)
6. Chakravarty, D., et al.: OncoKB: a precision oncology knowledge base. JCO Precis. Oncol. **1**, 1–16 (2017)
7. Suehnholz, S.P., et al.: Quantifying the expanding landscape of clinical actionability for patients with cancer. Cancer Discov. **14**(1), 49–65 (2024)
8. Banerji, S., et al.: Sequence analysis of mutations and translocations across breast cancer subtypes. Nature **486**(7403), 405–409 (2012)
9. Stephens, P., et al.: The landscape of cancer genes and mutational processes in breast cancer. Nature **486**(7403), 400–404 (2012)
10. Lefebvre, C., et al.: Mutational profile of metastatic breast cancers: a retrospective analysis. PLoS Med. **13**(12), 1436–1456 (2016)
11. Kan, Z., et al.: Multi-omics profiling of younger Asian breast cancers reveals distinctive molecular signatures. Nat. Commun. **9**(1), 1725 (2018)
12. Krug, K., et al.: Proteogenomic landscape of breast cancer tumorigenesis and targeted therapy. Cell **183**(5), 1436–1456 (2020)
13. Metastatic Breast Cancer Project. https://mbcproject.org/
14. Hoadley, K., et al.: Cell-of-origin patterns dominate the molecular classification of 10,000 tumors from 33 types of cancer. Cell **173**(2), 291–304 (2018)
15. Pallotta, S., Cascianelli, S., Masseroli, M.: RGMQL: scalable and interoperable computing of heterogeneous omics big data and metadata in R/Bioconductor. BMC Bioinform. **23**(1), 123 (2022)
16. Chicco, D., Jurman, G.: The Matthews Correlation Coefficient (MCC) should replace the ROC AUC as the standard metric for assessing binary classification. BioData Mining **16**(1), 4 (2023)
17. Meric-Bernstam, F., et al.: Survival outcomes by TP53 mutation status in metastatic breast cancer. JCO Precis. Oncol. **2**, 1–15 (2018)
18. Kotoula, V., et al.: Relapsed and de novo metastatic HER2-positive breast cancer treated with trastuzumab: tumor genotypes and clinical measures associated with patient outcome. Clin. Breast Cancer **19**(2), 113–125 (2019)
19. Hawsawi, Y.M., et al.: The role of BRCA1/2 in hereditary and familial breast and ovarian cancers. Mol. Genet. Genom. Med. **7**(9), e879 (2019)
20. Chen, J., et al.: Systemic deficiency of PTEN accelerates breast cancer growth and metastasis. Front. Oncol. **12**, 825484 (2022)
21. Wang, L., Zhang, S., Wang, X.: The metabolic mechanisms of breast cancer metastasis. Front. Oncol. **10**, 602416 (2021)
22. Fan, J., et al.: Clinical significance of FBXW7 loss of function in human cancers. Mol. Cancer **21**(1), 87 (2022)
23. Chen, S., Leng, P., Guo, J., Zhou, H.: FBXW7 in breast cancer: mechanism of action and therapeutic potential. J. Exp. Clin. Cancer Res. **42**(1), 226 (2023)
24. Gan, C., Zhou, K., Li, M., Shang, J., Liu, L., Zhao, Q.: NUP153 promotes HCC cells proliferation via c-Myc-mediated downregulation of P15INK4b. Dig. Liver Dis. **54**(12), 1706–1715 (2022)
25. Cerami, E., et al.: The cBio cancer genomics portal: an open platform for exploring multidimensional cancer genomics data. Cancer Discov. **2**(5), 401–404 (2012)

26. de Bruijn, I., et al.: Analysis and visualization of longitudinal genomic and clinical data from the AACR project GENIE biopharma collaborative in cBioPortal. Can. Res. **83**(23), 3861–3867 (2023)

27. Simpkins, F., et al.: Targeting CCNE1-amplified ovarian and endometrial cancers by combined inhibition of PKYMT1 and ATR. ESMO Open **9** (2024)

28. Stelzer, G., et al.: The GeneCards suite: from gene data mining to disease genome sequence analyses. Curr. Protoc. Bioinform. **54**(1), 1–30 (2016)

29. Yeh, C.H., Bellon, M., Nicot, C.: FBXW7: a critical tumor suppressor of human cancers. Mol. Cancer **17**, 1–19 (2018)

30. Ye, M., et al.: Targeting FBW7 as a strategy to overcome resistance to targeted therapy in non-small cell lung cancer. Can. Res. **77**(13), 3527–3539 (2017)

31. Roy, S.K., Shrivastava, A., Srivastav, S., Shankar, S., Srivastava, R.K.: SATB2 is a novel biomarker and therapeutic target for cancer. J. Cell Mol. Med. **24**(19), 11064–11069 (2020)

32. Jin, J., et al.: YAP-activated SATB2 is a coactivator of NRF2 that amplifies antioxidative capacity and promotes tumor progression in renal cell carcinoma. Can. Res. **83**(5), 786–803 (2023)

33. Ahmed, Z., Wan, S., Zhang, F., Zhong, W.: Artificial intelligence for omics data analysis. BMC Methods **1**(1), 4 (2024)

34. Wekesa, J.S., Kimwele, M.: A review of multi-omics data integration through deep learning approaches for disease diagnosis, prognosis, and treatment. Front. Genet. **14**, 1199087 (2023)

35. Gruca, A., Henzel, J., Kostorz, I., Stęclik, T., Wróbel, Ł, Sikora, M.: Maine: a web tool for multi-omics feature selection and rule-based data exploration. Bioinformatics **38**(6), 1773–1775 (2022)

36. Li, Y., Mansmann, U., Du, S., Hornung, R.: Benchmark study of feature selection strategies for multi-omics data. BMC Bioinform. **23**(1), 412 (2022)

37. Cascianelli, S., Galzerano, A., Masseroli, M.: Supervised relevance-redundancy assessments for feature selection in omics-based classification scenarios. J. Biomed. Inform. **144**, 104457 (2023)

38. Mongardi, S., Cascianelli, S., Masseroli, M.: Biologically weighted lasso: enhancing functional interpretability in gene expression data analysis. Bioinformatics **40**(10), btae605 (2024)

Functional Data Analysis and Clustering of Haematological Parameters in SARS-CoV-2 Patients

Daniela Volpatto[1,2,3], Marco Frattarola[2,3], Simone Pernice[2,3(✉)],
Francesca Cordero[2,3], Marco Beccuti[2,3], Giulio Mengozzi[4,5],
Stefano De Angelis[6], and Roberta Sirovich[1]

[1] Department of Mathematics "G. Peano", University of Turin, Turin, Italy
[2] Department of Computer Science, University of Turin, Turin, Italy
simone.pernice@unito.it
[3] CINI Infolife Laboratory, Roma, Italy
[4] Clinical Biochemistry Laboratory, Department of Medical Sciences, University of Turin, Turin, Italy
[5] Clinical Biochemistry Laboratory, City of Health and Science University Hospital, Turin, Italy
[6] UOC Laboratory Analysis ULSS2, Hospital of Castelfranco Veneto, Marca Trevigiana, Italy

Abstract. This study aims to evaluate haematological parameters of patients who have contracted the COVID-19 infection. In particular, we considered patients who were already hospitalised at the time of nasopharyngeal swab (NF) testing, and patients at the Emergency Department and/or who required hospitalisation following a positive result from the NF swab. The collected data are defined as longitudinal data (i.e. constituted by measurements accumulated sequentially over time), mainly characterised by observation times that are irregular, different for each patient, and distributed non-uniformly across the observation interval. In light of these considerations, we exploit CONNECTOR, a data-driven framework designed for longitudinal data, which returns a grouping of the curves of the haemochromocytometric parameters, based on a functional clustering algorithm. These clusters are analysed in terms of disease outcome and survival, finding a good correlation to mortality rates. Finally, the CONNECTOR clusters are exploited to stratify the patients based on profiles of the haemochromocytometric parameters evolution over time, showing that comorbidities appear to have an impact on mortality independently of the outcome of the monitoring carried out through the laboratory tests considered.

Keywords: Functional Data Analysis · Functional Clustering · Multivariate Longitudinal Data · COVID-19

1 Introduction

COVID-19, first identified in Wuhan, China, in late 2019, rapidly evolved into a global pandemic, officially declared by the World Health Organization (WHO)

L. Cerulo et al. (Eds.): CIBB 2024, LNBI 15276, pp. 165–179, 2025.
https://doi.org/10.1007/978-3-031-89704-7_13

on March 11, 2020. In Italy, the initial cases were detected on February 20, 2020, in Lombardy and Veneto, leading to a swift and severe outbreak across the country. SARS-CoV-2, classified as a class IV virus according to the Baltimore classification, is a single-stranded positive RNA virus (ssRNA+), closely related to SARS-CoV, the causative agent of the 2003 SARS epidemic. The virus primarily transmits via respiratory droplets and, to a lesser extent, via aerosols, with evidence suggesting possible transmission from contaminated surfaces.

The trends of haemochromocytometric parameters over time offer critical insights into the immune response and inflammatory status of COVID-19 patients. Monitoring these parameters is crucial for understanding disease progression, identifying patients at risk of severe complications, and optimising treatment strategies. Mathematically, this type of data is called *longitudinal data*, which provides a dynamic view of patient health over time, enabling the identification of trends and patterns that static measurements might miss. Classical methods, developed for finite-dimensional and independent observations, cannot be directly applied to longitudinal data (i.e., curves). The sampled variables are strongly correlated, and the problem becomes ill-conditioned in the context of multivariate linear models. By analysing data that vary over time, Functional Data Analysis (FDA) [1] provides the analytical ground to interpret longitudinal data. FDA approaches allow for a more nuanced understanding of disease progression and treatment efficacy, ultimately contributing to more precise and personalised medical care for COVID-19 patients. By leveraging longitudinal data, we aim to evaluate specific haemochromocytometric parameters in patients with COVID-19. We focus on patients who were already admitted at the time of their nasopharyngeal (NF) swab testing positive, as well as those who presented to the emergency department and/or required admission following a positive diagnosis. This analysis employs specialised functional clustering methods for sparse and irregular curves to address the complexity and variability inherent in the collected data.

2   Methods

2.1   The Cohort Description

We retrieved data from patients under the care of the *Azienda Ospedaliero Universitaria, Città della Salute e della Scienza* in Turin, from February 20th to April 30th, 2020. The patients were selected based on the COVID-19 positive results of a nasopharyngeal (NP) swab and were already hospitalised at the time of the test. The cohort consists of 558 patients (245 females, 313 males) with an average age of 63.6 years (minimum 13, maximum 97). The average length of hospitalisation was 15.5 days (minimum 1 day, maximum 222 days). The laboratory tests for these patients were collected using the DNLab software (Dedalus: Healthcare System Group), while clinical data were extracted from electronic medical records within the TrakCare software (Intersystems). The first positive swab test was conducted on average 10.8 days after the beginning of hospitalization. This analysis considers any admission from January 1,

2020, as the first day of hospitalisation. The first swab test resulting from the extractions was March 3, 2020, and the last swab test considered for this study was April 26, 2020. The collected laboratory tests are: blood count with leukocyte formula, reticulocytes and reticulocyte maturation indexes, erythroblasts, leukocyte cytofluorimetry, and nasopharyngeal swab testing. In this work we analyse all the parameters related to the blood counts: WBC (Leukocytes), Neutrophils, Lymphocytes, Monocytes, Eosinophils, Basophils, Thrombocytes (PLT) and Erythrocytes (RBC). Moreover, as suggested by the literature, we also analysed the ratio Neutrophils to Lymphocytes (NL ratio), see [2–4].

The clinical data were codified following the CIRS scale (Cumulative Illness Rating Scale) [5], commonly used in the geriatric field and useful in the context of this study to evaluate the chronic comorbidities present in the study population. The use of a geriatric scale is appropriate when considering that more than 75% of the subjects is over 57 years old. The CIRS scale provides a degree of severity (ranging from 0-"none" to 4-"extremely severe") for 13 relatively independent areas grouped under body systems. Since in the literature it has been showed that certain comorbidities are impactful factors for increasing the mortality rate in conjunction with the COVID-19 disease [6–8], we decided to consider only seven of the 13 body systems, namely: cardiac (heart only), hypertension (rating is based on severity; organ damage is rated separately), vascular (blood, blood vessels and cells, bone marrow, spleen, and lymphatics), respiratory (lungs, bronchi, and trachea below the larynx), renal (kidneys only), neurological (brain, spinal cord, and nerves; dementia is not included) and endocrine-metabolic (including diabetes, thyroid; breast; systemic infections; and toxicity). The results from the Cumulative Illness Rating Scale (CIRS) indicate that deceased patients exhibited greater severity of illnesses compared to those who survived, involving the cardiovascular system, neurological disorders, endocrine system disorders, and respiratory system disorders.

Data were anonymised and it is not possible to identify subjects based on the retained variables.

2.2   The Pipeline of Analysis

The analysis of the haematological curve profiles has been performed using two clustering techniques: the functional clustering model [9], implemented in the CONNECTOR tool [10], and Kmodes [11], implemented in the R package "klaR". A similar pipeline has been presented in [12] for the analysis of longitudinal data from patients affected by Multiple Sclerosis. In that case, instead of the Kmodes algorithm, the more classical Kmeans clustering algorithm was used. In this study, CONNECTOR is run on each of the cell populations counts over time, independently, obtaining a clusterisation for each haemochromocytometric parameter. Each patient is then described by an array of features each one coding for the cluster membership of an haemochromocytometric parameter. Then, Kmodes has been exploited to identify homogeneous patient groups based on the CONNECTOR cluster memberships of all the haemochromocytometric parameters, thus recovering the multivariate nature of the data.

CONNECTOR is an R package designed for the unsupervised analysis of longitudinal data, which exploits the model-based approach for clustering functional data [9], a functional clustering method designed for sparse and irregular curves. This approach utilises spline functions to fit smooth curves to the data, allowing for effective modelling and clustering of longitudinal data with varying lengths and measurement times. This tool is able to cluster longitudinal data with great flexibility and with such accuracy as to highlight differences in dynamics. The CONNECTOR pipeline consists of three main steps: (i) *Pre-processing*, (ii) *Model selection*, and (iii) *Cluster visualisation*. The first step involves the visualisation and inspection of the data. In the second step, various measures are computed to appropriately set the model's free parameters. Indeed, selecting the free parameters is a critical part of the analysis. The dimension of the spline basis vector is determined as the value with the highest cross-validated likelihood, using a ten-fold cross-validation as proposed by [9]. To aid in selecting the number of clusters, CONNECTOR provides two plots. The first one is the elbow plot which shows the total tightness (a measure of clustering dispersion) as the number of clusters varies. The second plot extends the well-known Davies-Bouldin (DB) index to the functional setting, termed the functional DB (fDB) index. By examining these plots, the optimal number of clusters can be inferred by looking for the elbow point in the first plot and the minimum value in the fDB index plot. Finally, in the last step, longitudinal data are sub-plotted into CONNECTOR clusters, and the discriminant power of each time point might be illustrated. Additionally, CONNECTOR provides a consensus matrix with a stability score, indicating the reproducibility of the clusters. For the sake of brevity we refer the reader to [10] for an exhaustive description of the CONNECTOR tool.

Finally, the Kmodes clustering algorithm is a variant of Kmeans clustering designed to handle categorical data, which are the CONNECTOR cluster memberships. In particular, the mode (the most frequent category) is used instead of the mean for cluster centroids, and the Hamming distance is used as distance metric, which measures the dissimilarity between two categorical objects by counting the number of mismatched attributes.

3   Results

The dataset has been pre-processed before being analyzed. All time points have been aligned to time $t_0 = 0$, which is the date of the first positive nasopharyngeal swab test. We considered only curves with at least three measurements, with positive observation time, hence subsequent to the positive nasopharyngeal swab test. Given that the 75% of the curves have less than twenty time observations, we decided to truncate the dataset at time $t_{max} = 20$. Finally, we filtered to keep only subjects with measurements for all the eight haemochromocytometric parameters: WBC (Leukocytes), Neutrophils, Lymphocytes, Monocytes, Eosinophils, Basophils, Thrombocytes (PLT) and Erythrocytes (RBC). We are left with 281 subjects measured for eight haemochromocytometric parameters.

The sample comprises 183 male subjects and 98 female subjects, with mean age equal to 68 years (minimum 19, maximum 97), balanced in the male and female subpopulations, with global death rate equal to 23.13%, with 95% confidence interval equal to (18.42%, 28.59%). The overall mortality in the observed population was confirmed to be no different from the mortality reported by literature in similar contexts in different nations and in different types of health services.

The curves of the trend of the haemochromocytometric parameters over time are analysed independently through CONNECTOR. For each cell population, CONNECTOR returned the clusters that collect subjects with the same trend of the parameter. For each cluster, we calculated the death event risk, and tested for homogeneity among clusters with a chi-square test. Additionally, we performed a Kaplan-Meier survival analysis to investigate whether the clusters could be associated to different evolution over time of the population at risk. This analysis allowed to assess the survival probabilities over time and determine if there were significant differences in the survival curves between the clusters, thereby providing insights into the potential prognostic value of the identified clusters.

The results of a complete analysis for WBC cell counts is reported in Fig. 1 with i) the four CONNECTOR clusters, coloured by death event with cluster labels colour-coded for increased and decreased death risks: clusters with increased death rate (larger than the upper extreme of the 95% confidence interval for the death rate on the whole sample) are red coloured, while clusters with reduced death risk (smaller than the lower extreme of the 95% confidence interval for the death rate on the whole sample) are green coloured, ii) the Kaplan-Meier curves plot, where x-axis represents time in days and the y-axis shows the survival function, and iii) the graphical representation of the residuals of the chi-square test for the homogeneity. The circles are colour-coded to represent the strength and direction of the residuals (violet for positive residuals, green for negative residuals). The frequency of death in the four clusters is significantly different (p-value < 0.0001) showing increased death rates in clusters C and D (36.6% and 51.5% respectively) and reduced death rates in clusters A and B (12% and 18% respectively). The survival analysis shows significantly different survival curves in the CONNECTOR clusters (p-value < 0.0001), which can be summarized in two different risk classes: good prognosis group (clusters A and B) and the bad prognosis group (clusters C and D). The CONNECTOR clusters account for different evolution of the WBC counts: almost constant in cluster A, slightly increasing around the day 10 in cluster B, decreasing monotonically from the value at time $t_0 = 0$ (which is larger than expected) in cluster C and increasing with large variance in cluster D. The rise of white blood cells (see cluster D) with the progress of time is supported by an increase in the peripheral circulation of cytokines and inflammatory chemokines that ultimately disrupt the inflammatory cascade triggering the mechanisms of the "cytokine storm", leading to multi-organ failure and finally leading to death, as already demonstrated in the literature also for other diseases of an infectious/inflammatory nature. In particular, the increase of WBC total count has been shown to be supported mainly by the increase in neutrophils but without an increase in lymphocytes.

The analysis also demonstrates how (see cluster B) an increase in total white blood cells followed by their normalization is an integral part not only of the development of inflammation but also of its resolution. This cluster reveals the importance of physiologically regulated inflammation as a protective mechanism against infectious disease. The analysis of WBC cluster A instead shows us how the absence of inflammation associated to the infectious pathology renders the course of the disease more benign and correlated to the minor mortality, that has been revealed to be the smaller of the four clusters. Cluster C patients instead show an apparently mixed population, the white blood cells at the beginning of the curves are for all increased to pathological levels and this seems to be due to two causes: delay in the diagnosis of COVID-19 (delay in performing nasopharyngeal swab), pre-existing pathological conditions. In both cases these have proven to be risk factors for mortality.

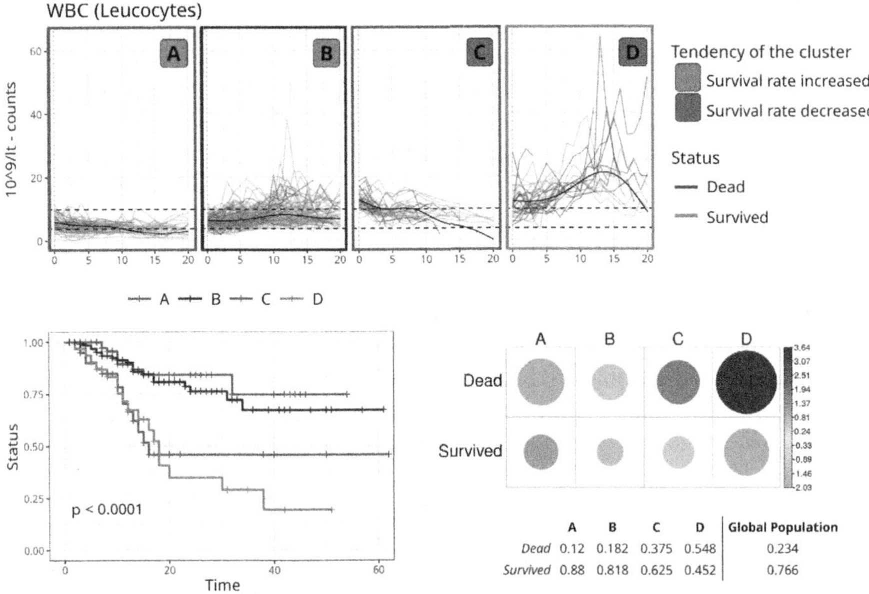

Fig. 1. WBC analysis. Comprehensive analysis of WBC cell counts, featuring: i) the four CONNECTOR clusters, ii) Kaplan-Meier survival curves, and iii) chi-square test for homogeneity. CONNECTOR free parameters are: spline basis dimension $p = 6$ and number of clusters $G = 4$.

The CONNECTOR clusters for Neutrophils, Lymphocytes and Monocytes are reported in Fig. 2, Fig. 3, and Fig. 4. The cluster labels are colour-coded for increased and decreased death risks: clusters with increased death rate (larger than the upper extreme of the 95% confidence interval for the death rate on the whole sample) are red coloured, while clusters with reduced death risk (smaller than the lower extreme of the 95% confidence interval for the death rate on the whole sample) are green coloured. The gray colour is associated to clusters with death risk within the range of the 95% confidence interval for the death rate on the whole sample, equal to (18.42%, 28.59%). The WBC sub-populations Neutrophils, Lymphocytes and Monocytes, are all grouped into clusters (respectively 4, 5 and 6 classes) with significantly different death rates and survival functions.

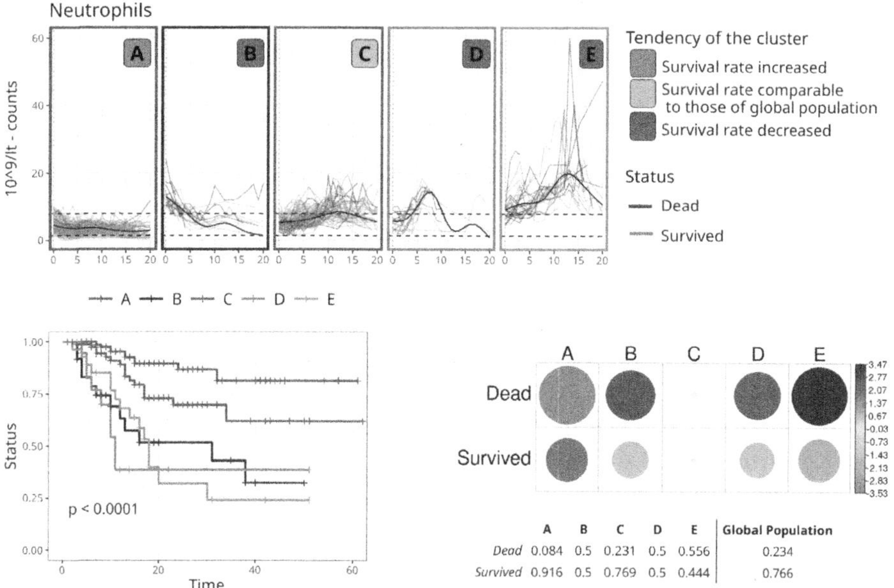

Fig. 2. Neutrophils analysis. Comprehensive analysis of Neutrophils cell counts, featuring: i) the five CONNECTOR clusters, ii) Kaplan-Meier survival curves, and iii) chi-square test for homogeneity. CONNECTOR free parameters are: spline basis dimension $p = 6$ and number of clusters $G = 5$.

Activated Neutrophils are crucial in the defense against a wide range of pathogens, but an improper hyperactivation can lead to severe collateral damage in form of immune-mediated target organ damage. The literature has now extensively demonstrated how Neutrophils are associated with an increased risk of mortality during infection with COVID-19, see [13]. This is shown in Fig. 2, where the absence of normalization in numerical terms of circulating Neutrophils in clusters D and E correlates with higher mortality. While in clusters A and C it is shown as an absence of increase of Neutrophils or their transient increase correlates with a lower mortality.

The Lymphocyte cell population shows, see Fig. 3, that there is a high risk of mortality in patients with lower lymphocyte count (see cluster A) compared to patients without lymphopenia. In addition, there was a significant correlation between the severity of lymphopenia and COVID-19 mortality. The other clusters show, in confirmation of cluster A, how instead a physiological or transiently

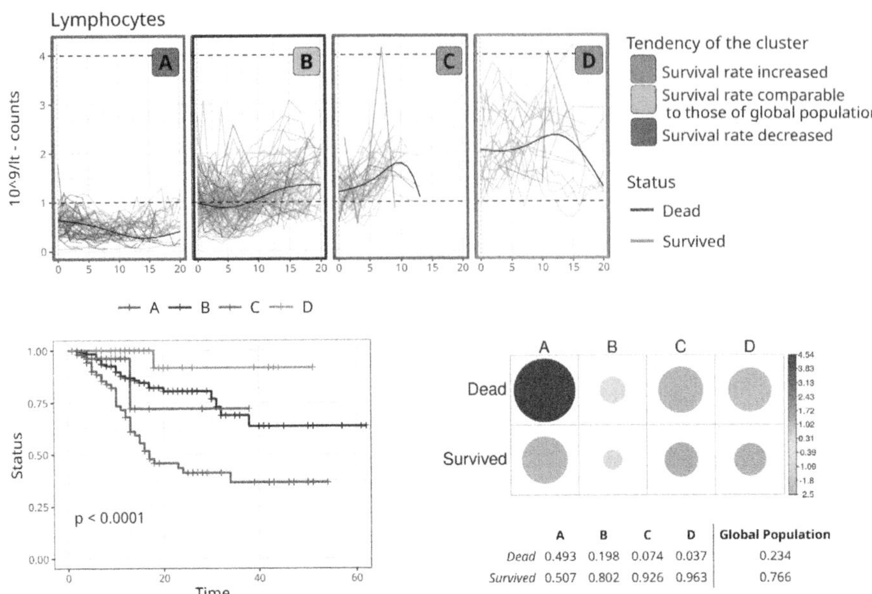

Fig. 3. **Lymphocytes analysis**. Comprehensive analysis of Lymphocytes cell counts, featuring: i) the four CONNECTOR clusters, ii) Kaplan-Meier survival curves, and iii) chi-square test for homogeneity. CONNECTOR free parameters are: spline basis dimension $p = 4$ and number of clusters $G = 4$.

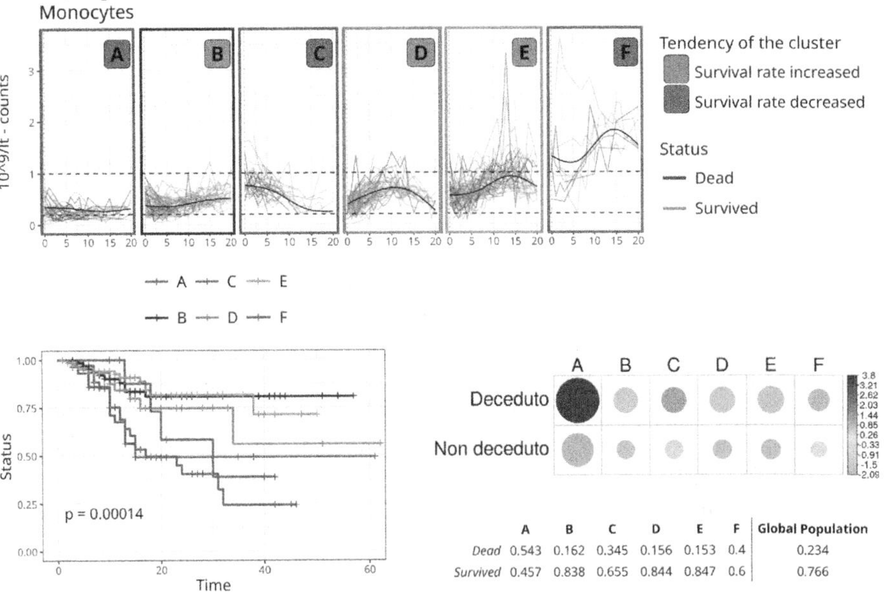

Fig. 4. **Monocytes analysis**. Comprehensive analysis of Monocytes cell counts, featuring: i) the six CONNECTOR clusters, ii) Kaplan-Meier survival curves, and iii) chi-square test for homogeneity. CONNECTOR free parameters are: spline basis dimension $p = 4$ and number of clusters $G = 6$.

increased lymphocytic count is not related to an increase in mortality or can also have a protective aspect against the risk of mortality.

Monocytes and macrophages are central players of the innate immune response and play a central role in the physiological regulation of inflammation, see [14]. They actively participate in all phases of the immune response, from the start of the inflammation process and through the adaptive immune response, to the clearance of cell debris and resolution of inflammation. The double role of Monocytes is very well highlighted in this analysis. Indeed, Fig. 4 clusters A and C, show how an absence of response in numerical terms of Monocytes correlates with a higher mortality as the uncontrolled response of cluster F.

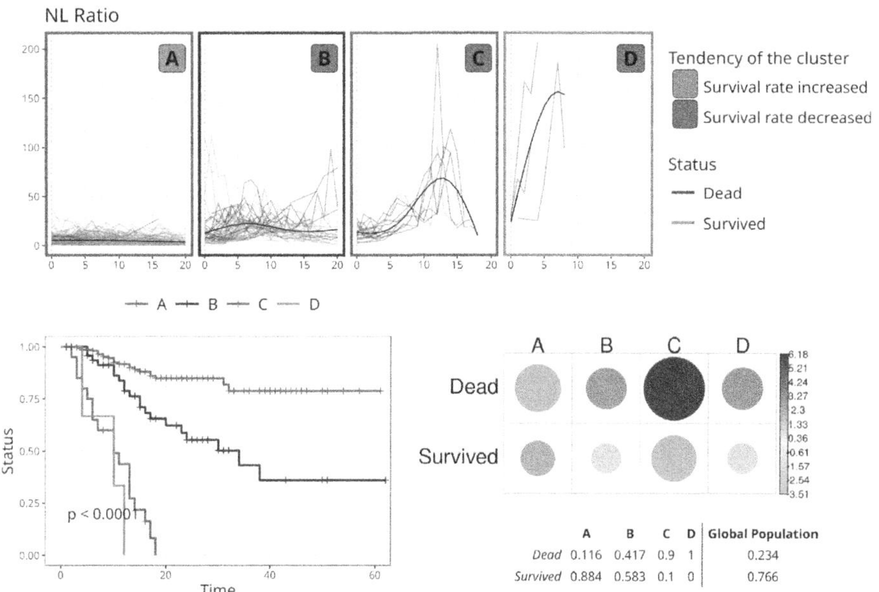

Fig. 5. Neutrophils to Lymphocytes ratio analysis. Comprehensive analysis of the Neutrophils to Lymphocytes ratio cell counts, featuring: i) the four CONNECTOR clusters, ii) Kaplan-Meier survival curves, and iii) chi-square test for homogeneity. CONNECTOR free parameters are: spline basis dimension $p = 4$ and number of clusters $G = 4$.

In the literature it has been suggested that particular significance is entailed by the ratio Neutrophils to Lymphocytes counts, see [2–4]. Hence we performed a clustering of the curves for this parameter. The results are illustrated in Fig. 5. One of the most recent meta-analysis which included 61 studies found that high Neutrophil-to-Lymphocyte Ratio (NLR) values on admission day were associated with progression towards severity and mortality [2]. The NLR showed a more significant prognostic performance towards severity than other markers such as white blood cell count, C-reactive protein, and neutrophil count [15]. NLR has already been proven to predict 30-day mortality in community acquired pneumonia (CAP) in elderly adults with a positive predictive value of 100% and a negative predictive value of 78% [16]. Our study showed findings coherent to the existing literature: B, C clusters and, with a reduced population number but with a 100% mortality, the D cluster show that higher NLRs are related to a higher risk of mortality even before the already cited 30 days of hospitalization with (CAP) or 28 days of hospitalization with COVID-19 disease [17]. As various cut-off have been proposed as a prognostic factor for mortality there is still there is no consensus regarding the optimal cut-off of NLR value for determining the elevated level, it may vary in different populations as may vary according with ethnicity, age, and sex [18].

The results on the Eosinophils and Basophils sub-populations are shown in Fig. 6 and Fig. 7. Although showing different trends so that they can be grouped into clusters by CONNECTOR, they are not significantly associated to the death risk, which cannot be rejected to be homogeneously distributed among the different clusters. The survival analysis as well produced Kaplan Meier estimated of the survival function that cannot be rejected to be similar. There is no evidence in the literature that correlates an increase or decrease in these two cell populations related to a higher or lower mortality during infection with COVID-19. The clusters shown in this study indeed show that there are no trends in time of these two cell populations that can clearly correlate with an increase or decrease in mortality.

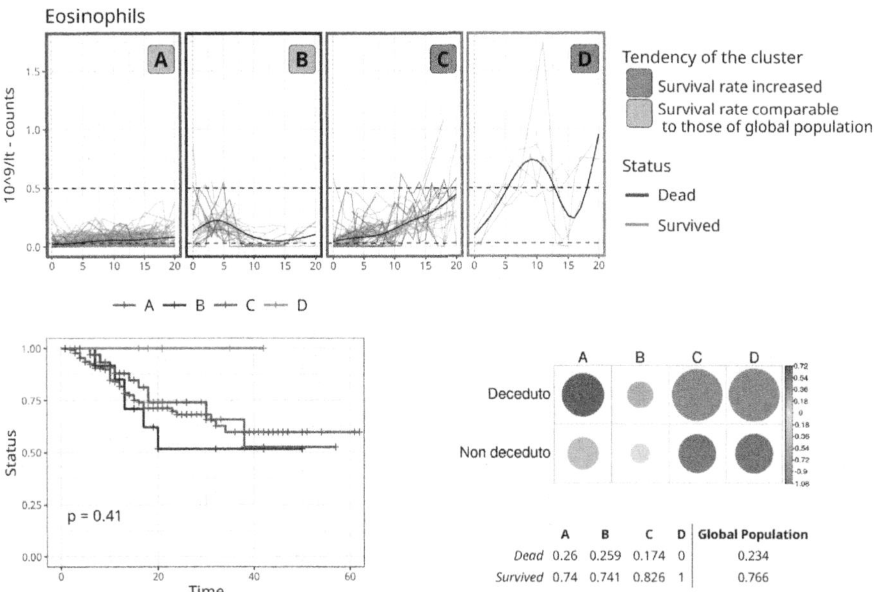

	A	B	C	D	Global Population
Dead	0.26	0.259	0.174	0	0.234
Survived	0.74	0.741	0.826	1	0.766

Fig. 6. Eosinophils analysis. Comprehensive analysis of the Eosinophils cell counts, featuring: i) the five CONNECTOR clusters, ii) Kaplan-Meier survival curves, and iii) chi-square test for homogeneity. CONNECTOR free parameters are: spline basis dimension $p = 6$ and number of clusters $G = 4$.

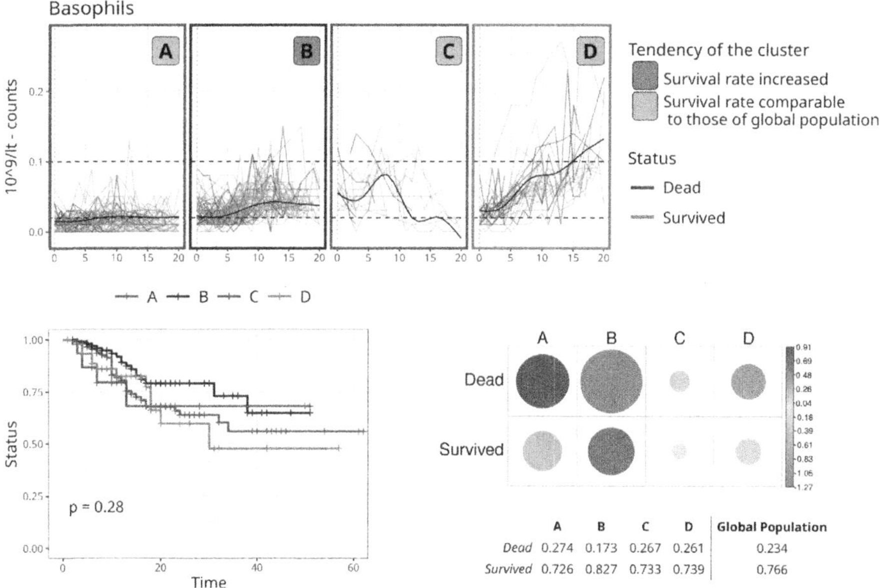

Fig. 7. **Basophils analysis**. Comprehensive analysis of the Basophils cell counts, featuring: i) the four CONNECTOR clusters, ii) Kaplan-Meier survival curves, and iii) chi-square test for homogeneity. CONNECTOR free parameters are: spline basis dimension $p = 6$ and number of clusters $G = 4$.

The second step of the analysis is designed to summarize the information coded by each haemochromocytometric parameter evolution over time into a single cluster membership for each patient. At the end of the first part of the analysis, each subject is described by a vector of features each reporting the CONNECTOR cluster label for each haemochromocytometric parameter. On this dataset, we used the Kmodes clustering algorithm to group patients with similar CONNECTOR cluster memberships. Since we cannot statistically claim that clusters based on eosinophil and basophil sub-populations trends differ in survival rates or risk of death, we decided to exclude the corresponding cluster memberships from the following analysis. Hence, we are left with seven parameters: WBC (Leukocytes), Neutrophils, Lymphocytes, Monocytes, Neutrophils to Lymphocytes ratio, Thrombocytes (PLT) and Erythrocytes (RBC). We ran Kmodes over different subsets of the cluster memberships features: 1. the full set, 2. the full set excluded PLT and RBC, 3. the smaller set composed by WBC, Neutrophils, Lymphocytes and Monocytes, 4. the smallest set composed by the Neutrophils to Lymphocytes ratio and Monocytes. The best clusterisation of patients results to be driven by the set number 3., composed by WBC, Neutrophils, Lymphocytes and Monocytes.

This result is not surprising as the rise of white blood cells (WBC) with the progress of time is supported by an increase in the peripheral circulation of cytokines and inflammatory chemokines that ultimately disrupt the inflammatory cascade triggering the mechanisms of the "cytokine storm", leading to multi-organ failure and finally leading to death, as already demonstrated in the literature also for other diseases of an infectious/inflammatory nature [8]. In particular the increase of WBC total count has been shown to be supported mainly by the increase in neutrophils but without an increase in lymphocytes as has also been previously showed by NLR data. This study results support previously reported findings that monitoring basic laboratory determinations (such as CBC and in particular WBC) could also serve as predictors of mortality. Activated neutrophils are crucial in the defence against a wide range of pathogens, but an improper hyper-activation can lead to severe collateral damage in form of immune mediated target organ damage. The literature has now widely demonstrated how neutrophil is associated with an increased risk of mortality during infection with COVID-19 [3]. Literature shows how a lower lymphocyte count is related with a higher risk of mortality. In addition, there was a significant correlation between the severity of lymphopenia and COVID-19 mortality [19].

Monocytes and macrophages are central players of the innate immune response and play a central role in the physiological regulation of inflammation. They actively participate in all phases of the immune response, from the start of the inflammation process and through the adaptive immune response, to the clearance of cell debris and resolution of inflammation. The clinical implication of this trend is faster and more effective resolution of inflammation. The immune mechanisms of this cell population are currently subject to careful study because a large part of their interaction with other cellular populations in circulation and tissues is being clarified and may lead to a better understanding of mechanisms of inflammation and resolution of inflammation, fully revealing their protective appearance against the response to an infection as well as their role in the uncontrolled inflammatory response [20]. Thrombocytes and Red Blood Cells have a more controversial role in the pathogenesis of COVID 19 syndrome, our data sets shows the same aspects: although coherent with the current literature, low platelet and RBC count are related with a higher mortality [21,22], still those elements have not found their respective roles inside a more broad point of view which contains aspects like comorbidity, age and sex. There is no evidence in the literature that correlates an increase or decrease in these two cell populations related to a higher or lower mortality during infection with COVID-19. The clusters shown in this study in fact show that there are no trends in time of these two cell populations that can clearly correlate with an increase or decrease in mortality.

Let us now describe this final case in detail. We ran Kmodes with different numbers of clusters and the withinnes measure indicated, using the Elbow method, $k = 4$ to be the optimal number of clusters. Figure 8 shows the four cell populations dynamics (rows) grouped by the Kmodes clusters (column). The final clusters, denoted by 1,2,3 and 4 in Fig. 8 are shown to have significantly different death rates (p-value < 0.0001) and significantly different survival functions (Kaplan Meyer curves are different with p-value < 0.0001). Cluster 1 is associated to the highest risk of death, equal to 62%, and cluster 3 as well has an increased mortality rate (33%), while clusters 2 and 4 have good prognosis with probability of death equal to 8% and 13% respectively. The survival analysis shows four (significantly) different survival scenarios, from the bad prognosis to the good prognosis.

We cross-referenced this final clustering with the CIRS total score, given by the sum of the severity of the chronic disease that involves of the apparatuses selected as potentially involved with the COVID-19 disease. From the two violin/box plots in Fig. 8 it is possible to infer that the mean CIRS score is not different among the Kmodes clusters (ANOVA pvalue = 0.09), while it is strongly different for the death/survived variable (pvalue < 0.0001). Indeed, it is well proven in the literature that comorbidities increase the risk of mortality during infection with COVID-19, see [23]. However, among the four CONNECTOR clusters both cluster number 1 and cluster number 3 are characterized by an increased mortality risk, even though only the cluster 1 CIRS scores are slightly increased. Those informations all together allow to claim that comorbidities may be associated with a bad prognosis for COVID-19 only for patients in cluster 1, while the bad prognosis of patients in cluster 3 cannot be associated with comorbidities. This defines a working hypothesis: this group of patients can be exploited to explore and find mechanisms of the disease which account for bad prognosis and are not associated to a general bad condition of the subject.

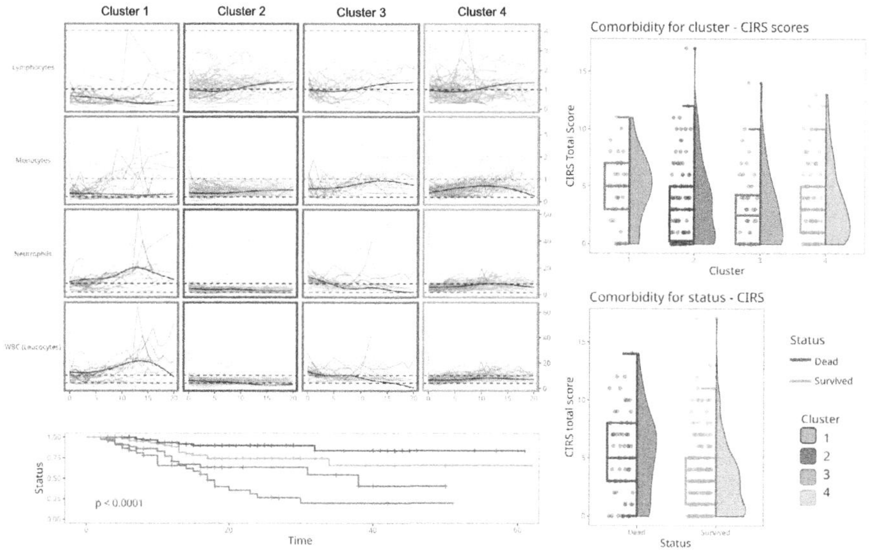

Fig. 8. Multivariate analysis. Final Kmodes clusters and CIRS scores. Death rates are: 62% in Cluster 1, 8% in Cluster 2, 33% in Cluster 3 and 13% in Cluster 4.

4   Conclusions

In this paper, we have introduced functional data analysis to study the evolution in time of the haemochromocytometric during COVID-19 infection disease. The results here presented are fully concordant with the evidences already present in the literature and confirm that the method proposed can be used to explore the disease from new perspectives.

References

1. Ramsay, J.O., Silverman, B.W.: Functional Data Analysis. Springer, New York (2005)
2. Ulloque-Badaracco, J.R., et al.: Prognostic value of neutrophil-to-lymphocyte ratio in COVID-19 patients: a systematic review and meta-analysis. Int. J. Clin. Pract. **75**(11), e14596 (2021)
3. Zinellu, A., Mangoni, A.A.: A systematic review and meta-analysis of the association between the neutrophil, lymphocyte, and platelet count, neutrophil-to-lymphocyte ratio, and platelet-to-lymphocyte ratio and COVID-19 progression and mortality. Expert Rev. Clin. Immunol. **18**(11), 1187–1202 (2022)
4. Sarkar, S., Khanna, P., Singh, A.K.: The impact of neutrophil-lymphocyte count ratio in COVID-19: a systematic review and meta-analysis. J. Intensive Care Med. **37**(7), 857–869 (2022)
5. Linn, B.S., Linn, M.W., Gurel, L.: Cumulative illness rating scale. J. Am. Geriatr. Soc. **16**(5), 622–626 (1968)
6. Chen, N., et al.: Epidemiological and clinical characteristics of 99 cases of 2019 novel coronavirus pneumonia in Wuhan, China: a descriptive study. The lancet **395**(10223), 507–513 (2020)
7. Huang, C., et al.: Clinical features of patients infected with 2019 novel coronavirus in Wuhan, China. The Lancet **395**(10223), 497–506 (2020)
8. Wang, D., et al.: Clinical characteristics of 138 hospitalized patients with 2019 novel coronavirus-infected pneumonia in Wuhan, China. JAMA **323**(11), 1061–1069 (2020)
9. James, G.M., Sugar, C.A.: Clustering for sparsely sampled functional data. J. Am. Stat. Assoc. **98**(462), 397–408 (2003)
10. Pernice, S., et al.: CONNECTOR, fitting and clustering of longitudinal data to reveal a new risk stratification system. Bioinformatics **39**(5), btad201 (2023)
11. Carreira-Perpinán, M.A., Wang, W.: The k-modes algorithm for clustering. arXiv preprint arXiv:1304.6478 (2013)
12. Pernice, S.: A new computational workflow to guide personalized drug therapy. J. Biomed. Inform. **148**, 104546 (2023)
13. Henry, B.M., et al.: Lymphopenia and neutrophilia at admission predicts severity and mortality in patients with COVID-19: a meta-analysis. Acta Bio Medica Atenei Parmensis **91**(3), e2020008 (2020)
14. Austermann, J., Roth, J., Barczyk-Kahlert, K.: The good and the bad: monocytes' and macrophages' diverse functions in inflammation. Cells **11**(12), 1979 (2022)
15. de Jager, C., et al.: The neutrophil-lymphocyte count ratio in patients with community-acquired pneumonia. PLoS ONE **7**(10), e46561 (2012)

16. Cataudella, E., et al.: Neutrophil-to-lymphocyte ratio: an emerging marker predicting prognosis in elderly adults with community-acquired pneumonia. J. Am. Geriatr. Soc. **65**(8), 1796–1801 (2017)

17. Zhang, S., et al.: Development and validation of a risk factor-based system to predict short-term survival in adult hospitalized patients with covid-19: a multicenter, retrospective, cohort study. Crit. Care **24**, 1–13 (2020)

18. Azab, B., Camacho-Rivera, M., Taioli, E.: Average values and racial differences of neutrophil lymphocyte ratio among a nationally representative sample of united states subjects. PLoS ONE **9**(11), e112361 (2014)

19. Chen, G., et al.: Clinical and immunological features of severe and moderate coronavirus disease 2019. J. Clin. Investig. **130**(5), 2620–2629 (2020)

20. Ligi, D., et al.: Deciphering the role of monocyte and monocyte distribution width (MDW) in COVID-19: an updated systematic review and meta-analysis. Clin. Chem. Lab. Med. (CCLM) **61**(6), 960–973 (2023)

21. Wang, Y., et al.: Significant association between anemia and higher risk for COVID-19 mortality: a meta-analysis of adjusted effect estimates. Am. J. Emerg. Med. **58**, 281–285 (2022)

22. Lippi, G., Plebani, M., Henry, B.M.: Thrombocytopenia is associated with severe coronavirus disease: (COVID-19) infections: a meta-analysis. Clin. Chim. Acta **506**(145–148), 2020 (2019)

23. Di Raimondo, D., et al.: The role of the cumulative illness rating scale (CIRS) in estimating the impact of comorbidities on chronic obstructive pulmonary disease (COPD) outcomes: a pilot study of the MACH (multidimensional approach for COPD and high complexity) study. J. Pers. Med. **13**(12), 1674 (2023)

Modeling and Simulation Methods for Computational Biology and Systems Medicine

Gene Set Optimization for Single Cell Transcriptomics

H. Robert Frost[(⊠)][iD]

Dartmouth College, Hanover, NH 03755, USA
hildreth.r.frost@dartmouth.edu

Abstract. Although single cell RNA-sequencing (scRNA-seq) provides unprecedented insights into the biology of complex tissues, analyzing such data on a gene-by-gene basis is challenging due to the large number of tested hypotheses and consequent low statistical power and difficult interpretation. These issues are magnified by the increased noise, significant sparsity and multi-modal distributions characteristic of single cell data. One promising approach for addressing these challenges is gene set testing, or pathway analysis. Unfortunately, statistical and biological differences between single cell and bulk transcriptomic data make it challenging to use existing gene set collections, which were developed for bulk tissue analysis, on scRNA-seq data. In this paper, we describe a procedure for customizing gene set collections originally created for bulk tissue analysis to reflect the structure of gene activity within specific cell types. Our approach leverages information about mean gene expression in the 81 human cell types profiled via scRNA-seq by the Human Protein Atlas (HPA) Single Cell Type Atlas. This HPA information is used to compute cell type-specific gene and gene set weights that can be used to filter or weight gene set collections. As demonstrated through the analysis of immune cell scRNA-seq data using gene sets from the Molecular Signatures Database (MSigDB), accounting for cell type-specificity can significantly improve gene set testing power and interpretability.

Keywords: gene set testing · pathway analysis · single cell transcriptomics · gene set optimization · cell type specificity

1 Introduction

1.1 Single Cell Analysis Challenges

Although single cell assays such as single cell RNA-sequencing (scRNA-seq) [15] are a powerful tool for studying complex tissues, technical and biological limitations make statistical analysis challenging [10,29]. Single cell methods profile very small amounts of genomic material, leading to significant amplification bias and sparsity relative to bulk tissue assays [2]. Single cell approaches for quality control, normalization and statistical analysis (e.g., zero-inflated models) only partially address these challenges [16,18]. In addition to the challenges of noise

L. Cerulo et al. (Eds.): CIBB 2024, LNBI 15276, pp. 183–195, 2025.
https://doi.org/10.1007/978-3-031-89704-7_14

and sparsity, important biological differences exist between bulk tissue and single cell data. As the average over a large number of cells, bulk tissue measurements are non-sparse, typically unimodal and, in many cases, approximately normally distributed. In contrast, single cell datasets reflect a heterogenous mixture of cell types and states resulting in multi-modal and non-normal distributions [2]. The diverse mixture of cell types and states found in complex tissues also leads to significant differences in gene expression patterns between bulk tissue and single cell data. As evidenced by projects such as the Human Protein Atlas (HPA) [26], gene activity in bulk tissue, as quantified via gene expression or protein abundance, can differ substantially from the activity occurring within the cell subpopulations comprising the tissue.

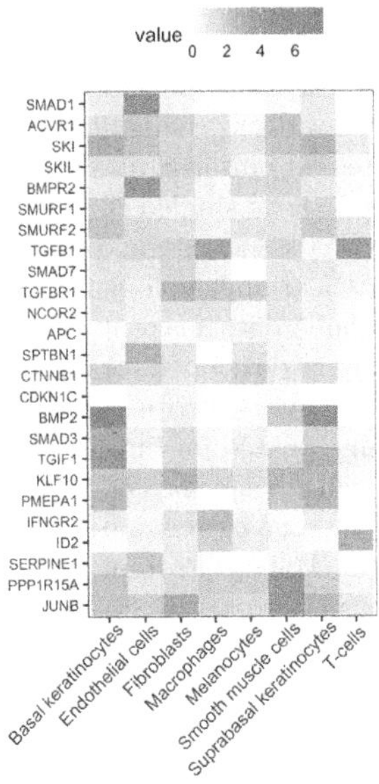

Fig. 1. Cell type-specific expression of genes in the MSigDB Hallmark TGF-β signaling pathway. Cells represent the fold-change in mean expression of the gene in a given cell type relative to the average across all 81 cell types profiled by the HPA Single Cell Type Atlas.

The HPA repository was recently updated with the Single Cell Type Atlas (SCTA) [25], which captures gene expression values for 81 common human cell types as measured by scRNA-seq on healthy tissue for 31 different tissue types. For the HPA SCTA, the source scRNA-seq data was obtained from the Single Cell Expression Atlas (SCEA) [17], the Human Cell Atlas (HCA) [11,21], the Gene Expression Omnibus (GEO) [3], and the European Genome-phenome Archive (EGA) [13]. The datasets from these source repositories were carefully curated to identify high-quality scRNA-seq data measured on 31 different tissue types where the samples were obtained from healthy individuals and processed without cell type enrichment. Using this data, cell type clusters where identified representing 81 distinct cell types. The mean expression profile of each cell type enables the quantification of the cell type-specificity of human protein coding genes. Importantly, mean gene expression differs not only between different cell types but also between bulk tissue samples and the cell types that comprise that tissue. Figure 1 illustrates these marginal differences for a subset of genes in the Molecular Signatures Database (MSigDB) [14] Hallmark TGF-β signaling pathway based on the cell types captured via scRNA-seq in human skin [23] as represented in the HPA SCTA. As illustrated by this figure, gene expression values measured via scRNA-seq on the cell types that comprise skin can look very different from the values computed via bulk RNA-seq on skin samples, which will be a weighted average of the cell type-specific measurements.

The pattern of co-expression can also vary significantly between single cells and bulk tissue. A comparison of gene co-expression in single cell and bulk glioblastoma samples performed by Wang et al. [27] found that over 90% of the gene co-expression pairs were unique to either the bulk or single cell data. This dramatic difference in the pattern of co-expression is due to the fact that co-expression at the bulk tissue level is often driven by variation in the proportion of cell types in a given tissue which can bear little resemblance to gene co-expression across cells of a specific type [4,27]. Genes that are uncorrelated at the single cell level can appear to be correlated at the bulk tissue level if the mean expression varies across cell types and cell type proportions vary across bulk samples. The inverse is also possible, i.e., genes whose expression is correlated at the single cell level can appear uncorrelated in bulk tissue.

A similar issue exists for the association of gene expression values with a given experimental condition, i.e., the experimental associations found at the bulk tissue level can be very dissimilar to those found for a specific cell type. Figure 2 provides a simplified illustration of the marginal and joint distribution characteristics of single cell and bulk tissue expression data. In this figure, the marginal distribution is represented by density plots for a single gene while the joint distribution is represented by covariance matrices. Collectively, the distributional differences between single cell and bulk tissue genomic data make it challenging to successfully analyze single cell expression data using biological models originally developed for bulk tissue, which represent the pattern of gene product abundance within an average cell.

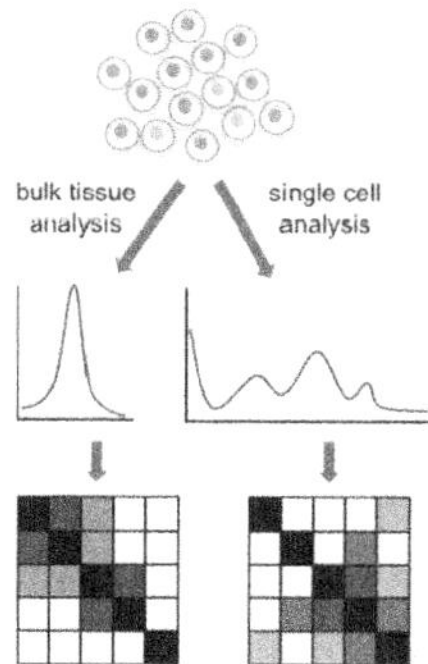

Fig. 2. Bulk tissue vs. single cell distributions. The middle and bottom rows illustrate approximate marginal and joint expression distributions.

1.2 Gene Set Testing for Single Cell Data

Although high-dimensional genomic data provides a molecular-level lens on biological systems, the gain in fidelity obtained by testing thousands of genomic variables comes at the price of impaired interpretation, loss of power due to multiple hypothesis correction and poor reproducibility [1,9]. To help address these challenges for bulk tissue data, researchers developed gene set testing, or pathway analysis, methods [12]. Gene set testing is an effective hypothesis aggregation technique that lets researchers step back from the level of individual genomic variables and explore associations for biologically meaningful groups of genes. Focusing on a small number of pathways can substantially improve power, interpretation and replication relative to an analysis focused on individual genomic variables [24]. The benefits that gene set-based hypothesis aggregation offers for the analysis of bulk tissue data are even more pronounced for single cell data given increased technical variance and sparsity. Although significant progress has been made developing gene set testing methods, including methods developed by us that are specifically optimized for scRNA-seq data [6,7], and building and maintaining gene set collections, existing collections were largely developed for the analysis of bulk tissue data. This is problematic since many gene sets in collections like the Molecular Signatures Database (MSigDB) [14] are defined to contain groups of genes whose expression in bulk tissue is correlated (e.g., MSigDB cancer modules [22]) or is associated with a specific experimental variable (e.g., MSigDB chemical and genetic perturbations). Such gene sets will often represent biological associations that do not hold at the single cell level [27]. It is important to note that new collections, e.g., the Human Cell Atlas-based MSigDB C8 collection [19], are being developed that contain gene sets derived from scRNA-seq data.

2   Data and Methods

To address the bulk tissue bias that exists in most public gene set collections, we have developed a procedure for customizing gene set collections to reflect the structure of gene activity within specific cell types as measured by single cell transcriptomic assays. Our approach leverages information about mean gene expression in the 81 human cell types profiled via scRNA-seq by the HPA SCTA. As detailed below, this SCTA information was used to compute cell type-specific gene and gene set weights that can be used to filter or weight gene set collections. An example vignette and gene and gene set weights for the 81 HPA SCTA cell types and MSigDB collections are available at https://hrfrost.host.dartmouth.edu/SCGeneSetOpt/.

2.1   Data Sources

The following data sources were leveraged to compute the cell type-specific gene and gene set weights and generate the paper results:

- Human Protein Atlas Single Cell Type Atlas (HPA SCTA): Information about the cell type-specific expression of human protein coding genes was obtained from the HPA SCTA via the downloadable file https://www.proteinatlas.org/download/rna_single_cell_type.tsv.zip.
- Molecular Signatures Database (MSigDB): Gene set definitions were obtained from the MSigDB v2024.1 via the downloadable files at https://www.gsea-msigdb.org/gsea/msigdb/index.jsp.
- 10k PBMC3k scRNA-seq data: The PBMC scRNA-seq dataset used to generated the results in Sect. 3 is also used in the Seurat Guided Clustering Tutorial (https://satijalab.org/seurat/articles/pbmc3k_tutorial.html), is included in the *SeuratData* R package and is freely accessible from 10x Genomics via a Creative Commons Attribution license at https://cf.10xgenomics.com/samples/cell/pbmc3k/pbmc3k_filtered_gene_bc_matrices.tar.gz. Processing of the PBMC3k dataset was performed using the same logic employed in the Seurat Guided Clustering Tutorial, which is also contained in the vignette available at https://hrfrost.host.dartmouth.edu/SCGeneSetOpt/.

2.2   Computation of Cell Type-Specific Gene and Gene Set Weights

Building on our prior work creating customized versions of MSigDB for different normal human tissue types [5], our method first computes cell type-specific weights for each protein coding human gene for all 81 normal human cell types profiled by the HPA SCTA ($w_{i,t}^g$ for gene i in cell type t). Specifically, $w_{i,t}^g$ is set to the fold-change between the mean normalized transcript abundance of gene i in cell type t as computed via scRNA-seq relative to the average across all 81 cell types. These gene-level weights are then leveraged to compute cell type-specific gene set weights for the sets in the MSigDB collections. Specifically, a weight $w_{j,t}^s$ for each gene set j in a target collection is computed as the -log of the p-value from a one-sided, two-sample test t test comparing the $w_{*,t}^g$ values for the genes in set j to the $w_{*,t}^g$ for genes not in j (this is similar to the competitive gene set test implemented by the *cameraPR* method in the R limma package [20]).

2.3   Using Gene Weights for Annotation Filtering and Weighting

The cell type-specific gene weights $w_{i,t}^g$ can be used to customize gene sets via either annotation filtering or weighting. Gene set annotations can be customized for cell type t by simply removing all annotations for each gene i if $w_{i,t}^g \leq T$, where T is a threshold ($T \geq 0$) that can be user specified or selected to optimize a gene set testing performance metric, e.g., replication of gene set testing results across related datasets. Filtering has the benefit of parsimony (it results in smaller and more easily intepreted sets), improved power (the remaining set members should be more likely to capture the relevant biological signal), and can work with any gene set testing method, however, it does require the specification of a threshold and ignores most of the information contained in the continuous weights. An alternate approach that uses the continuous gene weights and does not require a threshold is annotation weighting. In this scenario, the log of the gene weights is used to provide a directional weight that can be used with gene set testing methods like *fry* (see Sect. 2.6 for more details) that accept gene weights.

2.4   Using Gene Set Weights for Collection Filtering and Weighting

Similar to the application of gene-level weights, the cell type-specific gene set weights $w_{j,t}^s$ can be used for either filtering or weighting. Filtering can be performed by removing (or not using) all sets in a given collection where $w_{j,t}^s$ is less than some threshold. Collection filtering has the benefits of interpretation and statistical power since a smaller number of more biologically relevant gene sets are tested, which makes interpretation easier and reduces the multiple hypothesis correction penalty. Like annotation filtering, the downsides of collection filtering include the need for a specific threshold and fact that most of the information in the weights is not used. The weights can alternatively be used for p-value weighting (e.g., weighted FDR [8]) following gene set testing, which avoids the need for a specific threshold.

2.5 Choice of Cell Type Weights

The effectiveness of the filtering and weighting techniques detailed in Sects. 2.3 and 2.4 is strongly dependent on the what cell type weight is employed. Considerations for several common scenarios are discussed below:

- *Analysis of a single cell type in different experimental conditions*: For this scenario, gene set analysis is typically performed to identify sets that are differentially active between the experimental conditions. Using gene and/or gene set weights for type of cell in the dataset is usually appropriate since this will prioritize the pathways/functions most critical to the biology of that cell type.
- *Analysis of multiple cell types in different experimental conditions*: For this scenario, the goal of gene set analysis is usually to identify sets that are differentially active between the experimental conditions irrespective of the cell type. In this case, using an average (or weighted average) of the weights for all cell types present in the data can be effective. Similar to the single cell type case, this will priorize the pathways/functions most relevant to the function of those cell types.
- *Comparative analysis of different cell types in the same experimental condition*: For this scenario, gene set analysis aims to identify sets that are differentially active in cells of one type relative to cells of a different type. If the analysis is primarily focused on one of the cell types, then the weights for that type could be used. Alternatively, the average of the weights (or maximum weight) for the analyzed cell types could be used.

The results in Sect. 3 correspond to the last scenario and weights for just one of the cell types in the comparison are used (either B cell or T cell weights).

2.6 Gene Set Testing

The gene set testing results shown in Sect. 3 were generated using two techniques, both availble via the *limma* R package [20], for population-level gene set analysis:

- *Camera* [20]: This technique implements a competitive and population-level gene set test that accounts for inter-gene correlation.
- *Fry* [28]: This technique implementes a self-contained and population-level gene set test that can accept gene-level weights.

Both methods were executed using default parameters unless otherwise specified.

3   Results

To evaluate the cell type-specific weight model detailed in Sect. 2.2, we computed gene and gene set weights for all 81 normal human cell types included in the HPA SCTA. For each of these cell types, gene weights were generated for all human protein coding genes profiled by the HPA SCTA and gene set weights were calculated for all collections in v2024.1 of the MSigDB. To illustrate the utility of these weights and their application for the filtering and weighting approaches detailed in Sects. 2.3 and 2.4, we

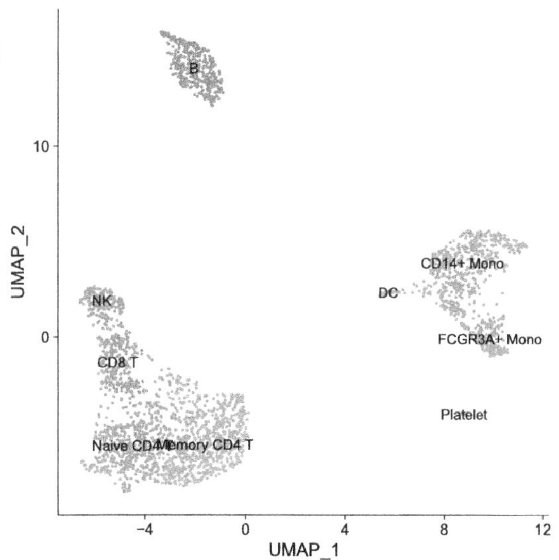

Fig. 3. Projection of PBMC scRNA-seq data onto the first two UMAP dimensions. Each point in the plot represents one cell.

performed various gene set analyses of the 10x PBMC3k example scRNA-seq dataset (visualized in Fig. 3) using both B and T cell weights and the sets in the Gene Ontology Biological Process collection (MSigDB collection C5.GO.BP). Results from these analyses are detailed in Sects. 3.1–3.4 below.

Both the cell type-specific weights and R logic for the B cell-based results are available on the paper website (https://hrfrost.host.dartmouth.edu/SCGeneSetOpt/).

3.1   Collection Filtering Using B Cell Gene Set Weights

As illustrated by Table 1, which lists the top ten MSigDB C5.GO.BP (Gene Ontology Biological Process) terms for B cells, the gene sets with the largest w^s values clearly reflect the known biological features of B cells.

Following the approach outlined in Sect. 2.4, we used the B cell-based gene set weights to filter the MSigDB C5.GO.BP collection with the goal of both improving statistical power by reducing the multiple hypothesis correction burden and improving the biological relevance and interpretability of the analysis by only testing sets specific to B cells. To perform this analysis, we created a filtered version of the C5.GO.BP collection that retained the sets with B cell weights above 10, which corresponds to 4.5% of the 5,777 C5.GO.BP sets retained after alignment with the PBMC scRNA-seq genes (annotations were removed for genes not captured in the PBMC data and then sets were eliminated if they had less

Table 1. Top 10 MSigDB C5.GO.BP gene sets based on B cell gene set weights.

Gene set	Weight (w^s)
GOBP_B_CELL_RECEPTOR_SIGNALING_PATHWAY	412
GOBP_B_CELL_ACTIVATION	270
GOBP_ANTIGEN_RECEPTOR_MEDIATED_SIGNALING_PATHWAY	206
GOBP_ADAPTIVE_IMMUNE_RESPONSE	205
GOBP_LYMPHOCYTE_ACTIVATION	200
GOBP_B_CELL_PROLIFERATION	186
GOBP_IMMUNE_RESPONSE_REGULATING_CELL_SURFACE_RECEPTOR_SIGNALING	185
GOBP_CELL_ACTIVATION	161
GOBP_POSITIVE_REGULATION_OF_IMMUNE_RESPONSE	155
GOBP_IMMUNE_RESPONSE_REGULATING_SIGNALING_PATHWAY	155

then 5 or greater than 200 members). We then performed a population-level and competitive gene set analysis using the *camera* method comparing set expression among B cells against expression in non-B cells. Weight-based collection filtering had the desired effect of improving the multiple correction-adjusted statistical significance of the results. Specifically, without filtering only 36 terms had FDR values < 0.1; with filtering this number increased to 67. This trend is visualized in Fig. 4.

3.2   Collection Filtering Using T Cell Gene Set Weights

Table 2 lists the top ten MSigDB C5.GO.BP (Gene Ontology Biological Process) terms according to the T cell-specific gene set weights. Similar to the top terms for B cells shown in Table 1, the Gene Ontology terms with the largest w^s values effectively capture the key aspects of T cell biology.

Table 2. Top 10 MSigDB C5.GO.BP gene sets based on T cell gene set weights.

Gene set	Weight (w^s)
GOBP_ADAPTIVE_IMMUNE_RESPONSE	500.0
GOBP_T_CELL_ACTIVATION	109
GOBP_T_CELL_RECEPTOR_SIGNALING_PATHWAY	101
GOBP_LYMPHOCYTE_ACTIVATION	91
GOBP_ANTIGEN_RECEPTOR_MEDIATED_SIGNALING_PATHWAY	81
GOBP_BIOLOGICAL_PROCESS_INVOLVED_IN_INTERSPECIES_INTERACTION	75
GOBP_T_CELL_DIFFERENTIATION	69
GOBP_ALPHA_BETA_T_CELL_ACTIVATION	68
GOBP_POSITIVE_REGULATION_OF_IMMUNE_SYSTEM_PROCESS	68
GOBP_CELL_ACTIVATION	67

Similar to the B cell analysis above, we created a filtered version of the C5.GO.BP collection that retained the sets with T cell weights above 10, which

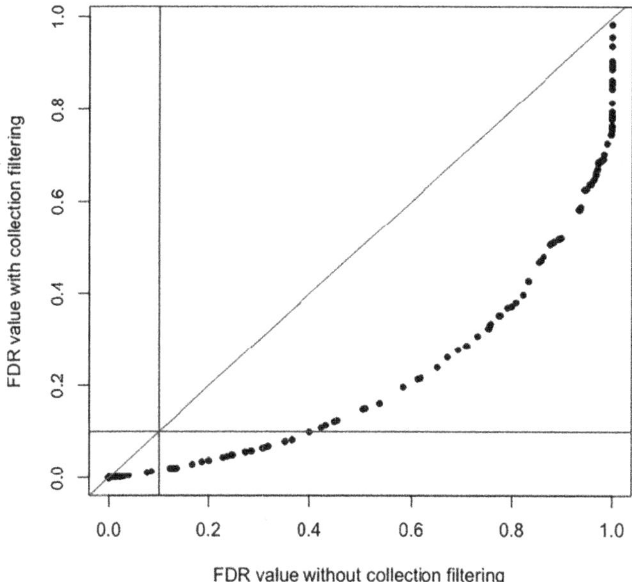

Fig. 4. Distribution of FDR values from a gene set analysis comparing B cells against other cell types in the PBMC data using the MSigDB C5.GO.BP collection. Each point captures the FDR values for one of the terms remaining after collection filtering with the x-axis representing the non-filtered FDR value and the y-axis representing the filtered FDR value. Blue lines reflect the FDR threshold of 0.1 and the red line reflects expected distribution for equal FDR values. (Color figure online)

corresponds to 4.4% of the 5,777 C5.GO.BP sets retained after alignment with the PBMC scRNA-seq genes. We then performed a gene set analysis using the *camera* method comparing set expression among T cells against expression in non-T cells. Weight-based collection filtering again had the desired effect of improving the multiple correction-adjusted statistical significance of the results. Specifically, without filtering 56 terms had FDR values < 0.1, with filtering this number increased to 68. This trend is visualized in Fig. 5.

3.3 Annotation Weighting Using B Cell Gene Weights

Following the approach in Sect. 2.3, we used the B cell-based gene weights to perform a weighted gene set analysis using the *fry* method. Specifically, set expression in B cells was compared to set expression in non-B cells and this analysis was performed both without weights and with gene weights set to the log2 of the B cell-based gene weight plus a pseudo-count of 1e-4 (the log transformation generates sign-based directional weights as need by *fry*). This weighting scheme prioritizes genes that are strongly up-regulated or down-regulated in B cells according to the HPA SCTA. The *fry* method (which is a fast approximation of the *roast* technique) generated unexpectedly small FDR values (much

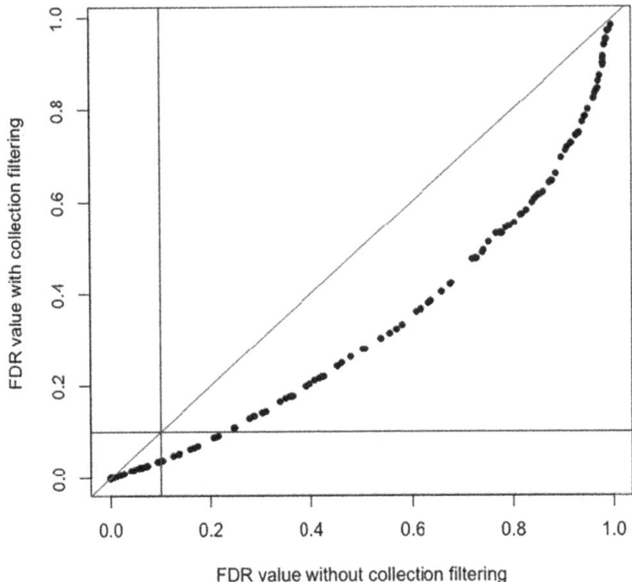

Fig. 5. Distribution of FDR values from a gene set analysis comparing T cells against other cell types in the PBMC data using the MSigDB C5.GO.BP collection. Each point captures the FDR values for one of the terms remaining after collection filtering with the x-axis representing the non-filtered FDR value and the y-axis representing the filtered FDR value. Blue lines reflect the FDR threshold of 0.1 and the red line reflects expected distribution for equal FDR values. (Color figure online)

smaller than *camera*), however, the rank ordering of the sets based on significance was similar to *camera* and matched the expected biology of B cells. Use of B cell gene weights improved the statistical power of the gene set analysis. Specifically, without weights 1,965 terms had FDR values $< 1e - 4$, with weights this number increased to 2,348.

3.4   Annotation Weighting Using T Cell Gene Weights

Similar to the B cell analysis in Sect. 3.3, we used T cell-based gene weights to perform a weighted gene set analysis comparing set expression in T cells to expression in non-T cells. Use of T cell gene weights also improved the statistical power of the gene set analysis. Specifically, without weights 2,920 terms had FDR values $< 1e - 4$, with weights this number increased to 3,590.

4   Conclusions

Gene set testing is a powerful tool for the analysis of scRNA-seq data that addresses the challenges of sparsity and noise. Unfortunately, the utility of gene set testing for single cell data is limited by the fact that most existing gene set collections were developed to capture gene activity within bulk tissue data, which can differ substantially from gene activity in specific cell types. In particular, the pattern of gene co-expression found among cells of a specific type is often significantly different from the pattern seen in bulk tissue samples, which is driven by cell type proportions. A similar issue exists for the association of gene expression values with a given experimental condition. To address this challenge, we explored methods for computing gene and gene set weights using information about the cell type-specificity of human protein coding genes from the HPA SCTA. These cell type-specific weights can be leveraged to improve the power and interpretability of gene set analyses through the filtering or weighting of gene set collections or gene set annotations. To support this type of analysis by other researchers, an example vignette along with gene and gene set weights for the 81 HPA SCTA cell types and MSigDB collections are available at https://hrfrost.host.dartmouth.edu/SCGeneSetOpt/.

Acknowledgments. This work was funded by National Institutes of Health grants R35GM146586, R21CA253408, and P30CA023108.

Disclosure of Interests. The authors have no conflicts of interest to declare.

References

1. Allison, D.B., Cui, X., Page, G.P., Sabripour, M.: Microarray data analysis: from disarray to consolidation and consensus. Nat. Rev. Genet. **7**(1), 55–65 (2006). https://doi.org/10.1038/nrg1749
2. Bacher, R., Kendziorski, C.: Design and computational analysis of single-cell rna-sequencing experiments. Genome Biol. **17**, 63 (2016). https://doi.org/10.1186/s13059-016-0927-y
3. Barrett, T., et al.: Ncbi geo: archive for functional genomics data sets–update. Nucleic Acids Res. **41**(Database issue), D991–5, January 2013. https://doi.org/10.1093/nar/gks1193
4. Crow, M., Paul, A., Ballouz, S., Huang, Z.J., Gillis, J.: Exploiting single-cell expression to characterize co-expression replicability. Genome Biol **17**, 101 (2016) https://doi.org/10.1186/s13059-016-0964-6
5. Frost, H.R.: Computation and application of tissue-specific gene set weights. Bioinformatics (2018). https://doi.org/10.1093/bioinformatics/bty217
6. Frost, H.R.: Reconstruction set test (reset): a computationally efficient method for single sample gene set testing based on randomized reduced rank reconstruction error. PLoS Comput. Biol. **20**(4), e1012084 (2024). https://doi.org/10.1371/journal.pcbi.1012084
7. Frost, H.R.: Variance-adjusted mahalanobis (vam): a fast and accurate method for cell-specific gene set scoring. Nucleic Acids Res. (2020). https://doi.org/10.1093/nar/gkaa582
8. Genovese, C.R., Roeder, K., Wasserman, L.: False discovery control with p-value weighting. Biometrika **93**(3), 509–524 (2006). https://doi.org/10.1093/biomet/93.3.509

9. Goeman, J.J., Buehlmann, P.: Analyzing gene expression data in terms of gene sets: methodological issues. Bioinformatics **23**(8), 980–987 (2007). https://doi.org/10.1093/bioinformatics/btm051

10. Heumos, L., et al.: Single-cell best practices consortium. In: Schiller, H.B., Theis, F.J.: Best practices for single-cell analysis across modalities. Nat. Rev. Genet. **24**(8), 550–572 (2023). https://doi.org/10.1038/s41576-023-00586-w

11. Hon, C.C., Shin, J.W., Carninci, P., Stubbington, M.: The human cell atlas: technical approaches and challenges. Brief. Funct. Genomics (2017). https://doi.org/10.1093/bfgp/elx029

12. Khatri, P., Sirota, M., Butte, A.J.: Ten years of pathway analysis: current approaches and outstanding challenges. PLoS Comput. Biol. **8**(2), e1002375 (2012). https://doi.org/10.1371/journal.pcbi.1002375

13. Lappalainen, I., et al.: The european genome-phenome archive of human data consented for biomedical research. Nat. Genet. **47**(7), 692–5 (2015). https://doi.org/10.1038/ng.3312

14. Liberzon, A., Subramanian, A., Pinchback, R., Thorvaldsdóttir, H., Tamayo, P., Mesirov, J.P.: Molecular signatures database (msigdb) 3.0. Bioinformatics **27**(12), 1739–40 (2011). https://doi.org/10.1093/bioinformatics/btr260

15. Macosko, E.Z., et al.: Highly parallel genome-wide expression profiling of individual cells using nanoliter droplets. Cell **161**(5), 1202–1214 (2015). https://doi.org/10.1016/j.cell.2015.05.002

16. McCarthy, D.J., Campbell, K.R., Lun, A., Wills, Q.F.: Scater: pre-processing, quality control, normalization and visualization of single-cell rna-seq data in r. Bioinformatics **33**(8), 1179–1186 (2017). https://doi.org/10.1093/bioinformatics/btw777

17. Papatheodorou, I., et al.: Expression atlas update: from tissues to single cells. Nucleic Acids Res. **48**(D1), D77–D83 (2020). https://doi.org/10.1093/nar/gkz947

18. Qiu, X., Hill, A., Packer, J., Lin, D., Ma, Y.A., Trapnell, C.: Single-cell mrna quantification and differential analysis with census. Nat. Methods **14**(3), 309–315 (2017). https://doi.org/10.1038/nmeth.4150

19. Regev, A., et al.: Human cell atlas meeting participants: the human cell atlas. Elife **6** (2017). https://doi.org/10.7554/eLife.27041

20. Ritchie, M.E., et al.: limma powers differential expression analyses for rna-sequencing and microarray studies. Nucleic Acids Res. **43**(7), e47 (2015). https://doi.org/10.1093/nar/gkv007

21. Rozenblatt-Rosen, O., Stubbington, M., Regev, A., Teichmann, S.A.: The human cell atlas: from vision to reality. Nature **550**(7677), 451–453 (2017). https://doi.org/10.1038/550451a

22. Segal, E., Friedman, N., Koller, D., Regev, A.: A module map showing conditional activity of expression modules in cancer. Nat. Genet. **36**(10), 1090–8 (2004). https://doi.org/10.1038/ng1434

23. Solé-Boldo, L., et al.: Single-cell transcriptomes of the human skin reveal age-related loss of fibroblast priming. Commun. Biol. **3**(1), 188 (2020). https://doi.org/10.1038/s42003-020-0922-4

24. Subramanian, A., et al.: Gene set enrichment analysis: a knowledge-based approach for interpreting genome-wide expression profiles. Proc. Natl. Acad. Sci. U S A **102**(43), 15545–15550 (2005). https://doi.org/10.1073/pnas.0506580102

25. Thul, P.J., et al.: A subcellular map of the human proteome. Science **356**(6340) (2017). https://doi.org/10.1126/science.aal3321

26. Uhlén, M., et al.: Proteomics. tissue-based map of the human proteome. Science **347**(6220), 1260419 (2015). https://doi.org/10.1126/science.1260419

27. Wang, J., et al.: Single-cell co-expression analysis reveals distinct functional modules, co-regulation mechanisms and clinical outcomes. PLoS Comput. Biol. **12**(4), e1004892 (2016). https://doi.org/10.1371/journal.pcbi.1004892
28. Wu, D., Lim, E., Vaillant, F., Asselin-Labat, M., Visvader, J.E., Smyth, G.K.: ROAST: rotation gene set tests for complex microarray experiments. Bioinformatics (Oxford, England) **26**(17), 2176–2182 (2010). https://doi.org/10.1093/bioinformatics/btq401. PMID: 20610611
29. Yuan, G.C., et al.: Challenges and emerging directions in single-cell analysis. Genome Biol. **18**(1), 84 (2017). https://doi.org/10.1186/s13059-017-1218-y

MicroRNAs as Biomarkers for Ulcerative Colitis

Flaminia Tani[1,2], Gloria Rita Bertoli[1,2],
and Bruno Giovanni Galuzzi[1,2(✉)]

[1] Institute of Bioimaging and Complex Biological Systems (IBSBC), Segrate, Italy
`brunogiovanni.galuzzi@cnr.it`
[2] National Biodiversity Future Center (NBFC), Palermo, Italy

Abstract. The integration of bioinformatics and molecular biology has advanced the search for biomarkers in several diseases. In this study, we searched for possible microRNA biomarkers related to Ulcerative Colitis, a chronic and progressive immune-mediated inflammatory condition characterized by inflammation of the gastrointestinal tract. Starting from a set of public datasets, we collect a set of miRNAs and analyze the possible interactions between miRNAs, their target genes, and the associated enriched pathways, related in particular to inflammation and oxidative stress.

Keywords: miRNA · ulcerative colitis · inflammation · oxidative stress

1 Introduction

The incessant interest in bioinformatics and molecular biology has made it possible to integrate the two disciplines to search for biomarkers for several diseases.

In this work, we considered Inflammatory bowel disease (IBD), a chronic and progressive immune-mediated inflammatory condition characterized by inflammation of the gastrointestinal tract. The two main clinical forms are Crohn's Disease (CD) and Ulcerative Colitis (UC), which have steadily increased in incidence and prevalence in recent decades. Currently, it is estimated that approximately 6.8 million people are affected by IBD worldwide, an increase of 85% over the last 30 years [8]. The etiology of IBD is multifactorial and depends on the complex interaction between mucosa microbiome, immunity, environmental factors, diet and human genetics. IBD affects large or small intestine tracts in the colon and rectum eliciting the formation of discontinuous areas of transmural inflammation of the mucosa.

As part of a research project developed for the National Biodiversity Future Center (NBFC) to study IBD, we aim to investigate how different biodiversity conditions impact IBD patients. To achieve this, we plan to select a small set of microRNAs (miRNAs), which serve as biomarkers for this disease and are

L. Cerulo et al. (Eds.): CIBB 2024, LNBI 15276, pp. 196–205, 2025.
https://doi.org/10.1007/978-3-031-89704-7_15

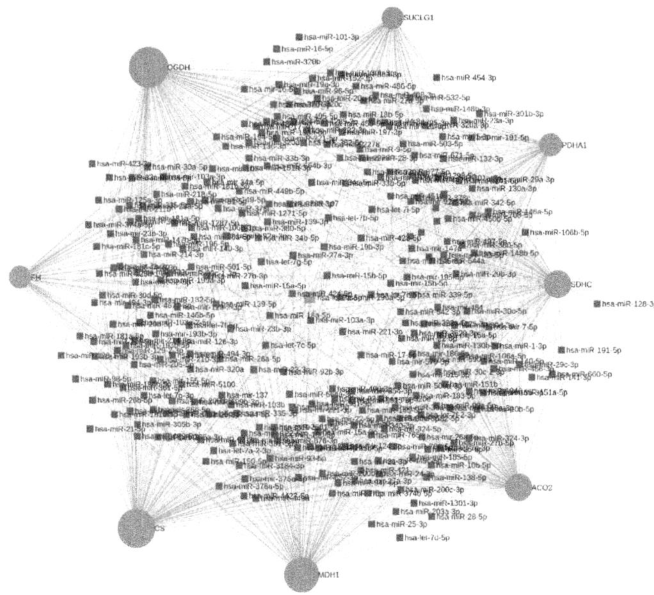

Fig. 1. Example of miRNA-gene association network.

potentially linked to key biological pathways, such as inflammation and oxidative stress. These miRNAs will be validated using the qPCR-TaqMan assay on biological samples (serum and plasma) of patients collected in the EXPOSITION cohort, part of the NBFC, whose recruitment is underway.

The epigenetic molecules that we consider, i.e. the miRNAs, are small noncoding RNAs that post-transcriptionally control gene expression through the inhibition of translation or degradation of their target mRNAs. Their role within organisms is multiple and complex. The complexity is related to the fact that a single miRNA can regulate the expression of several genes and a gene can be the target of multiple miRNAs (see Fig. 1). Their expression can impact the organism in different ways, including cellular differentiation, proliferation, apoptosis, and tumorigenesis. Some miRNAs have emerged as potentially useful biomarkers for several diseases [2]. In particular, their dysregulation was associated with the key signaling pathways involved in the pathogenesis of IBD [1] and to the inflammation and oxidative stress [9, 11]. Moreover, miRNAs can be isolated from a variety of non-invasive sources such as saliva, blood, and feces, making them relatively easy to extract, and more convenient compared to invasive tissue biopsies. In this work, we developed an *in silico* approach to identify a small subset of miRNAs biomarkers of the UC. To this aim, miRNAs differentially expressed in tissues from UC patients compared to healthy individuals were extracted from several public microarray datasets. Subsequently, we extracted the list of miRNA-gene interactions and performed an over-representation analysis to obtain a list of biological pathways controlled by these UC-miRNAs, and possible associated to inflammation and oxidative stress.

2   Data and Methods

2.1   Extracting Gene and miRNA Associations for IBD from MalaCards and HMDD Databases

We used the MalaCards and Human microRNA Disease Database (HMDD) to obtain biological associations with IBD. MalaCards is a comprehensive, human disease database that provides detailed information on various diseases, including associated genes, phenotypes, and pathways. It aggregates data from multiple sources to facilitate the understanding of disease mechanisms. The HMDD database contains curated experiment-supported associations between human precursor miRNAs and several diseases, making it a valuable resource for studying the regulatory functions of miRNAs in different pathological conditions. Since this database does not provide information on the maturation of miRNAs (i.e. "3p" or "5p" information), to compare a miRNA that includes the 3p and 5p annotations with another miRNA lacking this information, we exclude the 3p and 5p designations from the miRNA identifiers.

Using Malacard, we identified 70 genes for UC, 74 for CD, and 1117 for IBD. Using HMDD, we identified 39 miRNAs for UC, 50 for CD, and 116 for IBD, respectively.

2.2   Microarray Datasets

We searched for miRNA microarray datasets related to Ulcerative Colitis in the Gene Expression Omnibus (GEO) database. GEO was queried using the terms *Ulcerative Colitis*[All Fields] AND *microRNA*[All Fields], and, through manually-curated selection, four datasets containing miRNA expression profiles of disease samples of interest and healthy samples were obtained (Table 1). These datasets were acquired using different platforms and normalization processes, which makes difficult to combine them in a unique case/control dataset. Therefore, we decided to analyze the datasets independently of each other. Moreover, since dataset GSE32273 comprised three different assessments, namely microvesicles(M), platelets(P), and peripheral blood mononuclear cells (PBMC), we treated them as three separate datasets, each about a specific cell type and its corresponding controls. Finally, we extracted the common miRNAs among all the datasets, obtaining 590 miRNAs.

2.3   Statistical Analysis

We filtered out all non-human miRNAs from the analysis and standardized the notations to version v22 using the miRBaseConverter library [12]. We also verified that a log_2-transformation had been applied by checking if the 99th percentile is greater than 100 and the interquartile range is greater than 50. If not, we applied the log_2 transformation.

Table 1. Information about the GEO datasets related to Ulcerative Colitis.

Dataset	Platform	Tissue	Cases	Controls
GSE68306	GPL20111	Colon	9	16
GSE48957	GPL14613	Colon	10	10
GSE43009	GPL16384	Colon	5	5
GSE32273	GPL8786	Blood	66	66

Using GEOquery [3] and Limma [7] packages, we compute differentially expressed miRNAs for each dataset by comparing miRNA expression profiles between disease-affected samples and healthy normal samples. All the statistically significant miRNAs were obtained by checking that $p_{value} < 0.05$. In this analysis, we chose to not adjust the p_{value}, because we did not want to focus on multiple tests on the same samples but rather on the level of coherence between the statistical tests across multiple datasets.

2.4   miRNA-Gene Associations

We queried the database mirTarBase [5] to obtain direct interactions between the found miRNAs and the target genes. mirTarBase is a biological database containing more than 4.5 million of miRNA-target interactions (MTI). It employs an optimized scoring system capable of extracting MTIs efficiently from related articles downloaded from the PubMed literature database. Generally, the collected MTIs are validated experimentally and are divided between strong (Reporter assay, Western Blot, qPCR) and weak (Microarray, NGS, pSILAC) evidences. Subsequently, we used the collected MTIs to build the miRNA-gene network, where each miRNA and gene is a node of the network and a link is established between a miRNA and each controlled miRNA.

2.5   Enrichment Analysis

A functional enrichment analysis was conducted using Reactome [6] and ShinyGO [4], to predict the biological functions of a list of genes controlled by specific miRNAs. This analysis aims to identify and understand the biological processes or cellular functions significantly enriched in a specific set of genes or proteins. The goal is to identify whether there are functional categories such as metabolic pathways, cellular processes, or biological functions that are over-represented in the set of interest. All statistically significant pathways were filtered using the criterion of $q < 0.05$,where q is the False Discovery Rate (FDR). Therefore, results are statistically significant with a type I error rate of less than 5%.

3 Results

3.1 Few miRNAs Are Consistent Markers Across All the Datasets

From the six UC datasets, a total of 377 up-regulated miRNAs and 314 down-regulated miRNAs were identified. Details for each analyzed dataset are presented in Table 2. Surprisingly, no miRNA seems to have the same regulation direction (i.e. statistically significant up or down-regulation) in all the datasets. The intersection analysis, summarized in Fig. 2, revealed seven miRNAs that are consistently present in at least three datasets, all showing the same regulation direction (4 up and 3 down) without any inconsistencies, that is to say no one of these miRNAs is up-regulated in one dataset and down-regulated in another one. A total of 43 miRNAs (27 up-regulated, 16 down-regulated) has a specific regulation direction. The hsa-miR-194-5p is the only one consistently down-regulated in four datasets, but up-regulated in one dataset and without a significant difference in the last dataset. However, we considered it in the subsequent analysis since it is the most down-regulated miRNA across the six datasets.

For the subsequent analysis, we extracted the seven consistent miRNAs (hsa-miR-19b, hsa-miR-31, hsa-miR-141, hsa-miR-151a, hsa-miR-181b, hsa-miR-342, hsa-let-7c), and hsa-miR-194-5p.

Table 2. Description of miRNAs in datasets related to Ulcerative Colitis.

Dataset	Total	UP-regulated	DOWN-regulated
GSE68306	102	55	47
GSE48957	198	107	91
GSE43009	59	21	38
GSE32273(M)	174	115	59
GSE32273(P)	79	50	29
GSE32273(PBMC)	79	29	50

3.2 A Comparison with Previous miRNA-UC Associations

By comparing our miRNAs with the ones provided by the HMDD database (see Fig. 2), we obtain four miRNAs (hsa-miR-31, hsa-miR-342, hsa-miR-141, hsa-miR-194) associated with UC, two miRNAs with with CD (hsa-miR-181b, hsa-miR-19b), and one miRNA with IBD (hsa-let-7c). The hsa-miR-151a is not a marker of any of these diseases. However, from the HMDD, we found that this miRNA is associated with colorectal cancer, which is considered a potential complication of IBD. This analysis suggests that our list of miRNAs could serve as a valuable subset of biomarkers associated with UC.

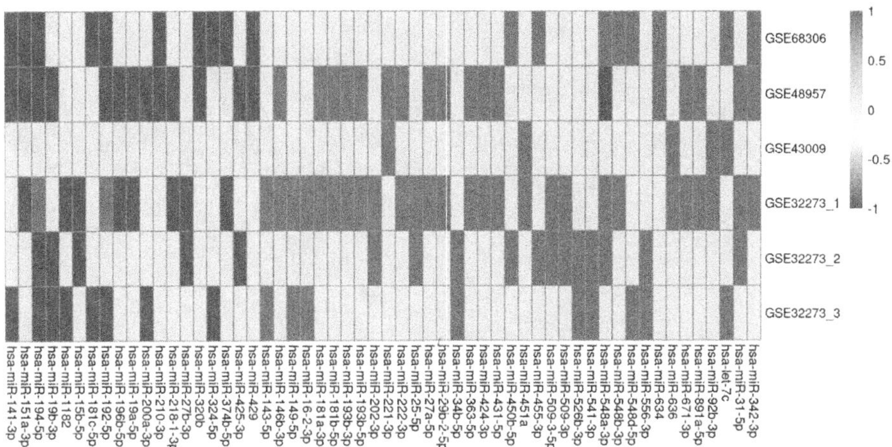

Fig. 2. The most common miRNAs across the datasets and their respective regulation. We coloured in red, blue and yellow, the up-regulation (positive log fold-change), down-regulation (negative fold change-change), and no statistical significance (i.e. $p_{value} \geq$ 0.05) of the miRNA in a specific dataset, respectively. (Color figure online)

3.3 miRNA-Genes Interaction Networks

Using the miRTarBase database, we identified the genes most directly affected by the activity of these miRNAs. We identified a network comprising 8 miRNAs and 2140 genes. From the total network, we extracted the sub-network that included only the genes having two strong (see Sect. 2.4) MTIs at least, for a total of 240 genes (see Fig. 3). From a preliminary analysis, most genes (212) are controlled by just one miRNA, miRNA hsa-miR-31-5p is the highest degree node regulating 64 gene targets, and 7 genes (L10, FOXP3, MLH1, SMAD4, TLR2, HSPA4, IL6) have previously been associated with UC according to the MalaCards database (Fig. 4).

3.4 Enrichment Analysis Unveils Many Pathways Involved in Inflammation and Oxidative Stress

Through Reactome, we imputed the 240 genes of the miRNA-gene interaction network and the 70 genes already associated with UC. The two lists are over-represented, with $q < 0.05$, in 429 and 234 pathways, respectively. We found that the two lists have 131 pathways in common, suggesting that both miRNAs and genes may operate in similar biological mechanisms, reinforcing the validity of their associations.

In Fig. 5, we reported a barplot depicting the 20 most significantly over-represented pathways. We can note that most pathways are part of a wide group of fundamental cellular and molecular processes in inflammation (e.g., Signaling by Interleukins) and oxidative stress (e.g. Cellular Senescences, Signaling by

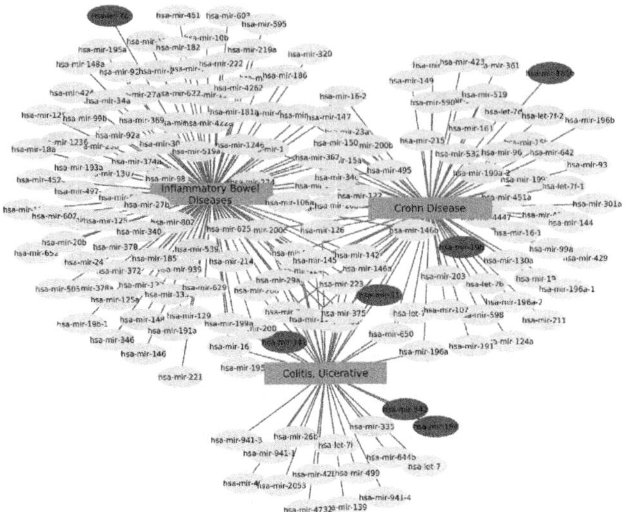

Fig. 3. The miRNA-disease network for UC, CD and IBD. We coloured in red the miRNAs extracted from the statistical analysis. (Color figure online)

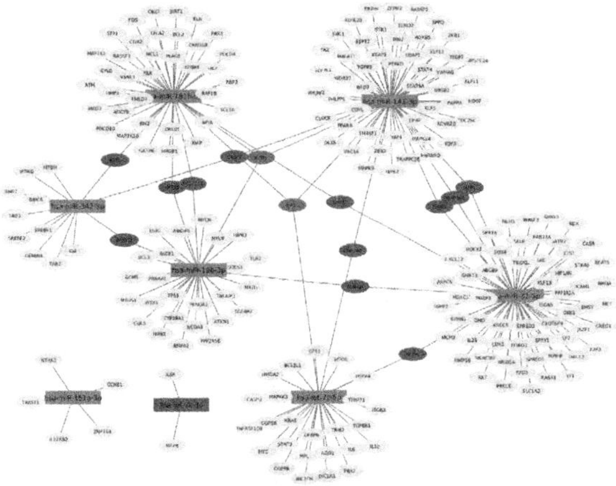

Fig. 4. The miRNA-gene interaction network, considering only the genes which have strong evidence of interaction with the associated miRNA. We coloured in green and yellow the miRNAs and genes, respectively. Moreover, we coloured in red and blue the nodes having degree 2 and 3, respectively. (Color figure online)

Receptor Tyrosine Kinases) or in both the processes (e.g. MAPK family signaling cascades, Cytokine Signaling in Immune system).

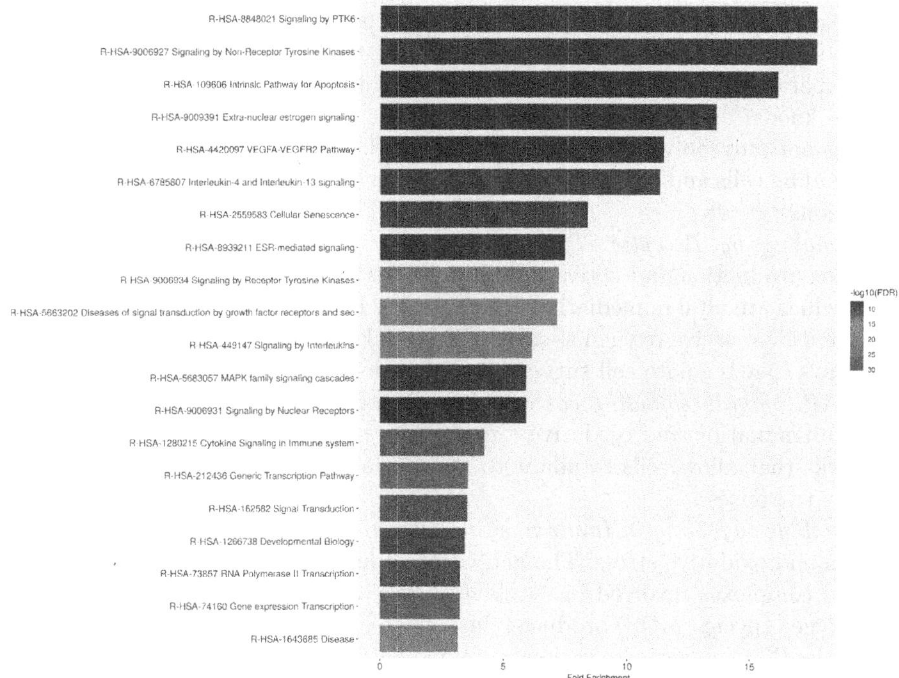

Fig. 5. Barplot of the most over-represented REACTOME pathways for the list of genes previously identified in Sect. 3.2.

4   Discussion

Among the possible pathway controlled by UC-associated miRNAs, we found the general *Signaling by Interleukins*, one of the main pathways involved in inflammation. Interleukins are a type of cytokines that regulate immune responses. Interleukin signaling typically begins when these cytokines bind to their specific receptors on target cells. This binding activates intracellular signaling pathways that result in the activation of transcription factors. These factors translocate to the nucleus and initiate the expression of pro-inflammatory genes, thereby amplifying the inflammatory response. Dysregulation of interleukin signaling is implicated in various inflammatory diseases, including autoimmune disorders, chronic inflammatory diseases, and even cancer. In particular, Interleukin-4 and Interleukin-13 signaling are primarily associated with Type 2 inflammation, exerting their effects by binding to the IL-4 receptor, leading to the activation of signaling pathways that influence various immune responses. They stimulate the differentiation of B cells producing plasma cells, promote eosinophil recruitment to inflamed tissues, and contribute to airway remodeling through effects on fibroblasts and smooth muscle cells.

Cellular senescence has been already found to be associated with IBD [10] as macrophages, activated during IBD, eliminates senescent cells, thus contributing to persistent inflammation. When cells experience stress, such as DNA damage or exposure to reactive oxygen species (ROS), they can enter a state of senescence characterized by permanent cell cycle arrest and a distinct secretory phenotype known as the senescence-associated secretory phenotype (SASP). This process not only prevents the proliferation of damaged cells but also influences surrounding cells and tissues, often contributing to inflammation and age-related conditions.

Signaling by Receptor Tyrosine Kinases can enhance pro-inflammatory cytokine production and activate pathways like MAPK and PI3K signaling cascade, which are vital in mediating inflammatory responses. Also, they can also be activated by reactive oxygen species (ROS), linking oxidative stress to signaling pathways that regulate cell survival, proliferation, and inflammation.

MAPK family signaling cascades are involved in the cellular response to both inflammation and oxidative stress. They participate in complex signaling networks that allow cells to adapt to stress conditions while mediating inflammatory responses.

Cytokine Signaling in Immune system is a pathway involved both in inflammation and oxidative stress. The activation of inflammasomes, which are multiprotein complexes involved in cytokine maturation, is also influenced by reactive oxygen species (ROS) produced during oxidative stress. ROS act as signaling molecules that can activate various cytokine pathways, linking oxidative stress to immune responses. For example, mitochondrial-derived ROS can trigger innate immune responses through pattern recognition receptors like Toll-like receptors, which subsequently lead to cytokine production. This connection emphasizes the dual role of ROS in promoting inflammation while simultaneously acting as a stress signal, contributing to chronic inflammatory diseases.

5 Conclusion

We present a work with the aim of assessing a possible small number of miRNA markers for UC. With this analysis, we found 7 IBD-associated miRNAs (hsa-miR-31, hsa-miR-342, hsa-miR-141, hsa-miR-194 associated to UC; hsa-miR–181b, hsa-miR-19b associated to CD; hsa-let-7c associated to IBD) and 1 miRNA, hsa-miR-151a, as a possible new biomarker of this pathology. The last miRNA is a well-established oncogenic miRNAs, deregulated in colorectal cancer. Our hypothesis is that, as the IBD pathology could degenerate into cancer, hsa-miR-151a could be considered an early detector of this transformation. Indeed, plasma hsa-miR-151a is proposed as a promising biomarker for early detection of colorectal cancer [13].

A possible limit of the results is related to the choice of the parameters analysis (e.g., the p_{value} threshold) which could provide possible alternative output results. Because of this, a parameter scan analysis will be done to improve the robustness of the results, to reduce also the possible false positive markers. Moreover, we will explore the level of expression of genes related to this subset of miRNAs in independent RNA-seq datasets to identify genes regulated at the post-transcriptional level, thereby reducing the number of potential false positive miRNA-gene interactions.

Funding. The work is Financed by the European Union - NextGenerationEU within the National Biodiversity Future Center project (NBFC, CN00000033, CUP B83C22002930006).

Availability of Data and Software Code. All miRNA datasets are available from the Gene Expression Omnibus database. We used MirTarBase to retrieve miRNA-gene associations, Reactome for pathway enrichment analysis, Cytoscape for network representation, and R scripts provided by GEO2R for statistical analysis.

Conflict of Interests. The authors declare that they have no competing interests.

References

1. Baumgart, D.C., Carding, S.R.: Inflammatory bowel disease: cause and immuno-biology. The Lancet **369**(9573), 1627–1640 (2007)
2. Cui, C., Zhong, B., Fan, R., Cui, Q.: HMDD V4. 0: a database for experimentally supported human microRNA-disease associations. Nucl. Acids Res. **52**(D1), D1327–D1332 (2024)
3. Davis, S., Meltzer, P.S.: GEOquery: a bridge between the gene expression omnibus (GEO) and bioconductor. Bioinformatics **23**(14), 1846–1847 (2007)
4. Ge, S.X., Jung, D., Yao, R.: ShinyGO: a graphical gene-set enrichment tool for animals and plants. Bioinformatics **36**(8), 2628–2629 (2020)
5. Huang, H.Y., et al.: miRTarBase update 2022: an informative resource for experimentally validated miRNA-target interactions. Nucleic Acids Res. **50**(D1), D222–D230 (2022)
6. Milacic, M., et al.: The reactome pathway knowledgebase 2024. Nucleic Acids Res. **52**(D1), D672–D678 (2024)
7. Ritchie, M.E., et al.: LIMMA powers differential expression analyses for RNA-sequencing and microarray studies. Nucleic Acids Res. **43**(7), e47–e47 (2015)
8. Roda, G., et al.: Crohns disease. Nat. Rev. Dis. Primers. **6**(1), 22 (2020)
9. Saliminejad, K., Khorram Khorshid, H.R., Soleymani Fard, S., Ghaffari, S.H.: An overview of microRNAs: biology, functions, therapeutics, and analysis methods. J. Cell. Physiol. **234**(5), 5451–5465 (2019)
10. Sienkiewicz, M., Sroka, K., Binienda, A., Jurk, D., Fichna, J.: A new face of old cells: an overview about the role of senescence and telomeres in inflammatory bowel diseases. Ageing Res. Rev. **91**, 102083 (2023)
11. Witwer, K.W., Halushka, M.K.: Toward the promise of microRNAs-enhancing reproducibility and rigor in microRNA research. RNA Biol. **13**(11), 1103–1116 (2016)
12. Xu, T., et al.: miRBaseConverter: an R/Bioconductor package for converting and retrieving miRNA name, accession, sequence and family information in different versions of miRBase. BMC Bioinform. **19**, 179–188 (2018)
13. Zhang, H., et al.: A panel of seven-miRNA signature in plasma as potential biomarker for colorectal cancer diagnosis. Gene **687**, 246–254 (2019)

PHeP: TrustAlert Open-Source Platform for Enhancing Predictive Healthcare with Deep Learning

Sandro Gepiro Contaldo[1,2], Emanuele Pietropaolo[3], Lorenzo Bosio[1],
Simone Pernice[1,2], Irene Terrone[1,2], Daniele Baccega[1,2], Yuting Wang[4],
Rahul Kumar Sahoo[4], Giuseppe Rizzo[5], Alessia Visconti[3(✉)],
Paola Berchialla[3], and Marco Beccuti[1,2]

[1] Department of Computer Science, University of Turin, Turin, Italy
[2] Laboratorio InfoLife, Consorzio Interuniversitario Nazionale per l'Informatica,
Roma, Italy
[3] Centre for Biostatistics, Epidemiology and Public Health, Department of Clinical
and Biological Sciences, University of Turin,Turin, Italy
`alessia.visconti@unito.it`
[4] Polytechnic of Turin, Turin, Italy
[5] LINKS Foundation,Turin, Italy

Abstract. Predictive healthcare, driven by the availability of large medical datasets and computational advancements, plays a crucial role in enhancing patient care and optimizing healthcare system performance. However, training and deploying predictive models in healthcare presents substantial challenges, such as specialized expertise, and the need for state-of-the-art and costly hardware. To address these barriers, we introduce the Predictive Healthcare Platform (PHeP), an open-source platform designed to simplify the use of pre-trained models. Through a real-world case study, we showcase PHeP functionality in making advanced predictive healthcare accessible to users without extensive computational skills, or expensive computational resources. This will foster a wider adoption of predictive healthcare in medical practice.

Keywords: predictive healthcare · deep learning · high performance computing · natural language processing · medical informatics

1 Introduction

Predictive healthcare, *i.e.*, the use of historical and/or real-time data to anticipate patients' medical needs [1], is crucial to improve both patients' quality of life and healthcare systems' performances. It has become possible thanks to the availability of large datasets of medical records, such as those collected in electronic health records (EHRs) and healthcare administrative databases (HADs),

---

S. Gepiro Contaldo and E. Pietropaolo—The authors contributed equally to the work.
P. Berchialla and M. Beccuti—The authors jointly supervised the work.

L. Cerulo et al. (Eds.): CIBB 2024, LNBI 15276, pp. 206–216, 2025.
https://doi.org/10.1007/978-3-031-89704-7_16

and to the recent development in deep learning approaches. EHRs and HADs track patients' medical history by collecting longitudinal information on medical diagnoses and procedures, as well as drug prescriptions [2,3]. Deep learning approaches are naturally able to deal with these very large amounts of longitudinal data, also when presenting with irregular time intervals between events [4]. These types of techniques have been successfully used to predict the risk of hospitalization, the onset of new diseases, and the worsening of existing ones [4–6].

Performances of deep learning approaches can be further improved by combining them with natural language processing techniques, which add contextual information by generating a compact representation of medical data (embedding). This line of research has been spurred by the development of the Bidirectional Encoder Representations from Transformers (BERT) [7]. BERT allows pre-training deep bidirectional representations from unlabelled text, which can be then fine-tuned to solve a wide range of tasks, including predictive healthcare tasks [8–10], without substantial architecture modifications.

Training BERT and its derived models, however, requires a very large amount of data, non-trivial expertise in data munging and deep learning technologies, state-of-the-art and costly hardware, and long processing time. For instance, training the large model (BERTLARGE) took 4 d on 16 Cloud TPUs (64 TPU chips total) [7] and training of Med-BERT [10] took a week on a single Nvidia Tesla V100 GPU of 32GB graphics memory capacity.

To address this and enhance the adoption of deep learning approaches in healthcare, we have developed a new open-source Predictive Healthcare Platform (PHeP). PHeP provides access to pre-trained models through an intuitive web-based graphical interface, aiming to simplify their usage for users without advanced computational skills.

Here, we describe PHeP and demonstrate its effectiveness by creating an online service which uses three pre-trained BERT models to predict hospitalization at three months from the last access to the national health service in a paediatric (under 14 years old) population.

2   Methods

This section outlines the main subcomponents of PHeP, developed under the TrustAlert project, details the hardware and software stack supporting it, and describes the pre-trained models used in our case study.

2.1   PHeP in a nutshell

PHeP sports a modular architecture with four main subcomponents: Collaborative Working Environment, Data Repository, Analysis Environment, and Prediction Environment (Figure 1**A**).

The *Collaborative Working Environment* (CWE, (https://www.trustalert. it) acts as the front-end platform, and serves as a centralized hub, allowing users to interact with various PHeP services, access resources, and engage with TrustAlert Project-related information [11]. Users can seamlessly navigate functionalities, collaborate with team members, access project documentation, monitor progress, and stay updated on developments. It is implemented through a WordPress multisite instance which allows for the creation of multiple sites within a single WordPress installation. Notably, the WordPress multisite instance was intentionally deployed in a separate virtual network from the other services to mitigate the potential impact of a security breach within WordPress on critical services. All connections between the CWE and the other primary services undergo a firewall service, thus effectively filtering malicious traffic.

The *Data Repository* is a centralized location providing a secure and structured environment, for storing, organizing, and managing data produced in the project, according to FAIR principles, including all pre-trained models developed for predictive purposes. It is implemented through Harvard Dataverse [12], a leading open-source data repository platform designed to facilitate the sharing, preservation, and discovery of research data. The repository offers persistent identifiers and version control features, ensuring the integrity and traceability of data over time. This facilitates the reuse and validation of data by other researchers, and supports reproducibility in scientific research. The Data Repository can interact with the other subcomponents through specifically developed REST APIs.

The *Analysis Environment* provides TrustAlert researchers with access to two powerful and versatile data analysis environments, *i.e.,* R and Python. The R environment is based on RStudio Server, which offers a robust and user-friendly interface for R programming and statistical analysis. The Python environment is hosted on JupyterHub, providing an interactive platform for Python coding, data visualization, and machine learning tasks. Thus, the *Analysis Environment* offers the necessary tools and resources for tasks such as data preprocessing, conducting exploratory data analysis, and building predictive models.

The *Prediction Environment* is the cornerstone of PHeP, embodying its cutting-edge Healthcare Predictive System (HPS). Through a seamlessly integrated web interface and an encrypted communication channel, this environment grants users intuitive access to a suite of advanced and secure predictive analytics tools. It can be easily instantiated with various pre-trained models, and customized to address different predictive healthcare tasks, such as the risk of short-term hospitalization (already implemented), the onset of new diseases, and the risk of non-urgent A&E accesses. Within the *Prediction Environment*, we developed a specialized web interface tailored for executing predictions via Python scripts. Harnessing the power of Next.js, a robust React framework renowned for its ability to construct high-quality web applications with server-side rendering and static generation, we streamlined the development process and expedited deployment, resulting in an intuitive and responsive platform. By using Next.js, we optimized code management and bolstered code reusability. The framework lazy loading strategy and automatic code-splitting efficiently render pages, loading only necessary content to enhance performance by fetching precisely what is required.

2.2   Hardware and Software Stack Supporting PHeP

The hardware infrastructure hosting PHeP relies on the open-access High-Performance Computing for Artificial Intelligence (HPC4AI) data centre (https://hpc4ai.unito.it), situated at the University of Turin. HPC4AI introduces an innovative HPC-cloud convergence architecture (Figure 1**B**), reshaping the traditional utilization of cloud and HPC systems to bolster AI applications. In this model, the cloud provides a modern interface for HPC, and HPC acts as an accelerator for the cloud. The HPC4AI cloud system is implemented using the open-source OpenStack cloud technology [13], and leverages different resources, including over 2400 physical cores, 60 TB RAM, 120 GPUs (NVidia T4/V100/A40), 25 Gb/s networking, and 4 storage classes with distinct characteristics. The use of OpenStack facilitates the implementation of a robust Deployment-as-a-Service (DaaS) feature, allowing the modelling of user-defined platforms (IaaS and PaaS) in a portable manner. Moreover, it automates their deployment on a virtualized infrastructure. The HPC subsystem is managed through SLURM software [14] and comprises 72 nodes with Intel architecture (68 nodes with 36 cores, 128GB RAM, OPA 100Gb/s and 4 nodes with 40 cores, 1 TB RAM, T4+V100 GPUs, eth 56Gb/s) and 8 nodes with Arm architecture (4 nodes with Ampere Altra 80 cores, 512GB RAM, 2xA100 GPUs, 2xBF2 DPUs, eth 100Gb/s and 4 nodes with 80 cores, Grace Hopper Superchips GH200). Additionally, there are 2 HPC storage systems utilizing BeeGFS [15] and LUSTRE [16], both in an all-flash configuration.

Figure 1**A** show an overview of the hardware and software stack supporting PHeP. Above the hardware layer, hosted within the HPC4AI data centre, lies the virtualization layer, powered by the OpenStack project, and which encompasses two Virtual Machines, collectively offering 20 Virtual Cores, 140GB RAM, and 1.4TB of storage. At the top is the application layer, which includes three of the integrated subcomponents (black boxes) described in the previous subsection, along with its corresponding microservices (orange boxes). The connections between these components are visually depicted by arrows, illustrating their interdependencies. This layered architecture ensures robust and scalable performance. The hardware provides a solid foundation, while virtualization efficiently facilitates resource management.

2.3   *Proof-of-Concept: BERT for Predicting Short-Term Hospitalization*

The data used in this study were collected by one of the local health services in the Piedmont region of Italy (ASL-CN2). The ASL-CN2 Ethics Committee and the Data Protection Officer authorized this study (Presa d'atto n. 4-2023).

Patients aged 14 and younger with at least one hospital admission were included in this study. Their medical history (*i.e.*, the set of medication dispensed or administered, and/or of medical diagnoses and/or procedures, hereafter: records) was collected from 2019-01-01 to 2024-02-29. Basic demographic data, *i.e.*, age group and living area, but not sex, were also available. We discarded patients whose medical history included more than 512 records due to limitations in the data encoding in the deep learning approach used. To exploit the

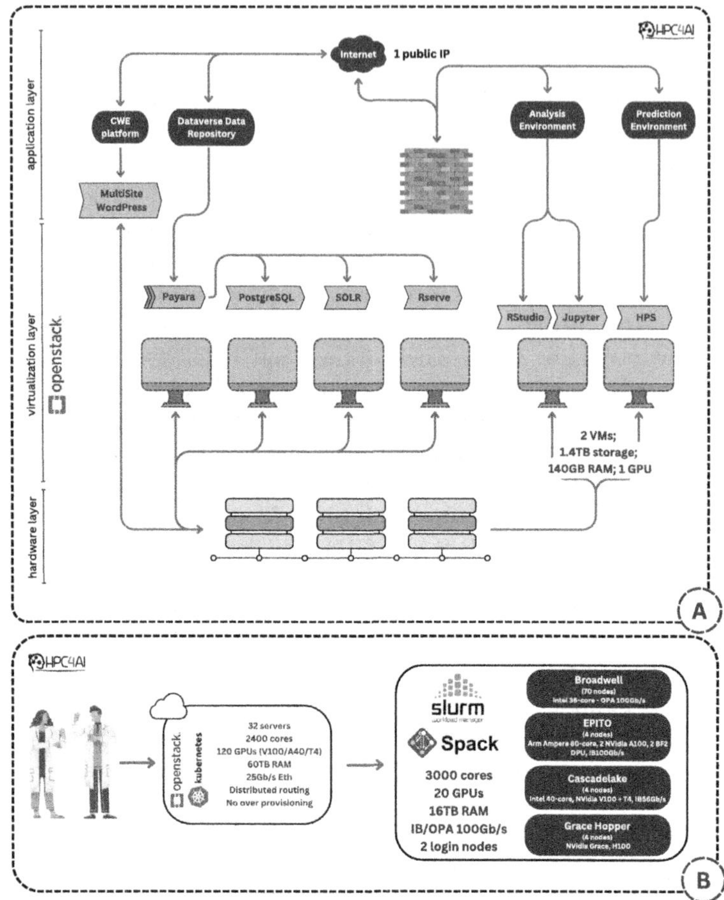

Fig. 1. (A) Overview of the hardware and software stack supporting PHeP. Black boxes represent PHeP subcomponents, orange boxes represent microservices for each subcomponent, and arrows illustrate their interdependencies. (B) The open-access High-Performance Computing for Artificial Intelligence (HPC4AI) architecture.

multi-head self-attention and positional encoding of transformer-based architectures of BERT, we built a chronologically ordered sequence of records describing the patients' medical history (hereafter: trajectory; the document), each including a variable number of records (the words) recorded during separate medical episode (the sentences). Then, we recorded the time between the last episode (not including medications) and the last hospitalization event. If the recorded time was smaller than 90 d, the patient was considered as a positive case of short-term hospitalization.

We used a tokenizer (*BertTokenizerFast* implementation) to map our dataset to a new vocabulary, which we then used to pre-train a BERT raw architecture both on the Masked Language Modeling (MLM) and Next Sentence Prediction (NSP) tasks originally proposed in [17] (*BertForMaskedLM* and *Bert-ForNextSentencePrediction* implementation, respectively), and on their combination (*BertForPreTraining* implementation), thus generating three models. The MLM task randomly masks a portion of the input sentences, and then learns to predict it based on the surrounding context, effectively capturing the relationships between same-episode diagnoses, procedures, and medications. The NSP tasks extract random pairs of sentences and classify whether these are consecutive (*i.e.*, one follows the other), thus learning temporal connections between medical episodes, and their associated diagnoses, procedures, and medications. Then, we fine-tuned these models to predict whether a patient will be hospitalized within 90 d from their last access to the national health service, given their available medical history (*BertForSequenceClassification* implementation).

Due to the proof-of-concept nature of this study, we did not perform any hyperparameters tuning, and the values selected from a literature review for both pre-training and fine-tuning are as follows: 8 epochs for pre-training, 1 epochs for fine-tuning, 512 as max sequence length, 10 as train batch size, 5×10^{-5} as learning rate, 768 as hidden size, 12 as number of hidden layers, 12 as number of attention layers, 3072 as intermediate size, GELU as hidden activation function, 0.1 as hidden dropout probability, and 0.02 as initializer range.

Patients were split 80:20 in training and testing set, respectively. For the fine-tuning task, we used two procedures to augment the information available in the training set. First, we generated $m_i - 1$ new trajectories (where m_i is the number of all the recorded medical events for the i-th patient) by removing, one at a time, medical events from the start of the trajectory and maintaining the original label. Then, we generated $n_i - 1$ new trajectories (where n_i is the number of hospitalization events for the i-th patient) by iteratively identifying the last recorded hospitalization event, truncating the trajectory at the previous episode, and generating a new label.

During model training, we used a 5-fold cross-validation (CV) strategy to test the stability of the learned models in reaching accurate results. Furthermore, to evaluate the impact of the augmentation procedure, we build and assessed three models, one for each task, using a non-augmented training set during fine-tuning.

This real world dataset was also used to generate four synthetic datasets of increasing size by reshuffling and recombining medical events, conditioned on age class, thus ensuring that the generated datasets maintained the statistical properties and, possibly, realistic patterns found in the original data. These synthetic datasets were then used to measure PHeP performances using a range of different GPUs, *i.e.*, NVIDIA Tesla T4 16GB, NVIDIA A100 40GB, and NVIDIA GraceHopper Superchip GH200 with H100 100GB.

3   Results

To validate the capabilities of the PHeP system, we conducted a comprehensive series of tests and experiments. First, using four synthetic datasets of increasing size and three different GPUs, we showed that PHeP can efficiently exploit the hardware resources available through the HPC-cloud convergence infrastructure, highlighting its ability to scale (Table 1).

Table 1. BERT running time. The table shows the time spent, per epoch, to pre-train (PT) and fine-tune (FT) the BERT raw model using different GPU as well as synthetic datasets of increasing size for the four implemented tasks. The number N of medical episodes is reported.

Task	GPU	N=16k	N=80k	N=160k	N=320k
MLM	T4	1min 57sec	9min 32sec	19min 07sec	37min 50sec
	A100	46sec	2min 26sec	4min 54sec	9min 48sec
	H100	11sec	57sec	1min 55sec	3min 48sec
NSP	T4	11min 30sec	56min 01sec	1h 52min	3h 44min
	A100	4min 40sec	14min 50sec	29min 35sec	59min 17sec
	H100	1min 12sec	6min 06sec	12min 11sec	24min 19sec
MLM+NSP	T4	14min 54sec	1h 15min	2h 30min	5h 01min
	A100	5min 44sec	18min 32sec	37min 15sec	1h 14min
	H100	1min 26sec	7min 22sec	14min 25sec	28min 38sec
fine-tuning	T4	12min 18sec	59min 53sec	1h 59min	3h 58min
	A100	4min 52sec	15min 54sec	30min 48sec	1h 01sec
	H100	1min 14sec	6min 13sec	12min 27sec	25min 36sec

We then tested the *Prediction Environment*, a central component of PHeP, by pre-training and storing six BERT models within the TrustAlert *Data Repository* to predict patients' hospitalization at three months from the last access to the national health service.

The models were trained on a cohort of 1,647 paediatrics ($\leq$ 14 years old) patients, with data collected over 5 years (Table 2). This dataset included 15,954 unevenly spaced medical episodes, summing up at 39,543 records, encompassing 2,083 unique codes, with frequencies ranging from 1 to 7,021 (median: 2, IQR: 1-6). The number of medical episodes per patient ranged from 2 to 148 (median: 5, IQR: 3-9). Among the most frequent records, there were outpatients appointments (N=7,021 for first referral, and N=2,814 for subsequent ones), injection or infusion of electrolytes (N=887) or antibiotics (N=827), dysphasia training (N=726), and interview with patients' parents (N=636). Among ICD-9 diagnosis, the most common involved respiratory diseases (N=404, *e.g.*, acute bronchitis, bronchopneumonia, acute bronchiolitis due to respiratory syncytial virus), obstructive sleep apnoea syndromes (N=105), dehydration (N=96), anti-neoplastic chemotherapy (N=96), fever (N=74), and epilepsy (N=44), again in line with this very young population.

Since many records were either unique or present at very low frequency, with, for instance, 98% of records showing a relative frequency lower than 0.05%, we used a 5-fold CV strategy to assess performance stability. We noticed that,

Table 2. Sample characteristics. Categorical variables are presented as number (percentage), and the comparison between training and testing set was performed *via* χ^2 test. Count variables are presented as median [interquartile range] and compared using the Wilcoxon test.

		Overall	Training set	Testing set	P-value
N		1,647	1,317	330	-
Age group	< 1	160 (9.7)	135 (10.3)	25 (7.6)	0.50
	1-4	567 (34.4)	454 (34.5)	113 (34.2)	
	5-9	588 (35.7)	465 (35.3)	123 (37.3)	
	10-14	332 (20.2)	263 (20.0)	69 (20.9)	
N of unique diagnoses/procedures		4 [1, 9]	4 [1, 9]	4 [2, 9]	0.37
N of unique medications[a]		0 [0, 0]	0 [0, 0]	0 [0, 0]	0.62
N of hospitalizations		6 [3, 10]	6 [3, 10]	6 [4, 11]	0.55

[a]Zero values for this distribution are due to the fact that > 90% of the children and adolescents in this dataset do not assume any medication.

when using data augmentation during model fine-tuning, we obtained very stable accuracies (standard deviation < 0.06; Table 3, Scenario 1) suggesting that all trained models were robust. However, performances dropped between 18 and 32% during testing. We hypothesized that this was due to the augmentation procedure, which generated, on average, nine very similar trajectories for each patient. Therefore, we performed a new experiment in which no augmentation was applied during fine-tuning, observing comparable accuracies in the 5-fold CV and in the testing set (Table 3, Scenario 2). We also noted that, apart from the model generated using the NSP task, whose performance remained unchanged, accuracies were better when no augmentation was performed. suggesting that such a procedure may bias the training sample and should be avoided, even though more extensive experimentations are needed to confirm this observation. We also observed that the accuracies of the models that were pre-trained on both the MLM and the NSP task were slightly higher, suggesting that both the relationships between events recorded within the same episode (as learn by MLM) and the temporal connections between medical episodes (as learn by NSP) are important for the prediction of short-term hospitalization, with the latter playing a smaller role, as suggested by the lowest performances reached by this task alone (Table 3). Matthews correlation coefficients confirmed the observed trends in accuracies.

These pre-trained models can be used throughout an intuitive and reactive web interface (Figure 2, http://trustalert.hpc4ai.unito.it:3000/), where users should only upload a CSV file containing the patients' hospitalization history. The results of the computation is a text file whose first column contains the patient ID, as provided in the input file, and the second column contains the prediction label, *i.e.*, '1' if the patient is expected to be hospitalized within three months, '0' otherwise. In this way, the prediction tasks can be performed in min-

Table 3. Accuracy in predicting short-term hospitalization in a cohort of 1,647 paediatrics patients. The Table shows, for each task used for pre-training, the results of the stability test (5-fold CV) as well as the accuracies obtained on the testing set by the models trained on the complete training set. Accuracies obtained during 5-fold CV are presented as mean $\pm$ standard deviation. Scenario 1 and 2 perform fine-tuning using data with or without augmentation, respectively.

		MLM	NSP	MLM+NSP
Scenario 1	5-fold CV	0.848 ± 0.015	0.825 ± 0.005	0.824 ± 0.053
	testing set	0.578	0.579	0.673
Scenario 2	5-fold CV	0.765 ± 0.024	0.546 ± 0.027	0.729 ± 0.048
	testing set	0.724	0.579	0.766

utes by non-experts without requiring any technical knowledge, software installation or parameter tuning, making predictive healthcare accessible to healthcare providers which often lack these capabilities. Notably, input files are not stored on our servers, thus respecting GDPR privacy regulations.

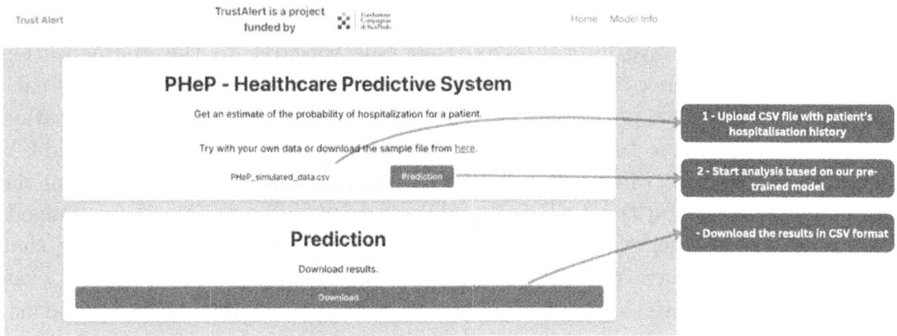

Fig. 2. The PHeP user interface for our case study. The figure shows how to (1) load and (2) process the data, and (3) to download the prediction results.

4   Conclusion

In this work, we introduce the Predictive Healthcare Platform (PHeP), an innovative open-source platform developed under the TrustAlert project. PHeP is designed to simplify the use of pre-trained models in predictive healthcare, providing an intuitive web-based graphical interface. This makes it accessible to users without extensive computational skills or resources, thus widening its adoption.

Through a real-world case study, PHeP has demonstrated its efficiency and adaptability in handling large-scale medical datasets. Our experiments, conducted with varying dataset sizes and GPUs, have shown robust performance. The platform houses six pre-trained BERT models that predict paediatric patients' risk of short-term hospitalization, trained on a large cohort of real patients. The platform's user-friendly interface allows non-experts to perform these tasks in minutes, making predictive healthcare more accessible to healthcare providers. Looking ahead, future enhancements to PHeP will focus on expanding its range of capabilities, allowing, for instance, the prediction of the onset of new diseases and of the risk of non-urgent A&E accesses. We will also ensure that models are transparent and explainable, further improving its user-friendliness and accessibility. This positions PHeP as a valuable tool in fostering the ongoing revolution of predictive healthcare.

Acknowledgments. D.B. is a Ph.D. student enrolled in the National Ph.D. in Artificial Intelligence, XXXVII cycle, health and life sciences course organized by Università Campus Bio-Medico di Roma.

Funding Information. This work is part of the TrustAlert project, which was supported by the Fondazione Compagnia San Paolo and Fondazione CDP under the "Artificial Intelligence" call. HPC4AI (https://hpc4ai.unito.it) was set up thanks to an initial investment of 4.5M€ from a competitive funding call from the Piedmont Region via EU POR-FESR 2014-2020.

Availability of Data and Software Code. Data on study participants are available to *bona fide* researchers under managed access due to governance and ethical constraints. The raw data should be requested *via* contact with the corresponding author, and will be shared only after approval from the ASL-CN2 Ethics Committee and the Data Protection Officer.

The source code of the service interface is available at: https://github.com/qBioTurin/TrustalertPredictionService. The source code for our case study, *i.e.*, the prediction of short term hospitalization using healthcare administrative data using BERT, is available at: https://github.com/edoppiap/bert_medical_records.

Conflict of interests. The authors declare no conflict of interests.

References

1. Rajkomar, A., Dean, J., Kohane, I.: Machine learning in medicine. N. Engl. J. Med. **380**(14), 1347–1358 (2019)
2. Jensen, P.B., Jensen, L.J., Brunak, S.: Mining electronic health records: towards better research applications and clinical care. Nature Rev. Genet. **13**(6), 95–405 (2012)
3. Boncyk, C.S., Jelly, C.A., Freundlich, R.E.: The blessing and the curse of the administrative database (2020)
4. Nguyen, P., Tran, T., Wickramasinghe, N., Venkatesh, S.: *Deepr*: a convolutional net for medical records. IEEE J. Biomed. Health Inform. **21**(1), 22–30 (2017)

5. Pham, T., Tran, T., Phung, D., Venkatesh, S.: Predicting healthcare trajectories from medical records: a deep learning approach. J. Biomed. Inform. **69**, 218–229 (2017)
6. Choi, E., Bahadori, M.T., Sun, J., Kulas, J., Schuetz, A., Stewart, W.: Retain: an interpretable predictive model for healthcare using reverse time attention mechanism. In: Advances in Neural Information Processing Systems, vol. 29 (2016)
7. Devlin, J., Chang, M.-W., Lee, K., Toutanova, K.: BERT: pre-training of deep bidirectional transformers for language understanding. arXiv preprint arXiv:1810.04805 (2018)
8. Shang, J., Ma, T., Xiao, C., Sun, J.: Pre-training of graph augmented transformers for medication recommendation. In: Proceedings of the Twenty-Eighth International Joint Conference on Artificial Intelligence, IJCAI-19, pp. 5953–5959. International Joint Conferences on Artificial Intelligence Organization, vol. 7 (2019)
9. Li, Y., et al.: BEHRT: transformer for electronic health records. Sci. Rep. **10**(1), 7155 (2020)
10. Rasmy, L., Xiang, Y., Xie, Z., Tao, C., Zhi, D.: MED-BERT: pretrained contextualized embeddings on large-scale structured electronic health records for disease prediction. NPJ Digit. Med. **4**(1), 86 (2021)
11. Trustalert project (2023). https://www.trustalert.it. Accessed 22 May 2024
12. Brooke, D.: Community built infrastructure: the dataverse project (2020)
13. Horizon official site (2023). https://docs.openstack.org/horizon/latest/. Accessed 22 May 2024
14. Slurm documentation (2023). https://slurm.schedmd.com. Accessed 22 May 2024
15. BeeGFS documentation (2023). https://doc.beegfs.io/latest/. Accessed 22 May 2024
16. Lustre official site (2023). https://www.lustre.org. Accessed 22 May 2024
17. Vaswani, A., et al.: Attention is all you need. CoRR, abs/1706.03762 (2017)

Cutting Slices of Complexity in Cancer Therapy Design: An Agent-Based Model of Dabrafenib in Melanoma

Stefano Maestri[(✉)] [iD]

University of Camerino, Camerino 62032, Italy
`stefano.maestri@acm.org`

Abstract. Agent-based modelling has emerged as a promising approach for studying cell signalling pathways due to its ability to grasp the critical role of local interactions within biochemical reactions. The flexibility of agent-based models allows for the emergence of complex behaviours from simple local rules, reducing the reliance on extensive experimental data. This potential can be harnessed to investigate the behaviour of mutated pathways in cancer and identify therapeutic targets for developing personalised treatments. In this study, we define an agent-based model to simulate the inhibitory effect of the targeted therapy drug dabrafenib on the BRAFV600E-MEK-ERK signalling cascade in melanomas. We tested an abstraction method that sets up the simulation in a three-dimensional environment but takes a slice of its volume to carry out the simulations. The model's reliability is supported by validation against a clinical study, providing encouraging evidence for the use of agent-based models as in silico support in cancer therapy design.

Keywords: Agent-based Models · Targeted Therapies · Dabrafenib · Personalised Medicine · Melanoma · Repast4Py

1 Introduction

The study of new therapies for cancer treatment has evolved significantly with advancements in computational modelling techniques. Among the most promising approaches, agent-based modelling has gained attention, especially in the last two decades, for its ability to incorporate a wide range of biological properties into the model, including genetic mutations, cellular behaviours, and environmental influences [1].

In the context of cancer therapy design, an agent-based model (ABM) offers a framework for integrating patient-specific data into simulations, making it well-suited for developing personalised treatment strategies [19].

ABMs simulate individual entities (*agents*) and their interactions within a system to study emergent behaviours. The system evolves as each agent perceives and responds to its unpredictable environment populated by other agents; the agent's perception leads to appropriate actions to modify the environment

L. Cerulo et al. (Eds.): CIBB 2024, LNBI 15276, pp. 217–229, 2025.
https://doi.org/10.1007/978-3-031-89704-7_17

(Fig. 1). Global behaviour emerges from local interactions that are dynamically generated at each simulation step, a property that forms the basis of the bottom-up modelling approach enabled by ABMs. This represents a defining feature of the agent-based methodology: modelling complex systems does not require extensive reliance on experimental data, as some of the system's properties emerge from local agent interactions. As a consequence, when experimental data are difficult to obtain—such as for kinetic modelling—ABMs may be preferable to data-intensive machine-learning approaches or network-based models that heavily rely on curated interaction datasets [8,20].

By representing molecules, cells, or other biological entities as agents, ABMs can simulate how variations in genetic mutations, signalling pathways, and microenvironmental conditions influence tumour progression and response to therapy [3,13].

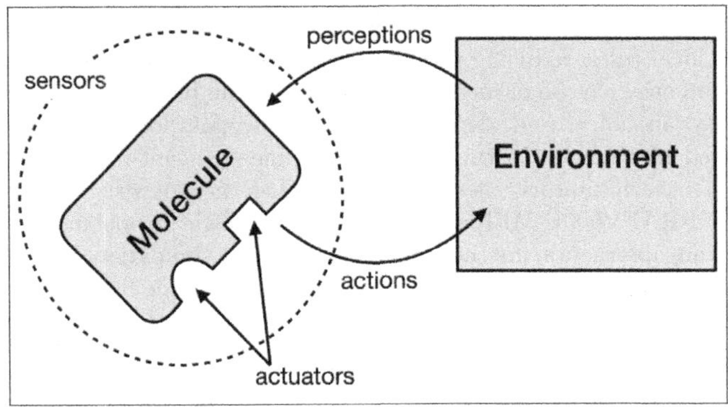

Fig. 1. Molecule behaviour modelled through an agent-based approach. The figure illustrates the main features of the ABM framework described in this work: an agent reproduces a molecular behaviour by interacting with its environment through actuators, which represent the molecule's binding domains, and sensors, which abstract the effect of co-resonating electrodynamic forces that facilitate molecular interactions (see [12] and [14] for more details on this property of the model).

We have already successfully applied this approach to show the role played by electrodynamic forces in promoting biomolecular interactions [14]. In this paper, we test the capability of our ABM to support cancer treatment.

Specifically, we simulated the inhibitory effect of the targeted therapy drug dabrafenib (DBF) on the BRAFV600E-MEK-ERK signalling cascade (see Fig. 2 for more details on the part of the pathway considered in this study). The BRAFV600 mutation is present in approximately 50% of melanomas [2]; by explicitly targeting this mutation, the development of BRAF inhibitors, such as DBF, has significantly impacted melanoma treatment. DBF inhibits the activity of the mutant BRAF protein, thereby reducing the aberrant signalling and inducing apoptosis in melanoma cells [10].

MAPK/ERK is a well-known signalling pathway that has already been used as a case study for computational modelling, including ABMs [1,17]; for this reason, different works provide the quantitative data we need to simulate our model [9]. Nonetheless, to our knowledge, this is the first attempt to employ agent-based modelling to simulate the DBF inhibition of the pathway.

For this aim, we improved and implemented our modelling approach into a recently released framework, Repast4Py, since it is built on Repast HPC (High-Performance Computing) and hence optimised to build large, distributed ABMs, which should be able to manage the complexity of thousands of molecular interactions [4]. This study, however, tests an abstraction approach that allows simulations to run on a mid-range multi-core system, such as a 12-core workstation (see Data and Methods), without fully exploiting the framework's HPC potential. Still, the validation we provide for our results against a clinical study supports the model's reliability, even at such a level of abstraction.

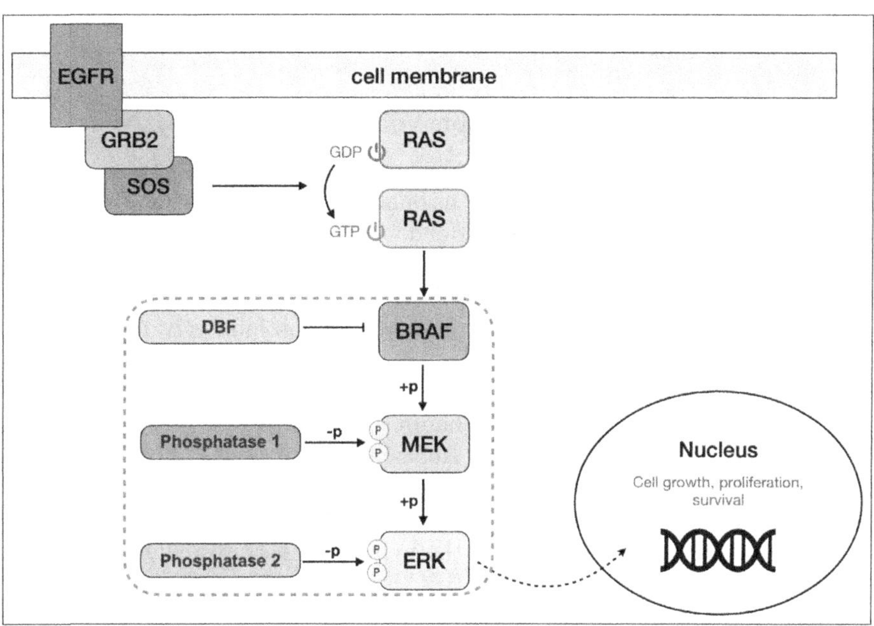

Fig. 2. Simplified representation of the MAPK/ERK pathway and the effect of DBF. The figure illustrates key molecules in the MAPK/ERK cascade and the pathway's role in fundamental cellular processes such as growth, proliferation, and survival. The dashed green square highlights the portion of the pathway considered in this study, which includes the inhibitory effect of DBF on BRAF. Downstream in the pathway, MEK and ERK are activated through double phosphorylation, a process counteracted by phosphatase 1 (for MEK) and phosphatase 2 (for ERK).

2   Data and Methods

In our ABM, each molecule is represented by a *reactive agent*, an autonomous system able to perceive changes in its environment and react to them based on pre-defined rules or conditions (Fig. 1).

Formally, a *reactive agent* is defined by a 6-tuple $\langle E, Per, Ac, see, action, do \rangle$ where:

- E is the set of all possible states of the environment.
- Per is a partition of E representing the agent's perception of the environment.
- Ac is a set of possible actions.
- *see*: $E \rightarrow Per$ maps the environment to the agent's perception.
- *action*: $Per \rightarrow Ac$ maps perceptions to actions.
- *do*: $Ac \times E \rightarrow E$ updates the environment based on the agent's actions.

An agent observes the environment based on its perception (*see*), selects an appropriate action (*action*), and acts upon the environment (*do*) to modify it [7]. For further details on the implementation of these formal definitions to model the behaviour of molecules as reactive agents, refer to Maestri et al. [14].

2.1   The *Slicing* of the Simulation Volume

In the model that underlies our original agent-based simulator, the environment perceived by the agents is a three-dimensional space representing a portion of the cytoplasm [14]. This representation allowed us to model each molecule as a sphere whose radius is estimated from its molecular weight and the average value of the molar-specific volume of a protein in solution [5,16]. However, that choice came with the drawback of a high computational cost, prohibitively impacting the simulation length (one second is simulated in about 24 h on an 8-core CPU). That made it almost impossible to validate our results against empirical experiments, which often span from one to several hours (if not days).

Leveraging HPC systems, such as Repast4Py, can significantly reduce simulation time, making it a viable solution to the high computational demands of our models. However, in the current preliminary study, we set aside the HPC capabilities of Repast4Py. We tested the hypothesis that to reduce the computational cost of the simulations when HPC is unavailable, a *slice* of a simulation cube can be taken as a rough—but effective—approximation of the simulation volume (Fig. 3). Thus, we set up the simulation based on the initial configuration of the modelled molecules, including properties such as molecule radius and initial concentration, within a three-dimensional space. We then *sliced* the simulation volume and constrained the molecules to move in two dimensions, specifically on a square with a thickness of one simulation unit (1 nanometre in this study).

2.2   Experimental Data for the Simulation Setup

Only the following experimental parameters are required to set up the signalling pathway simulation:

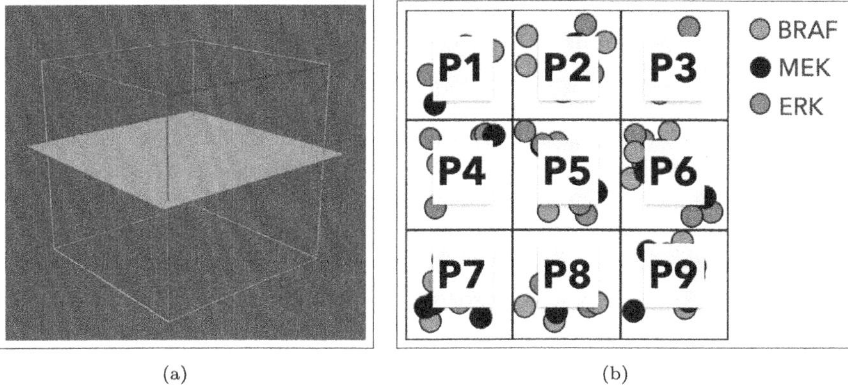

(a)   (b)

Fig. 3. The cube *slicing* process. To reduce the computational cost of the simulations when HPC is unavailable, the simulation cube is *sliced* (a), and the molecule concentrations are scaled accordingly (the exact position of the *slice* within the cube volume is not relevant). This process makes the molecules move on a square managed as a grid by Repast4Py (b). Molecule movements and interactions are distributed across different space portions, each of them controlled by a dedicated process (or rank), allowing for parallel computation. A space portion is made of several grid cells (not shown). This figure, representing a scenario involving nine processes (P1, P2, ..., P9), is only for illustrative purposes and is not representative of the actual complexity or diversity of the full simulation. For more details on how Repast4Py manages rank computations, see the Repast4Py User Guide.

–   The initial concentration and molecular weight of the species involved.
–   The kcat (turnover number) of the enzymatic reactions.
–   The volume of the simulation environment.

The molecular weight can be retrieved from databases such as UniProt and PubChem [11,18], and is needed to estimate the molecule radius, as mentioned above. The latter is then used to calculate the diffusion coefficient through the Stokes-Einstein equation for the Brownian motion of a spherical particle (Eq. 1).

$$D = \frac{k_B T}{6\pi \eta r} \tag{1}$$

where k_B is the Boltzmann constant, T the temperature, η the viscosity of the environment, and r the radius of the molecule. For our simulations, we set $T = 298.15$ kelvin and $\eta = 0.0011$ pascal-second.

The movement vector magnitude is finally obtained as the average value of the square of the molecule displacement in a time t corresponding to a tick of the simulation clock (Eq. 2). Refer to [14] for more details on the physics at the basis of the model.

$$<x^2> = 2Dt \tag{2}$$

In the current study, the measuring unit of space is the nanometre (10^{-9} metres), while the time unit is a tenth of a second (10^{-1} seconds), corresponding to one tick of the simulation clock.

The turnover number is required because an enzyme-substrate complex must wait a time equal to its reciprocal before releasing the reaction product. To simulate the effect of DBF on the MAPK/ERK pathway, the kcat values, along with the species' molar concentrations, have been retrieved from Hamis et al. [9].

The species involved and the related concentrations have been structured in a Python dictionary to be imported into the Repast4Py ABM model (which, as the framework's name suggests, is written in Python language) [4].

The model comprises the molecular species listed in Table 1 along with their initial concentrations.

Table 1. Molecular species represented in the ABM. The initial concentration of (unphosphorylated) ERK is 0 because we analysed the DBF effects by measuring the percentage decrease of phosphorylated ERK (pERK + ppERK) starting from the concentrations reached at a steady state when no drug was administered (all ERK in the simulation environment is phosphorylated). pMEK and ppMEK are also formed in the course of the simulation.

species	initial concentration (mol/L)
BRAF	10e-9
DBF	various concentrations (see Results)
ERK	0
pERK	8.0074e-8
ppERK	1.1185e-6
MEK	1.2e-6
pMEK	0
ppMEK	0
phosphatase 1	3e-10
phosphatase 2	1.2e-7

3   Results

We ran the simulation setup on a volume of 300 pl, corresponding to 3e14 cubic nanometres, meaning that the actual simulations have been performed on a square of 66943 × 66943 nanometres. This volume allows us to maintain the concentration values provided in Hamis et al. [9]; as reported in Table 1, these are significantly disproportioned, especially regarding the concentration of BRAF.

Converting molar concentrations to the corresponding number of molecules and subsequently *slicing* the simulation cube results in too few BRAF molecules being represented at smaller simulation volumes, potentially compromising accuracy.

As a first step, we tested *eleven scenarios*, in which we simulated the administration of the following DBF concentrations (expressed in nanomoles per litre): 0 (i.e., a cytoplasm completely free of DBF), 192.4, 384.9, 577.3, 769.8, 962.2, 1154.7, 1347.1, 1539.6, 1732.1, and 1924.6 nM.

They represent reference DBF dosages of the clinical study conducted by Falchook et al. [6], whose patient data points are also reported in Fig. 6 of Hamis et al. [9] (reproduced here in Fig. 5b). We adopted a conversion from ng/mL to nanomolar concentration based on the latter (which considers the molecular weight of DBF as 519.6 g/mol); thus, the dabrefenib nanomolar concentrations listed above correspond to intervals of 100 ng/mL, from 0 to 1000. Falchook et al. measured phosphorylated ERK levels in BRAF-positive melanoma biopsy samples before and five days after DBF monotherapy administration (doses between 70 and 200 mg were administered twice daily).

To achieve an accurate and detailed comparison with the plots shown in the two articles mentioned above, we simulated each scenario for 8 h (completed with a mid-range 12-core CPU and 32 GB of RAM in about 12 h each).

We observed the effect of DBF by measuring the percentage decrease of phosphorylated ERK (pERK + ppERK) from baseline concentrations, determined by simulating a system with no drugs and no phosphorylated ERK until it reached a steady state. The concentrations measured at this point are 80.074 nM for pERK and 1118.5 nM for ppERK (after the *slicing* process, these are reduced to 0.012 nM and 0.167 nM, respectively).

In a system in which no DBF was administered (Fig. 4a), the agent-based simulation shows an initial 39% drop in phosphorylated ERK caused by the effect of phosphatase 2. However, its concentration grows again as the result of the kinase cascade activation and then oscillates at values between 0.15 and 0.16 nM, corresponding to a decrease of just about 13%. When the administration of DBF is simulated, the drug clearly produces its effects on the pathway inhibition: a concentration of 192.4 nM (Fig. 4b) shows a drop in phosphorylated ERK of about 70% from its baseline levels; concentrations of 962.2 nM and 1924.6 nM (Fig. 4c and 4d, respectively) push the phosphorylated ERK reduction to more than 90%.

3.1 Similarities and Discrepancies with Other Studies

The values described above are in fair accordance with the results provided by Hamis et al. and Falchook et al., even if the 1924.6 nM of DBF scenario is more faithful to the prediction of the Hamis et al. model than the clinical study proposed by Falchook et al. Indeed, with the same initial concentration of BRAF as our model (10 nM), the latter observed a phosphorylated ERK concentration

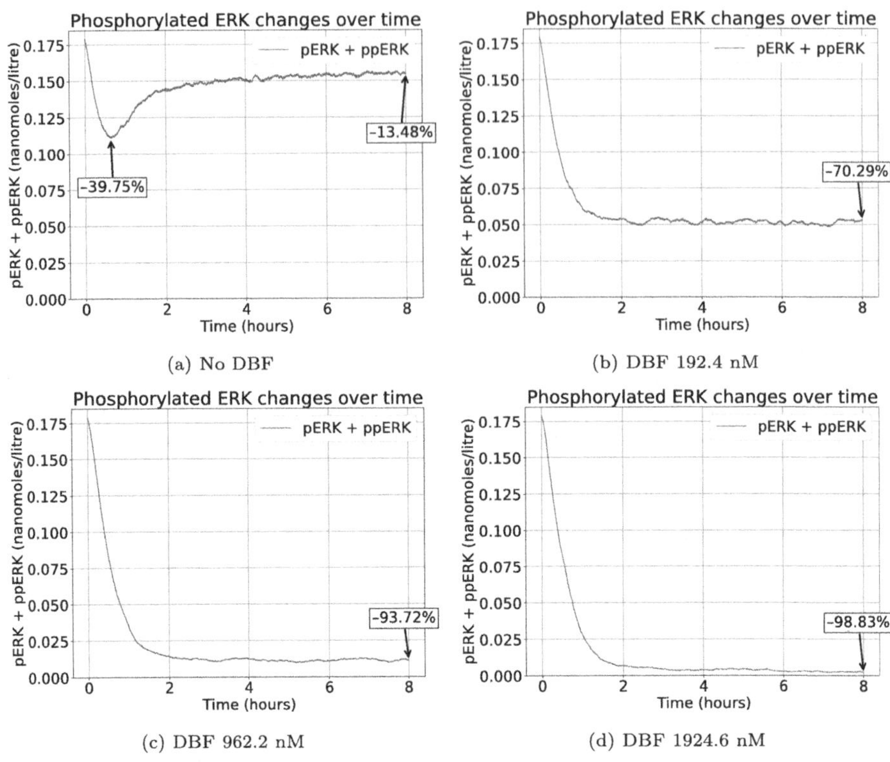

(a) No DBF

(b) DBF 192.4 nM

(c) DBF 962.2 nM

(d) DBF 1924.6 nM

Fig. 4. Concentration changes over time of phosphorylated ERK during 8-hour simulations. The plots illustrate four representative scenarios from those described in the Results: a cytoplasm environment free of DBF, along with the minimum, mid-range, and maximum DBF concentrations from the dosing interval tested, based on the clinical study by Falchook et al. [6]. With no DBF in the environment (a), after an initial 39% drop due to the effect of phosphatase 2, the phosphorylated ERK decrease stabilises around a -13% of the initial pERK + ppERK concentration. Simulating the administration of progressively higher concentrations of DBF, (b), (c), and (d), the phosphorylated ERK decrease is significantly more evident.

change similar to that found in drug-resistant patient-derived xenografts [21]. The causes underlying this discrepancy will be investigated in future works.

To facilitate a precise comparison between our ABM, the ordinary differential equation (ODE) model from [9], and the clinical data in [6], we simulated—in addition to the DBF concentrations described above—the same eight drug concentrations used in the clinical trials, that is: 138.5, 179, 205, 404, 560, 572.5, 726, and 908 ng/mL[1]. This approach allowed us to directly evaluate how well each model captures the dynamics of phosphorylated ERK reduction at varying DBF levels, providing a clear benchmark across the three scenarios.

---

[1] The concentrations of the clinical data were approximated from Fig. 2c of Falchook et al. by visually estimating the values from the plot.

Table 2. Comparison of RMSE and MAE between our agent-based and the ODE simulations from Hamis et al.. The agent-based simulation shows a slightly higher RMSE (21.29) compared to the ODE simulation (20.77), indicating that the ODE model produces fewer large outliers in error. However, the ABM has a lower MAE (15.40) than the ODE model (16.83), suggesting that, on average, the ABM's errors are more consistent and smaller than those in the ODE model despite occasional larger deviations.

Simulations	RMSE	MAE
Agent-based	21.29	15.40
ODE	20.77	16.83

In comparing our agent-based simulation against the ODE simulations and the clinical data, both models show a reasonable fit at higher concentrations of DBF (Fig. 6).

Our ABM demonstrates more consistent residuals across the entire concentration range despite showing larger initial errors at concentrations below 200 ng/mL. In contrast, the ODE model, while exhibiting a slightly lower Root Mean Squared Error (RMSE) (Table 2), shows greater variability in its residuals, particularly at lower concentrations. RMSE measures the square root of the average squared differences between predicted and actual values, emphasising larger errors. This makes it particularly useful when larger deviations are more concerning, but it can also amplify the effect of outliers.

On the other hand, the ABM shows a lower Mean Absolute Error (MAE), which calculates the average magnitude of prediction errors without squaring them. MAE provides a more straightforward interpretation by measuring errors on the same scale as the data, giving equal weight to all deviations.

These results suggest that while the ODE model achieves a marginally better overall error, as the RMSE indicates, the ABM provides more stable and reliable predictions across the concentration range, especially at lower DBF concentrations, where the ODE model exhibits more variability.

These results show that the *slicing* process does not significantly affect our ABM's ability to allow faithful biological behaviour to emerge from biomolecular interactions.

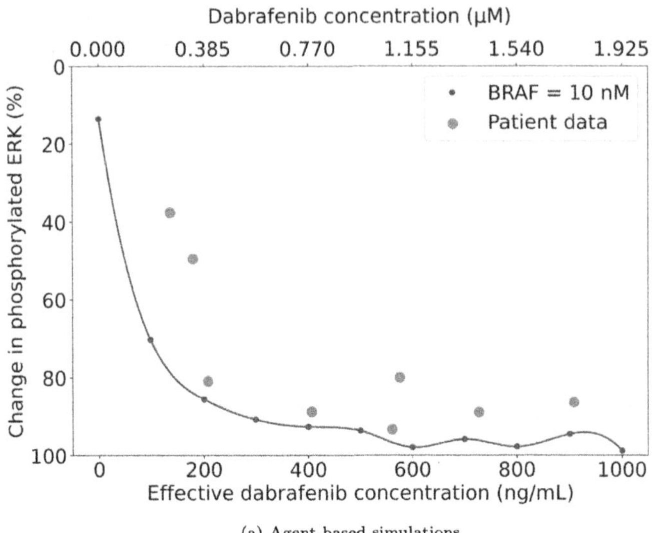

(a) Agent-based simulations

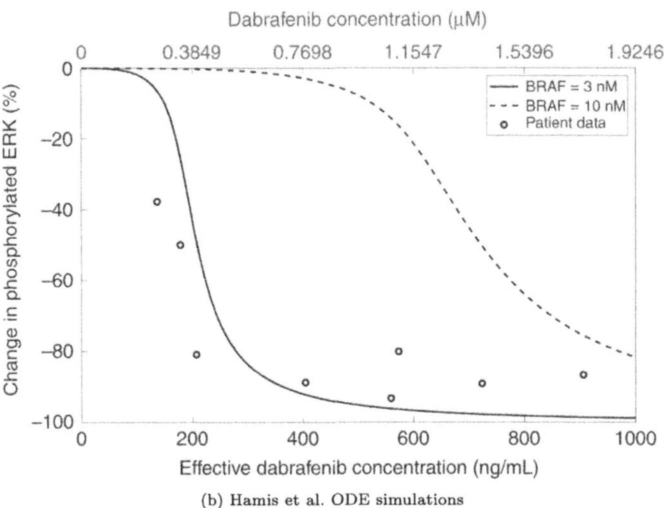

(b) Hamis et al. ODE simulations

Fig. 5. Comparison of the results of the agent-based simulations with other works. Plot **(a)** puts together the percentage decreases of phosphorylated ERK of the eleven scenarios described in the Results along with the patient data points (in light blue) shown in Fig. 2c of Falchook et al. [6]. DBF concentrations are expressed in μM for a more accurate comparison. Plot **(b)** is adapted from Fig. 6 of Hamis et al. © CC BY 4.0 DEED [9]. To better convey the trend in our simulations and mitigate segmentation caused by the limited number of points, we interpolated the data in Plot **(a)** using a cubic spline. The outcome closely matches the behaviour predicted by the computational model from Hamis et al. (with an initial BRAF concentration of 3 nM) and aligns well with the patient data observed in Falchook et al.

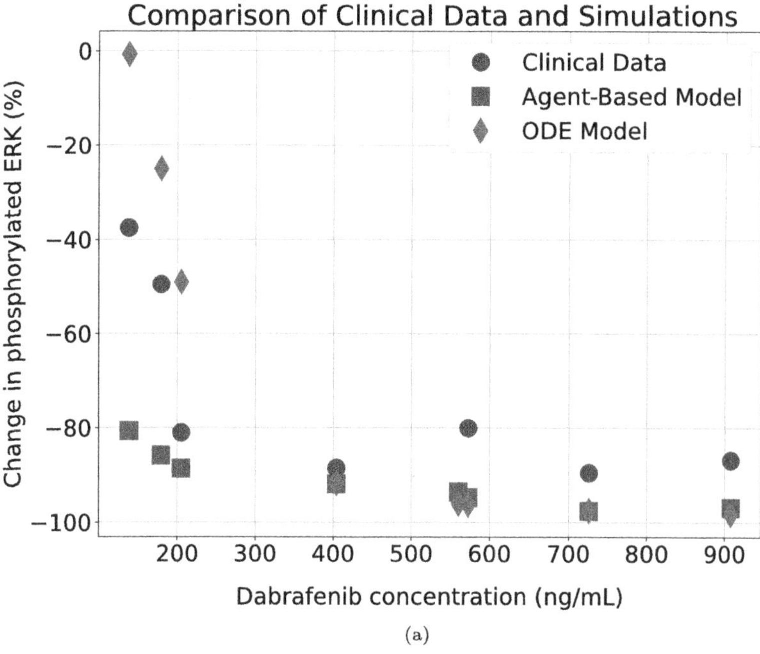

(a)

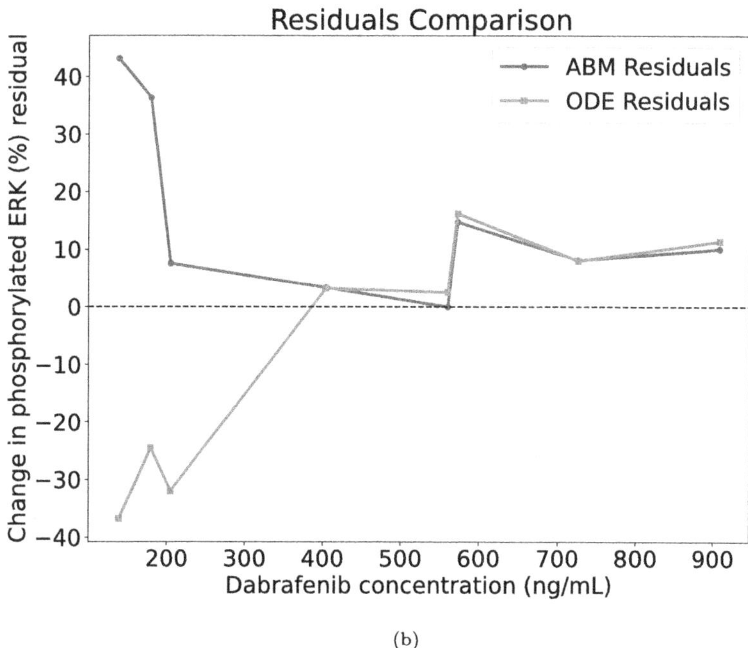

(b)

Fig. 6. Comparison of clinical data with simulation results. Plot **(a)** shows clinical data alongside the ABM and ODE model predictions, highlighting the changes in phosphorylated ERK at varying DBF concentrations. Plot **(b)** is a residuals plot comparing the discrepancies between observed clinical data and predictions from both the ABM and ODE models across the same concentration range.

4   Conclusion

This work evaluated an ABM of signalling pathways using Repast4Py, a recently released simulation framework. At this stage, rather than fully leveraging Repast4Py's HPC capabilities, we focused on testing the hypothesis that abstracting the simulation cube as a *slice* of its volume can faithfully reproduce the behaviour of DBF as an inhibitor of the MAPK/ERK cascade. The comparison with a clinical study gives us a promising indication of the validity of our approach. The next step in our study is to fully exploit Repast4Py and run the simulations in a distributed environment to observe the agents' behaviour in three-dimensional space. We are also working on increasing the simulation's performance by running the numerical computations on a GPU using Python packages such as Numba.

These choices will allow us to test the simulation on a more realistic cell size, i.e., 3–10 pl volume, without the reduction in the number of molecules caused by the *slicing* process. The current volume of 300 pl is indeed achievable just by human fatty cells or oocytes and may be the cause at the basis of the discrepancy with the Hamis et al. ODE simulations (see Results).

Nonetheless, we believe this paper's results already provide a clear clue to ABM's reliability in reproducing the effects of cancer therapy drugs and lay promising groundwork for future developments. Among them, integrating topological data analysis techniques, such as those we applied to visualise molecular interactions in metabolic reactions [15], could offer additional insights into the spatial dynamics of the agent-based model and further enhance the understanding of emergent behaviours in cancer therapy simulations.

Acknowledgments. This study was funded by Diatech Pharmacogenetics S.R.L.

Disclosure of Interests. The simulations presented in this article were conducted using proprietary software developed as part of a company-funded project. The source code cannot be shared publicly due to confidentiality agreements. Apart from this, the author has no competing interests to declare that are relevant to the content of this article.

References

1. An, G., Mi, Q., Dutta-Moscato, J., Vodovotz, Y.: Agent-based models in translational systems biology. WIREs Syst. Biol. Med. **1**(2), 159–171 (2009). https://doi.org/10.1002/wsbm.45
2. Ascierto, P.A., et al.: The role of BRAF V600 mutation in melanoma. J. Transl. Med. **10**, 85 (2012). https://doi.org/10.1186/1479-5876-10-85
3. Belenchia, M., et al.: Agent-based learning model for the obesity paradox in RCC. Front. Bioeng. Biotechnol. **9** (2021). https://doi.org/10.3389/fbioe.2021.642760
4. Collier, N., Ozik, J.: Distributed agent-based simulation with Repast4Py. In: 2022 Winter Simulation Conference (WSC), pp. 192–206 (2022). https://doi.org/10.1109/WSC57314.2022.10015389

5. Erickson, H.P.: Size and shape of protein molecules at the nanometer level determined by sedimentation, gel filtration, and electron microscopy. Biol. Proc. Online **11**(1), 32–51 (2009). https://doi.org/10.1007/s12575-009-9008-x

6. Falchook, G.S., et al.: Dose selection, pharmacokinetics, and pharmacodynamics of BRAF inhibitor dabrafenib (GSK2118436). Clin. Cancer Res. **20**(17), 4449–4458 (2014). https://doi.org/10.1158/1078-0432.CCR-14-0887

7. Genesereth, M.R., Nilsson, N.J.: Logical Foundations of Artificial Intelligence. Morgan Kaufmann, Los Altos, Calif (1987)

8. Guo, M.G., Sosa, D.N., Altman, R.B.: Challenges and opportunities in network-based solutions for biological questions. Briefings in Bioinformatics **23**(1), bbab437 (2022). https://doi.org/10.1093/bib/bbab437

9. Hamis, S.J., Kapelyukh, Y., McLaren, A., Henderson, C.J., Roland Wolf, C., Chaplain, M.A.J.: Quantifying ERK activity in response to inhibition of the BRAFV600E-MEK-ERK cascade using mathematical modelling. Br. J. Cancer **125**(11), 1552–1560 (2021). https://doi.org/10.1038/s41416-021-01565-w

10. Hauschild, A., et al.: Dabrafenib in BRAF-mutated metastatic melanoma: a multicentre, open-label, phase 3 randomised controlled trial. The Lancet **380**(9839), 358–365 (2012). https://doi.org/10.1016/S0140-6736(12)60868-X

11. Kim, S., et al.: PubChem 2023 update. Nucleic Acids Res. **51**(D1), D1373–D1380 (2023). https://doi.org/10.1093/nar/gkac956

12. Lechelon, M., et al.: Experimental evidence for long-distance electrodynamic intermolecular forces. Sci. Adv. **8**(7), eabl5855 (2022). https://doi.org/10.1126/sciadv.abl5855

13. Leighow, S.M., Landry, B., Lee, M.J., Peyton, S.R., Pritchard, J.R.: Agent-based models help interpret patterns of clinical drug resistance by contextualizing competition between distinct drug failure modes. Cel. Mol. Bioeng. **15**(5), 521–533 (2022). https://doi.org/10.1007/s12195-022-00748-6

14. Maestri, S., Merelli, E., Pettini, M.: Agent-based models for detecting the driving forces of biomolecular interactions. Sci. Rep. **12**(1), 1878 (2022). https://doi.org/10.1038/s41598-021-04205-8

15. Piangerelli, M., Maestri, S., Merelli, E.: Visualising 2-simplex formation in metabolic reactions. J. Mol. Graph. Model. **97**, 107576 (2020). https://doi.org/10.1016/j.jmgm.2020.107576

16. Richards, F.M.: Areas, volumes, packing and protein structure. Annu. Rev. Biophys. Bioeng. **6**, 151–176 (1977). https://doi.org/10.1146/annurev.bb.06.060177.001055

17. Shuaib, A., Hartwell, A., Kiss-Toth, E., Holcombe, M.: Multi-compartmentalisation in the MAPK signalling pathway contributes to the emergence of oscillatory behaviour and to ultrasensitivity. PLoS ONE **11**, e0156139 (2016). https://doi.org/10.1371/journal.pone.0156139

18. The UniProt Consortium: UniProt: The Universal Protein Knowledgebase in 2023. Nucleic Acids Res. **51**(D1), D523–D531 (2023). https://doi.org/10.1093/nar/gkac1052

19. Wang, Z., Bordas, V., Sagotsky, J., Deisboeck, T.S.: Identifying therapeutic targets in a combined EGFR-TGFβR signalling cascade using a multiscale agent-based cancer model. Math. Med. Biol. J. IMA **29**(1), 95–108 (2012). https://doi.org/10.1093/imammb/dqq023

20. Xu, C., Jackson, S.A.: Machine learning and complex biological data. Genome Biol. **20**(1), 76 (2019). https://doi.org/10.1186/s13059-019-1689-0

21. Xue, Y., et al.: An approach to suppress the evolution of resistance in BRAFV600E-mutant cancer. Nat. Med. **23**(8), 929–937 (2017). https://doi.org/10.1038/nm.4369

Machine Learning for Structured Data in Clinical Informatics and Medical Biology

Forward and Backward Feature Selection Guided by Prior Biological Knowledge for Enhanced Interpretability

Sofia Mongardi[(✉)] [iD], Silvia Cascianelli [iD], and Marco Masseroli [iD]

Department of Electronics, Information and Bioengineering (DEIB), Politecnico di Milano, Milan, Italy
{sofia.mongardi,silvia.cascianelli,marco.masseroli}@polimi.it

Abstract. In gene expression analysis, the high dimensionality and limited sample size often lead to instability and overfitting of predictive models. While feature selection algorithms are commonly used to identify the most predictive genes, traditional approaches tend to focus solely on quantitative contributions, which can limit the discovery of deeper biological insights. To address this, we propose a novel wrapper-based approach that integrates prior biological knowledge into the feature selection process. Our approach extends standard forward feature selection by iteratively adding the most promising gene while ensuring it provides biological value, computed from prior knowledge derived from publicly available data sources. Additionally, we apply the same concept to backward selection, iteratively removing features that contribute the least to the predictive performance while providing limited additional biological information.

Keywords: Machine learning · feature selection · biological interpretability · genomics · classification

1 Introduction

Analyzing and deriving meaningful insights from large-scale transcriptomics data presents significant challenges. Indeed, such gene expression datasets significantly suffer from the "large p, small n" problem, as the number of features (genes, p) far exceeds the number of available samples (n). This imbalance frequently results in issues like overfitting and instability when applying and training predictive models on such data. To address these challenges, feature selection algorithms are commonly employed to identify genes having the highest predictive and discriminative power. Traditional feature selection methods tend to focus exclusively on the quantitative contribution of genes to the predictive task, which may limit the biological interpretability of the selected genes. Incorporating prior biological knowledge, such as known gene functions and relationships, is crucial for understanding the underlying biological mechanisms, and can help

L. Cerulo et al. (Eds.): CIBB 2024, LNBI 15276, pp. 233–247, 2025.
https://doi.org/10.1007/978-3-031-89704-7_18

minimize the risk of spurious correlations whilst better linking findings to existing knowledge [3]. Feature selection algorithms that integrate such knowledge can enhance the selection process by considering both statistical significance and biological relevance, leading to improved predictivity and better biological insights. This approach may accelerate discovery as well as enable more meaningful downstream analysis.

Many existing knowledge-based feature selection methods in the literature are filter-based [6,14], where genes are ranked according to a specific metric before being used as input for predictive models. These methods can sometimes be overly restrictive or ineffective, ultimately prioritizing biological relevance over predictive accuracy. Instead, wrapper-based feature selection approaches, while computationally more intensive, offer a valid alternative. Indeed, wrapper methods have the advantage of implicitly considering feature dependencies, including interactions and redundancies, often yielding improved performance compared to filter methods.

In this paper, we present a novel wrapper method for feature selection that integrates prior biological knowledge into the forward and backward feature selection frameworks. For forward selection, our method incrementally selects the most promising features at each step, ensuring that they not only improve the predictive performance but also provide additional biological insights. Alternatively, for the backward algorithm, our approach iteratively removes features that contribute the least to the predictive performance while providing limited additional biological information. We achieve this selection by evaluating the similarity between the feature embeddings of prior knowledge and by considering the overall biological relevance of each feature/gene, based on information extracted from publicly available knowledge bases. By combining these aspects, our approaches can identify predictive and biologically informative features. We validate the effectiveness of our methods through extensive experiments on breast cancer (BRCA) gene expression data, specifically focusing on the classification of patients into their intrinsic BRCA subtypes, obtained using the PAM50 test [15].

2   Materials and Methods

2.1   Data and Pre-processing

We considered 1,053 RNA-seq profiles of BRCA patients from The Cancer Genome Atlas (TCGA) [4,12,21], including 25,150 genes per sample. We discarded any miRNA gene (1,650), all the non-coding genes (4,155) according to RefSeq [11] and the genes (1,421) not expressed (raw counts ≤ 4) in at least 80% of the samples. We removed 10 samples with the top-five expressed genes accounting for at least 20% of the total sample raw counts. We then normalized gene expressions in log_2 transformed *reads per million* (RPM) for each gene g_i in each sample s_j: $RPM_{ji} = \log_2 \frac{\# \; reads \; mapped \; to \; g_i}{total \; \# \; reads \; for \; sample \; s_j} * 10^6$. A total of 17,924 genes were retained after pre-processing. We retrieved from cBioPortal[1]

---

[1] https://www.cbioportal.org/.

the BRCA intrinsic subtypes of all the samples, assigned by the state-of-the-art PAM50 test [15]. We considered *Basal* (175), *HER2-enriched* - Her2 (81), *Luminal A* - LumA (543), and *Luminal B* - LumB (207) subtypes for a total of 1,006 patients, discarding the *Normal-like* (37) class whose clinical significance is controversial.

2.2 Prior Knowledge Score

We used the *Gene Information Score* (*GIS*), describing each gene's overall biological relevance as proposed in [10], to summarize the prior existing information for each gene across one or multiple different biological knowledge bases. Considering a dataset with q genes and l knowledge bases, where each knowledge base is represented as a directed acyclic graph (DAG) containing n_l terms, we can retrieve the most specific terms associated with each gene for each knowledge base. By exploiting the DAG structure, we can trace the ancestors of these specific terms at any depth, considering all or a subset of the relationships. This generates an expanded list of terms (annotations) linked to each gene in the dataset that can be subsequently used to build a binary annotation matrix $B \in \mathbb{R}^{q \times \sum_{k=1}^{l} n_k}$, with genes along the rows, annotation terms along the columns, and binary values indicating whether a gene is annotated with a given term or not. From the binary matrix B, a weighted annotation matrix $W \in \mathbb{R}^{q \times \sum_{k=1}^{l} n_k}$ can be obtained considering the depth and number of descendants of each term in each knowledge base of interest. See Appendix A, Sect. A.1, for further details on the construction of W. Starting from W, the *GIS* can be then computed as:

$$GIS(g) = \frac{\sum_{m:W_{g,t}>0} W_{g,t}}{\sum_{m:W_{g,t}>0} 1} \tag{1}$$

where $W_{g,t}$ is the weight for the term t associated with the gene g in weighted annotation matrix W. Given this formula, the obtained *GIS* values are in the range [0,1]. A *GIS* value of 0 indicates that there is no available prior biological information in the considered knowledge, while higher *GIS* values tend to suggest a greater number of specific annotations.

Eventually, we transformed the *GIS* into w_{GIS} as follows:

$$w_{GIS}(g) = \frac{1}{1 + GIS(g)} \tag{2}$$

The higher the *GIS* and biological relevance, the lower the corresponding w_{GIS}, which, by design, is always a non-negative value bounded in the range [0.5,1].

2.3 Prior Knowledge Embeddings

The Non-negative Matrix Factorization (NMF) was used to obtain embeddings of prior knowledge from the weighted annotation matrix W (defined in Sect. 2.2) that summarize each gene's existing biological information across different knowledge bases. The NMF algorithm aims to learn matrices U and H

such that $W \approx UH$. Further details on the NMF algorithm and its implementation are reported in Sect. A.2 of Appendix A. We run the algorithm for several iterations and, after convergence, we directly extracted the embeddings of prior knowledge from U.

2.4   Feature Selection Algorithm

Our proposed approaches are an extension of the traditional forward and backward feature selection algorithms that also consider each feature's prior biological relevance in addition to its predictive power. In this section, we briefly introduce the two algorithms as well as their corresponding extensions.

Forward Feature Selection. Given a predefined model and an initially empty set of selected features, forward feature selection [7] iteratively evaluates the performance impact of adding each remaining candidate feature to this set. The feature that leads to the greatest improvement in a specified performance metric is then added to the selected features. This process repeats until a predefined stopping criterion is met, such as when adding more features no longer significantly enhances the model's performance, or when the maximum number of iterations is reached. Our proposed extension builds upon this framework: at each iteration k, it selects the feature that improves a predefined performance metric p as well as incorporates additional biological information. This is done by evaluating the distance between the embeddings of the prior knowledge associated with the features already selected $(FS^{(k)})$ and the embedding of the candidate feature (f_c). Depending on the chosen criterion c, the algorithm may use the maximum (d_{max}) or average (d_{mean}) distance between the embeddings of the selected and candidate features. This distance is adjusted based on the candidate feature's overall biological relevance, represented by w_{GIS}, to limit scenarios where features without prior information are selected. Additionally, a hyperparameter α controls the extent to which biological information influences the selection process. When $\alpha = 0$, the algorithm performs standard feature selection. The total number of selected features is capped by the maximum number of iterations k_{max}. The detailed steps of our proposed feature selection algorithm are outlined in Algorithm 1.

Backward Feature Selection. Given a predefined model, backward feature selection [7] starts with a predefined initial set containing all the features and, at each iteration k, it evaluates the impact on the performance of removing each feature from this set. The feature whose removal causes the greatest improvement (or smallest decrease) in a specified performance metric is removed from the set. This process is repeated until a stopping criterion is met, such as when removing additional features significantly degrades the model's performance or when the minimum number of features has been reached. Similarly, our proposed extension removes the features that improve a predefined metric p the most, that are also providing limited additional biological information. We do

---

Algorithm 1: Proposed Forward Feature Selection

Input: Iteration k, number of iteration k_{max}, set of candidate features FC, set of selected features FS, performance metric p, distance metric d, criterion c, prior knowledge embedding e, hyperparamter α

Output: Best performance metric p_{best}, best iteration k_{best} and final feature subset FS_{best}

$FC^{(0)} = \{f_1, f_2, \ldots, f_p\}, \quad FS^{(0)} \leftarrow \emptyset, \; FS_{best} \leftarrow \emptyset$

if $k = 0$ **then**

$\quad f^* = \arg\max_{f_c \in FC^{(0)}} p(FS^{(0)} \cup \{f_c\})$

$\quad FS^{(1)} \leftarrow FS^{(0)} \cup \{f^*\}$

$\quad FC^{(1)} \leftarrow FC^{(0)} \setminus \{f^*\}$

$\quad p_{best} = p(FS^{(1)})$

$\quad k_{best} = 1$

end

while $k < k_{max}$ **do**

$\quad f^* = \arg\max_{f_c \in FC^{(k)}} (1 - \alpha)p(FS^{(k)} \cup \{f_c\}) + \alpha(d_c(e_{f_c}, e_{FS^{(k)}})/2w_{GIS}(f_c))$

$\quad FS^{(k+1)} \leftarrow FS^{(k)} \cup \{f^*\}$

$\quad FC^{(k+1)} \leftarrow FC^{(k)} \setminus \{f^*\}$

$\quad$ **if** $p(FS^{(k+1)}) > p_{best}$ **then**

$\quad\quad p_{best} = p(FS^{(k+1)})$

$\quad\quad k_{best} = k + 1$

$\quad$ **end**

end

$FS_{best} \leftarrow FS^{(k_{best})}$

---

this by measuring the similarity between the embeddings of the prior knowledge of the remaining features and the embedding of the discarded one at each iteration k of the algorithm. Based on the selected criterion c, the minimum (s_{min}) or average (s_{mean}) similarity between embeddings is used. As before, we also adjust the computed similarity using the w_{GIS} to avoid degenerate scenarios, and include a hyperparameter α to control how much prior biological information contributes to the feature selection process. If $\alpha = 0$, the proposed algorithm converges to standard backward feature selection. The maximum number of iterations k_{max} determines the minimum number of features that the algorithm can select. The pseudo-code describing our proposed backward extension is reported in Algorithm 2.

2.5   Experiments

We considered the dataset described in Sect. 2.1 for an application use case, and classified the BRCA samples according to their intrinsic subtypes. We split the data into training (80%) and test (20%) sets, and standardized the data prior to model fitting (and testing) using the training set statistics (mean and standard deviation). We built the weighted annotation matrix W by using publicly available biological knowledge bases, specifically Reactome (Number of terms

---

Algorithm 2: Proposed Backward Feature Selection

Input: Iteration k, number of iteration k_{max}, set of candidate features FC, set of selected features FS, performance metric p, similarity metric s, criterion c, prior knowledge embedding e, hyperparamter α

Output: Best performance metric p_{best}, best iteration k_{best} and final feature subset FS_{best}

$FC^{(0)} = \emptyset, \quad FS^{(0)} \leftarrow \{f_1, f_2, \ldots, f_p\}, \quad FS_{best} \leftarrow \emptyset$

if $k = 0$ **then**

$\quad f^* = \arg\max_{f_c \in FC^{(0)}} p(FS^{(0)} \cup \{f_c\})$

$\quad FS^{(1)} \leftarrow FS^{(0)} \setminus \{f^*\}$

$\quad FC^{(1)} \leftarrow FC^{(0)} \cup \{f^*\}$

$\quad p_{best} = p(FS^{(1)})$

$\quad k_{best} = 1$

end

while $k < k_{max}$ **do**

$\quad f^* = \arg\max_{f_c \in FC^{(k)}} (1 - \alpha)p(FS^{(k)} \setminus \{f_c\}) + \alpha(s_c(e_{f_c}, e_{FS^{(k)}}) \times w_{GIS}(f_c))$

$\quad FS^{(k+1)} \leftarrow FS^{(k)} \setminus \{f^*\}$

$\quad FC^{(k+1)} \leftarrow FC^{(k)} \cup \{f^*\}$

$\quad$ **if** $p(FS^{(k+1)}) > p_{best}$ **then**

$\quad\quad p_{best} = p(FS^{(k+1)})$

$\quad\quad k_{best} = k + 1$

$\quad$ **end**

end

$FS_{best} \leftarrow FS^{(k_{best})}$

$= 2{,}629$; maximum depth $= 12$) [8], the Human Phenotype Ontology (HPO) (17,895; 18) [16] and the three Gene Ontology (GO) [2] sub-ontologies: Biological Processes (30,540; 19), Cellular Components (12,471; 13) and Molecular Functions (4,473; 15). Since forward and backward feature selection requires a predefined model, we assessed and compared the performance of multi-class logistic regression (LR) and support vector machine (SVM) classifiers. As the focus of this study was not on optimizing performance, we utilized the default parameters defined in the scikit-learn Python package[2]. To enhance generalization and prevent overfitting, we employed 5-fold cross-validation to evaluate the performance improvement of each candidate feature for both algorithms. For the forward selection approach, we tested two different maximum iteration values ($k_{max} = 30, 50$) and two different values for the new hyperparameter α ($\alpha = 0.3, 0.5$). We also tested α values greater than 0.5 but observed significant decreases in performance for these settings, indicating that such values may overly prioritize the biological aspect at the expense of the statistical one. Due to the high dimensionality of the feature space and the limited number of samples, we could not consider the complete feature space in backward feature selection for stability reasons. Therefore, we performed an initial filtering using

---

[2] https://scikit-learn.org/stable/.

the Fisher's scores [5], to select the $k_{init} = 200$ most predictive features. Then, we applied our proposed backward extension to refine the selection towards a final set, setting $k_{max} = 150$. We used the F1-score as the performance metric p, the cosine distance as the distance metric d in Algorithm 1, and the cosine similarity as the similarity metric s in Algorithm 2. We evaluated the final model performance on the test set using also accuracy, recall, and precision.

2.6 Biological Validation

To assess the effectiveness of our proposed approach in improving the biological interpretability of the results, we conducted several evaluations comparing the feature subsets selected by standard feature selection algorithms with those selected by our proposed method. For each selected feature subset, we used Fisher's Exact Test [20] to evaluate the statistically significant enrichment of GO terms and Reactome biological pathways. The resulting p-values were adjusted for multiple testing using the Benjamini-Hochberg false discovery rate (FDR) correction, with a significance threshold set at 0.05. Additionally, we retrieved cancer-specific and disease-specific pathways from the KEGG database [9] and calculated the overlap between the selected feature subsets and the gene signatures associated with these pathways. Identifying biologically enriched gene subsets can provide valuable insights into disease-related biological mechanisms, potentially deepening our understanding of the underlying biological context.

3   Results

Here, we report all the results from the performed experiments, comparing standard forward and backward feature selection algorithms and our proposed approaches.

3.1   Forward Feature Selection

Classification performance and biological validation findings are shown in Table 1 and Table 2 for $k_{max} = 50$. The results for $k_{max} = 30$ are available in Appendix B, Sect. B.1. Our proposed approach outperformed the standard feature selection algorithm for both the LR and SVM models. Indeed, the $LR_{d_{max}}$ and $SVM_{d_{max}}$ models achieved respectively a 6% and 4% gain in performance compared to the standard algorithm. For both classifiers, the d_{max} criterion and $\alpha = 0.3$ appeared to work the best in terms of classification performance (Table 1). Similar findings were obtained in terms of biological validation. Indeed, our approach retrieved a substantially higher number of GO term and Reactome pathway annotations (Table 2), with an increased number of statistically significant terms. Furthermore, the number of selected genes included in cancer-specific and BRCA-specific KEGG pathway signatures was also higher, on average, for both classifiers when selecting features using our proposed forward feature selection algorithm. Overall, it appears that, for both the LR and SVM classifiers,

Table 1. Classification performance on the test set for $k_{max} = 50$ for the forward algorithm. Best performance in bold.

Method	N[1]	α^2	Accuracy	Precision[3]	Recall[3]	F1-score[3]
LR[4]	41	-	0.856	0.851	0.848	0.856
LR$_{d_{max}}$	21	0.3	**0.916**	**0.894**	**0.939**	**0.914**
LR$_{d_{mean}}$	48	0.3	0.856	0.827	0.899	0.855
LR$_{d_{max}}$	33	0.5	0.891	0.844	0.917	0.870
LR$_{d_{mean}}$	48	0.5	0.861	0.808	0.865	0.824
SVM[3]	47	-	0.886	0.841	0.885	0.860
SVM$_{d_{max}}$	38	0.3	**0.926**	0.894	**0.941**	0.914
SVM$_{d_{mean}}$	48	0.3	0.896	0.855	0.900	0.874
SVM$_{d_{max}}$	47	0.5	0.916	0.888	0.932	0.907
SVM$_{d_{mean}}$	31	0.5	0.916	**0.902**	0.933	**0.915**

[1] Number of selected features.
[2] Hyperparameter for the prior knowledge in the feature selection process.
[3] Macro-average metrics.
[4] Standard feature selection.

the combination of d_{max} and $\alpha = 0.3$ returned the best results in terms of classification performance and biological validation. Specifically, using the d_{max} as the distance criterion c (Algorithm 1) returned not only a higher number of significant GO terms and Reactome pathways, but also a higher number of disease-specific genes across different k_{max} values (see KEGG BRCA in Table 2 and Table 6). Comparing the results in Table 1 and Table 5, is clear that increasing the number of features may not always lead to improved performance results, while it increases the number of significant GO terms and Reactome pathways (see Table 2 and Table 6). Furthermore, analyzing some of the key selected genes, we found that the majority of the feature sets obtained through our proposed methods included the *LGALS9* gene, whose elevated expression enhances breast cancer cell invasiveness [13], the *SLC38A5* gene, an amino acid transporter which has been recently associated with BRCA tumor progression and poor prognosis [17], and the *TREM2*, which has a critical role in tumor-associated macrophages (TAMs) as it drives macrophage polarization to an immune-suppressive state, promoting tumor growth and leading to T cell exhaustion in breast cancer [18]. All these genes were never selected by the standard algorithm.

It is worth noting that the maximum iteration value k_{max}, α, and the criterion c are all hyperparameters that can be fine-tuned using k-fold cross-validation, allowing for the selection of the optimal feature set and classifier, based on the best cross-validation performance. We did not perform such an optimization as the primary goal of this study was to evaluate the impact of such hyperparameters on the feature selection process and to assess the effectiveness of our approach in selecting the most promising genes both in terms of predictivity and biological relevance.

Table 2. Biological validation results for $k_{max} = 50$ for the forward algorithm.

Method	N	α	GO[1]	Reactome[1]	KEGG Cancer[2]	KEGG BRCA[3]
LR[4]	41	-	1,659 (0.0%)	241 (0.0%)	5	4
$LR_{d_{max}}$	21	0.3	3,175 (64.7%)	207 (7.7%)	6	4
$LR_{d_{mean}}$	48	0.3	3,587 (5.9%)	371 (3.8%)	7	1
$LR_{d_{max}}$	33	0.5	4,456 (62.6%)	303 (16.5%)	8	4
$LR_{d_{mean}}$	48	0.5	3,508 (12.5%)	399 (3.5%)	6	2
SVM[4]	47	-	1,987 (0.0%)	284 (0.0%)	3	4
$SVM_{d_{max}}$	38	0.3	4,911 (64.0%)	438 (11.9%)	13	8
$SVM_{d_{mean}}$	48	0.3	3,733 (15.1%)	284 (0.0%)	7	1
$SVM_{d_{max}}$	47	0.5	3,294 (64.7%)	369 (14.4%)	11	6
$SVM_{d_{mean}}$	31	0.5	2,959 (0.4%)	315 (0.3%)	5	1

[1] Number of annotation terms retrieved by the selected N features with the specified α (and percentage of significantly enriched terms, with FDR p-value < 0.05)
[2] Matches between the selected gene sets and cancer-related KEGG signatures.
[3] Matches between the selected gene sets and BRCA-related KEGG signatures.
[4] Standard feature selection.

3.2   Backward Feature Selection

Classification results from the experiments performed (Sect. 2.5) are reported in Table 3, while findings from the biological validation are shown in Table 4. Similar to forward selection, our proposed backward extension outperformed the standard backward method for both the LR and SVM models. However, the performance gain was less pronounced compared to forward selection, largely due to the initial filtering we had to apply. Although our approach selected a larger number of features, we achieved comparable performance when evaluating feature sets of similar size. Using the mean as the criterion c for the similarity measure (s_{mean}) led to improved performance with respect to using the minimum criterion (s_{min}). For both classifiers, increasing the hyperparameter α resulted in the selection of larger feature sets, with minimal feature selection refinement following the initial filtering step. While using the mean criterion s_{mean} improved the classification performance, it produced poor results in terms of biological validation. Specifically, the number of retrieved GO and Reactome terms (Table 4), as well as the corresponding number of significant terms, were substantially

Table 3. Classification performance on the test set for $k_{max} = 150$ for the backward algorithm. Best performance in bold.

Method	N[1]	α[2]	Accuracy	Precision[3]	Recall[3]	F1-score[3]
LR[4]	72	-	0.926	0.915	0.944	0.928
LR$_{s_{min}}$	117	0.3	0.916	0.895	0.946	0.917
LR$_{s_{mean}}$	62	0.3	**0.941**	**0.925**	**0.962**	**0.941**
LR$_{s_{min}}$	160	0.5	0.921	0.893	0.905	0.898
LR$_{s_{mean}}$	173	0.5	0.931	0.913	0.910	0.911
SVM[3]	52	-	0.901	0.884	0.936	0.906
SVM$_{s_{min}}$	50	0.3	0.896	0.856	0.934	0.885
SVM$_{s_{mean}}$	81	0.3	0.926	**0.912**	0.958	**0.931**
SVM$_{s_{min}}$	53	0.5	0.906	0.864	0.938	0.892
SVM$_{s_{mean}}$	198	0.5	**0.941**	0.905	**0.962**	0.928

[1] Number of selected features.
[2] Hyperparameter for the prior knowledge in the feature selection process.
[3] Macro-average metrics.
[4] Standard feature selection.

Table 4. Biological validation results for $k_{max} = 150$ for the backward algorithm.

Method	N	α	GO[1]	Reactome[1]	KEGG Cancer[2]	KEGG BRCA[3]
LR[4]	72	-	2,761 (0.0%)	273 (0.4%)	6	3
LR$_{s_{min}}$	117	0.3	4,095 (17.0%)	441 (15.2%)	17	7
LR$_{s_{mean}}$	62	0.3	2,051 (0.0%)	219 (0.5%)	2	3
LR$_{s_{min}}$	160	0.5	4,299 (7.8%)	481 (10.4%)	18	8
LR$_{s_{mean}}$	173	0.5	3,995 (2.5%)	441 (5.2%)	17	8
SVM[4]	52	-	2,517 (5.5%)	251 (9.2%)	4	2
SVM$_{s_{min}}$	50	0.3	3,336 (30.8%)	310 (19.3%)	8	1
SVM$_{s_{mean}}$	81	0.3	2.503 (0.0%)	251 (1.2%)	3	3
SVM$_{s_{min}}$	53	0.5	3,430 (30.1%)	299 (17.4%)	8	1
SVM$_{s_{mean}}$	198	0.5	4,344 (4.2%)	479 (9.8%)	19	8

[1] Number of annotation terms retrieved by the selected N features with the specified α (and percentage of significantly enriched terms, with FDR p-value < 0.05).
[2] Matches between the selected gene sets and cancer-related KEGG signatures.
[3] Matches between the selected gene sets and BRCA-related KEGG signatures.
[4] Standard feature selection.

lower. In contrast, employing the minimum criterion s_{min} significantly increased these metrics, particularly for feature sets of a size comparable to those obtained with the standard backward algorithm. When evaluating the number of selected genes included in cancer-specific and BRCA-specific KEGG pathway signatures, the values were mostly lower than those achieved with the standard algorithm for similarly sized feature sets. Although the backward selection extension achieved strong classification performance, it only slightly improved over the standard approach in terms of the metrics used for biological validation, differently from the proposed forward extension. These less favorable biological validation outcomes may be attributed to the initial filtering step, which likely influences the selection process and limits the consideration of other feature interactions that could enhance the biological interpretability of the results.

4   Conclusions

In this work, we introduced two extensions to the forward and backward feature selection algorithms that incorporate prior biological knowledge in the feature selection process. This is achieved by leveraging embeddings of such knowledge and considering the overall biological relevance of features. Experimental results demonstrate that our algorithms select highly predictive genes based on their discriminative power, while also ensuring that these genes are biologically informative, mainly for the forward selection extension, significantly enhancing the interpretability of the results. Our approach provides a general and flexible strategy for knowledge-guided feature selection that is applicable across various domains. Furthermore, the proposed framework can be readily adapted to only consider domain-specific information, if available, and to integrate new information as it becomes available over time. Our proposed algorithms have a minimal computational overhead compared to the standard feature selection methods, with the only computationally intense part being the generation of the embeddings prior to the execution of the feature selection algorithm. The paper limits its comparison to standard feature selection approaches. To provide a more comprehensive evaluation, future work would include comparison with specific state-of-the-art methods, as well as an analysis of different distance metrics and embedding algorithms, focusing on developing more efficient and accurate methods for generating these embeddings, to reduce the time and computational resources required while generating semantically rich representations. Additionally, we plan to explore the integration of more complex forms of prior knowledge, including known gene interactions, to potentially further enhance the selection process and improve the biological relevance of the results.

Funding Information. This work was supported by the MUSA - Multilayered Urban Sustainability Action - project, funded by the European Union - NextGenerationEU, under the National Recovery and Resilience Plan (NRRP) Mission 4 Component 2 Investment Line 1.5: Strengthening of research structures and creation of R&D "innovation ecosystems", set up of "territorial leaders in R&D".

Conflict of Interests. The authors declare to have no conflicts of interest.

A   Prior Knowledge

A.1   Construction of Matrix W

Given a dataset with q genes and l knowledge bases, where each knowledge base is structured as a directed acyclic graph (DAG) containing n_l terms, we can identify the most specific terms linked to each gene within each knowledge base and build a binary matrix B. From this matrix B, we can obtain a weighted annotation matrix $W \in \mathbb{R}^{q \times \sum_{k=1}^{l} n_k}$ by considering the depth and number of descendants of each term in each knowledge base of interest [1]. For each term t, we can compute its *structure-based Information Content* (IC_{struct}) [19] as follows:

$$IC_{struct}(t_{j,k}) = \frac{depth(t_{j,k})}{max\_depth_k} \times \left(1 - \frac{log(desc(t_{j,k}) + 1)}{log(total\_terms_k)}\right) \tag{3}$$

where $t_{j,k}$ is the jth term in the kth knowledge base, $depth(t_{j,k})$ and $desc(t_{j,k})$ are the maximum depth and the number of descendants of the term t_j in k, and max_depth_k and $total\_terms_k$ are the maximum depth and the total number of terms of k.

A.2 NMF Algorithm

NMF is an unsupervised learning algorithm used to decompose a positive-defined matrix W into the product of two lower-rank non-negative matrices U and H. The objective of NMF is to approximate W as the product of two non-negative matrices U and H, minimizing the Frobenius norm of the difference:

$$\|W - UH\|_F^2 = \sum_{i=1}^{m}\sum_{j=1}^{n}(W_{ij} - (UH)_{ij})^2 \tag{4}$$

where W is a $p \times m$ matrix, U is a $p \times q$ matrix, and H is a $q \times m$ matrix. After randomly initializing U and H with non-negative values, the NMF algorithm iteratively updates the two matrices using a multiplicative update rule:

$$H_{ij}^n \leftarrow H_{ij}^{n-1} \frac{((U^{n-1})^T W)_{ij}}{((U^{n-1})^T U^{n-1} H^{n-1})_{ij}} \tag{5}$$

$$U_{ij}^n \leftarrow U_{ij}^{n-1} \frac{(W(U^{n-1})^T)_{ij}}{(U^{n-1} H^n (H^n)^T)_{ij}} \tag{6}$$

where n is the index for the iteration. The algorithm runs iteratively until a predefined number of iterations or convergence is reached, which generally occurs when the change in the matrices U and H is smaller than a specific threshold.

To obtain the prior knowledge embeddings, we run the NMF algorithm on the weighted annotation matrix W for 3,000 iterations, checking that the non-negative matrices U and H were stable, and then extracted the embeddings from U. We used an embedding size of 64. We also evaluated different embedding sizes (i.e., 128 and 256) and noticed that increasing the dimension did not significantly improve performance.

B   Additional Results

In this section, we report additional findings from the performed experiments.

B.1   Forward Feature Selection

Here, the classification performance and biological validation results are shown in Table 5 and Table 6 for $k_{max} = 30$.

Table 5. Classification performance on the test set for $k_{max} = 30$. Best performance in bold.

Method	N^1	α^2	Accuracy	Precision[3]	Recall[3]	F1-score[3]
LR^4	30	-	0.876	0.861	0.879	0.868
$LR_{d_{max}}$	21	0.3	**0.916**	**0.894**	**0.939**	**0.914**
$LR_{d_{mean}}$	27	0.3	0.856	0.828	0.903	0.856
$LR_{d_{max}}$	27	0.5	0.886	0.914	0.844	0.870
$LR_{d_{mean}}$	30	0.5	0.881	0.832	0.894	0.856
SVM^3	30	-	0.876	0.829	0.890	0.855
$SVM_{d_{max}}$	27	0.3	**0.936**	**0.904**	**0.956**	**0.926**
$SVM_{d_{mean}}$	29	0.3	0.881	0.853	0.893	0.871
$SVM_{d_{max}}$	29	0.5	0.891	0.850	0.920	0.877
$SVM_{d_{mean}}$	23	0.5	0.906	0.869	0.897	0.881

[1] Number of selected features.
[2] Hyperparameter for the prior knowledge in the feature selection process.
[3] Macro-average metrics.
[4] Standard feature selection.

Table 6. Biological validation results for $k_{max} = 30$.

Method	N	α	GO[1]	Reactome[1]	KEGG Cancer[2]	KEGG BRCA[3]
LR^4	30	-	1,393 (0.0%)	162 (0.0%)	3	3
$LR_{d_{max}}$	21	0.3	3,175 (64.7%)	207 (7.7%)	6	4
$LR_{d_{mean}}$	27	0.3	2,129 (0.0%)	244 (0.0%)	3	1
$LR_{d_{max}}$	27	0.5	3,962 (57.6%)	241 (21.6%)	6	3
$LR_{d_{mean}}$	30	0.5	2,925 (1.0%)	251 (1.6%)	4	1
SVM^4	30	-	1,737 (0.12%)	249 (0.0%)	2	1
$SVM_{d_{max}}$	27	0.3	3,845 (57.8%)	339 (2.1%)	8	5
$SVM_{d_{mean}}$	29	0.3	2,744 (2.0%)	198 (0.0%)	4	1
$SVM_{d_{max}}$	29	0.5	4,045 (65.1%)	247 (10.1%)	8	3
$SVM_{d_{mean}}$	23	0.5	2,330 (0.0%)	269 (0.0%)	4	0

[1] Number of annotation terms retrieved by the selected N features with the specified α (and percentage of significantly enriched terms, with FDR p-value < 0.05)
[2] Matches between the selected gene sets and cancer-related KEGG signatures.
[3] Matches between the selected gene sets and BRCA-related KEGG signatures.
[4] Standard feature selection.

References

1. Acharya, S., Saha, S., Nikhil, N.: Unsupervised gene selection using biological knowledge: application in sample clustering. BMC Bioinform. **18**(1), 513 (2017)
2. Ashburner, M., Ball, C.A., Blake, J.A., et al.: Gene Ontology: tool for the unification of biology. Gene Ontol. Consortium. Nature Genetics **25**(1), 25–29 (2000)
3. Bellazzi, R., Zupan, B.: Methodological review: towards knowledge-based gene expression data mining. J. Biomed. Inform. **40**(6), 787–802 (2007)
4. Network, C., et al.: Comprehensive molecular portraits of human breast tumours. Nature **490**(7418), 61–70 (2012)
5. Duda, R.O., Hart, P.E., Stork, D.G.: Pattern Classification, 2nd edn. Wiley, New York (2001)
6. Fang, O.H., Mustapha, N., Sulaiman, N.: An integrative gene selection with association analysis for microarray data classification. Intell. Data Anal. **18**, 739–758 (2014)
7. Ferri, F., Pudil, P., Hatef, M., Kittler, J.: Comparative study of techniques for large-scale feature selection. In: Gelsema, E.S., Kanal, L.S. (eds.) Pattern Recognition in Practice IV, Machine Intelligence and Pattern Recognition, vol. 16, pp. 403–413. North-Holland (1994)
8. Jassal, B., Matthews, L., Viteri, G., et al.: The Reactome pathway knowledgebase. Nucleic Acids Res. **48**(D1), D498–D503 (2019)
9. Kanehisa, M., Goto, S.: KEGG: kyoto encyclopedia of genes and genomes. Nucleic Acids Res. **28**(1), 27–30 (2000)
10. Mongardi, S., Cascianelli, S., Masseroli, M.: Biologically weighted LASSO: enhancing functional interpretability in gene expression data analysis. Bioinformatics, btae605 (2024)
11. O'Leary, N.A., Wright, M.W., Brister, J.R., et al.: Reference sequence (RefSeq) database at NCBI: current status, taxonomic expansion, and functional annotation. Nucleic Acids Res. **44**(D1), D733–D745 (2016)
12. Pallotta, S., Cascianelli, S., Masseroli, M.: RGMQL: scalable and interoperable computing of heterogeneous omics big data and metadata in R/Bioconductor. BMC Bioinform. **23**(1), 123 (2022)
13. Pally, D., et al.: Galectin-9 signaling drives breast cancer invasion through extracellular matrix. ACS Chem. Biol. **17**(6), 1376–1386 (2022)
14. Papachristoudis, G., Diplaris, S., Mitkas, P.A.: SoFoCles: feature filtering for microarray classification based on Gene Ontology. J. Biomed. Inform. **43**(1), 1–14 (2010)
15. Parker, J.S., Mullins, M., Cheang, M., et al.: Supervised risk predictor of breast cancer based on intrinsic subtypes. J. Clin. Oncol. **27**(8), 1160–1167 (2009)
16. Robinson, P.N., Köhler, S., Bauer, S., et al.: The Human Phenotype Ontology: a tool for annotating and analyzing human hereditary disease. Am. J. Hum. Genetics **83**(5), 610–615 (2008)
17. Shen, X., et al.: SLC38A5 promotes glutamine metabolism and inhibits cisplatin chemosensitivity in breast cancer. Breast Cancer **31**(1), 96–104 (2024)
18. Sun, R., et al.: Neutral ceramidase regulates breast cancer progression by metabolic programming of TREM2-associated macrophages. Nat. Commun. **15**(1), 966 (2024)
19. Teng, Z., Guo, M., Liu, X., et al.: Measuring gene functional similarity based on group-wise comparison of GO terms. Bioinformatics **29**, 1424–1432 (2013)

20. Upton, G.: Fisher's Exact Test. J. R. Stat. Soc. Ser. A Stat. Soc. **155**(3), 395–402 (2018)
21. Weinstein, J.N., Collisson, E.A., Mills, G.B., et al.: The cancer genome atlas pan-cancer analysis project. Nat. Genet. **45**(10), 1113–1120 (2013)

The Impact of Mis-Labeled Artefacts on Deep Learning Models for EEG Analysis: a Case Study

Alberto Zancanaro[1] and Giulia Cisotto[2(✉)]

[1] Department of Information Engineering, University of Padova, Padova, Italy
`alberto.zancanaro.1@phd.unipd.it`
[2] Department of Mathematics, Informatics and Geosciences, University of Trieste,
Trieste, Italy
`giulia.cisotto@units.it`

Abstract. Electroencephalography (EEG) is largely used for its very informative value of brain activity, high temporal resolution, but also portability and relatively low cost. However, its modeling is a challenging task due to the unavailability of large datasets and its very low signal-to-noise ratio. Deep learning (DL) has the potential to cope with some of these weaknesses. Unfortunately, DL models are very sensitive, not only to the size, but also to the quality of the input, and learning from clean EEG is not always guaranteed. In this work, we show how *hvEEGNet*, a DL model that provides high-fidelity reconstruction of multi-channel EEG data, can handle different amounts and types of artefacts. Specifically, we show how mis-labeled artefacts in the benchmark *dataset 2a* from the BCI competition IV lead to reconstruction failures and we investigate the relationship between the quality of the input and the model's learning ability. This work shows the effectiveness of *hvEEGNet* as an anomaly detector, but also opens new critical directions for future investigations towards the development of more reliable and fair DL models for noisy EEG data.

Keywords: deep learning · anomaly detection · classification · EEG · motor imagery · EEGNet · autoencoder · VAE · BCI · dataset 2a

1 Introduction

Electroencephalography (EEG) is a biological signal used in different clinical and research applications, e.g. emotions recognition and brain-computer interface (BCI) due to its non-invasive nature, relatively low cost, and high temporal resolution. The last decade has seen a considerable increase in the use of deep learning (DL) models to analyze EEG data for many different tasks. According to [6] motor imagery (MI) classification is the most common one (22% of studies), followed by emotions recognition (16%), and mental workload assessment (16%). Unfortunately, unlike other research areas such as image classification or natural

L. Cerulo et al. (Eds.): CIBB 2024, LNBI 15276, pp. 248–255, 2025.
https://doi.org/10.1007/978-3-031-89704-7_19

language processing, a few large public EEG datasets are available. Nonetheless, some benchmark datasets are very widely used in the community: for example, *dataset 2a* and *2b*, collected for the BCI Competition IV [4], are present in about two third studies applying DL models for classifying different MI in EEG recordings [2]. On the other hand, EEG is known to be particularly prone to artefacts and interferences from other sources of electrical activity, like facial muscles contraction or power line noise. Artefacts and interferences have the effect of strongly altering the distribution of the actual EEG data and this, in turn, might significantly affect the ability of the DL models to learn the true patterns in the EEG data, possibly leading to learning biases [9]. This problem can be further exacerbated by the often small size of the datasets.

In this work, we show how the presence of mis-labeled artefacts in the *dataset 2a* has an impact on the learning and the performance of our *hvEEGNet*, a DL model previously shown to reconstruct multi-channel EEG data with very high-fidelity [5]. Also, we investigate its reconstruction failures due to un-labeled artefacts, while suggesting *hvEEGNet* as an effective anomaly detector.

The rest of the paper is organized as follows. Section 2 describes the *dataset 2a* from the BCI competition IV, the expert labeling map, and shortly introduces the *hvEEGNet* model. Section 3 presents and discusses the main results. Lastly, Sect. 4 concludes the work, while highlighting the most relevant future perspectives.

2   Materials and Methods

2.1   Dataset and Expert Labels

The *dataset 2a* is a popular and well-known dataset, largely used to assess the effectiveness of machine learning (ML) and DL models to classify movements from EEG [4]. It consists of 22-channel EEG recordings from 9 subjects (identified as S1, S2, etc.) that performed four different MI tasks: the imagination of the right hand, left hand, tongue, and feet movements, respectively. The signals were sampled at 250 Hz, and then filtered using a band-pass filter with passband 0.5-100 Hz, and a notch filter at 50 Hz. For every subject, two *sessions* were recorded (in two different days). Each one consists of 288 MI *repetitions*, with one fourth repetitions for each movement class, presented in a random order. Each recording was, in turn, split into six *runs* of 48 repetitions. During each repetition, the subject was asked to wait for 2 s and then perform the MI for 4 s. Then, each repetition is 6 s long. This dataset comes with a set of labels, given by an expert, which defined each repetition as *clean* or *artefactual*. In Fig. 1, we report a visual representation of the presence of artefacts for each subject, session, and repetition. To note, the expert assigned the label by considering the whole multi-channel recording at each repetition: thus, labels do not distinguish between clean or corrupted channels. Neither, labels are provided with a comment that supports the expert's choice. The vast majority of the previous works used the EEG data collected during the first session as the training set for their DL models, while those collected during the second session as the test set.

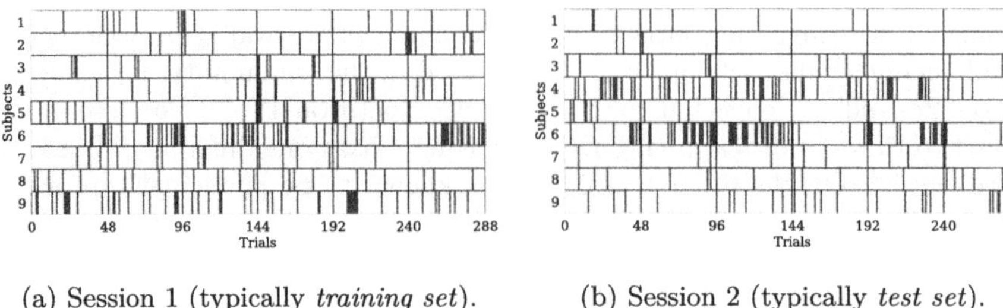

(a) Session 1 (typically *training set*). (b) Session 2 (typically *test set*).

Fig. 1. Expert's labels for *dataset 2a*. Each dark red vertical line represents a repetition labeled as *artefactual* by the expert. Runs are divided by black vertical lines (one every 48 repetitions).

2.2 HvEEGNet

In this work, we used the unsupervised DL model named *hvEEGNet* [8] proposed in [5]. *hvEEGNet* is a hierarchical variational autoencoder (VAE) [10] with the encoder and the decoder designed using the blocks of the popular *EEGNet* architecture [8]. Specifically, temporal, spatial, and separable convolutional neural networks are used to form the three main blocks of our architecture. As any other autoencoder, the VAE has an encoder-decoder architecture that learns the distribution of the data during the training through a process of compression and reconstruction. Furthermore, from its latent space, it is possible to extract a compressed representation of the input. In [5], we showed that *hvEEGNet* is able to provide a high-fidelity reconstruction of multi-channel EEG data. A complete description of the model can be found in [5]. Note that we adopted a *within-subject* learning by training one model for each subject, separately: more precisely, data from session 1 were used in the training phase, while data from session 2 for the test. A cross-subject extension of this work needs further investigations that will be addressed in the future. Finally, we measured the performance of *hvEEGNet* in terms of reconstruction error, using the soft dynamic time warping (SDTW) similarity metric (see references included in [5]) between every input repetition and its corresponding reconstructed one. In mathematical terms, for every subject, we built a matrix $\overline{\mathbf{E}} = [e_{c,r}] \in \mathbb{R}^{C \times R}$, where $e_{c,r}$ is the reconstruction error of the c-th channel of the r-th repetition, with $c = 1, 2, ..., C$ and $r = 1, 2, ..., R$. $e_{c,r}$ is computed as the SDTW between the original input $x_{c,r}(t)$ and the reconstructed $\hat{x}_{c,r}(t)$, with t representing the discrete time dimension.

3 Results and Discussion

In this section, we present the results of our analysis. First, we show how *hvEEG-Net* can tolerate the presence of artefacts producing very high-fidelity reconstructions in most cases. Second, we provide some examples of poor reconstructions and put forward hypotheses on the possible causes behind these failures. Lastly, we open the discussion to the most relevant directions we intend to develop in the near future. To note, in the following, we report results of *hvEEGNet* applied to *dataset 2a* with $C = 22$ and $R = 288$. Also, we highlight that no pre-processing steps were applied on the dataset, as the latter was supposed to be already cleaned and made ready to be input in any ML model.

3.1   Robustness of *hvEEGNet* against artefacts

Our model revealed a good degree of robustness to artefactual data. Despite the presence of numerous artefacts detected by the domain expert, both in the training set and in the test set of all subjects, our model is able to perform generally very well. An example is reported in Fig. 2, where the reconstruction ability of *hvEEGNet* in S9 is clearly visible both in the training set (session no.1 of *dataset 2a*) and in the test set (session no.2 of *dataset 2a*). This result is in line with an unfortunately limited literature [1], where auto-encoders were effectively trained using a mix of clean and artefactual data. This represents a more realistic scenario where auto-encoders might help: especially in EEG analysis, it is challenging to define and obtain a clean dataset [7].

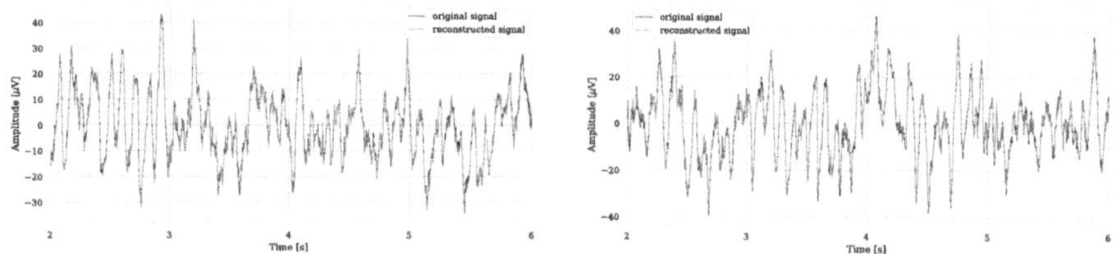

(a) Repetition 253 - session 1 (ch. "Cz").   (b) Repetition 102 - session 2 (ch. "C3").

Fig. 2. An example of high-fidelity reconstruction using *hvEEGNet* in the case of a large number of artefacts for S9. Black line is for the original signal, red for the reconstructed one. Repetitions were randomly chosen among all repetitions of S9.

3.2   Reconstruction Performance in Relation to Input Statistical Distribution

When the number and the type of artefacts significantly change between the training and the test sets, the performance of *hvEEGNet* strongly decreases. Table 1 gives an intuition of this: for some subjects in the dataset, the amount of artefacts (as detected by the domain expert) is larger in the test set compared to the training set. This can be seen, for example, for S4 where the number of artefacts is more than doubled in the test set. Figure 3 shows the matrix $\overline{\mathbf{E}}$ of the reconstruction errors (with reference to Sect. 2), both for the training and the test set of S4. Here, we can observe the presence of a few large errors in session 1 (training), but several ones in session 2 (test). We verified (not reported for space constraints) that this does not happen with the other subjects. Thus, we can infer that our *hvEEGNet* model is able to handle a certain number of artefacts during its training. However, as the number of artefacts significantly increases w.r.t. the training set, the model failed more frequently. Besides, we also verified that the errors in the test set correspond to actual artefacts in the data.

Table 1. Number of artefacts per subject, as identified by the expert.

Session no.	S1	S2	S3	S4	S5	S6	S7	S8	S9
1 (training set)	15	18	18	26	26	69	17	24	51
2 (test set)	7	5	15	60	12	73	11	17	24

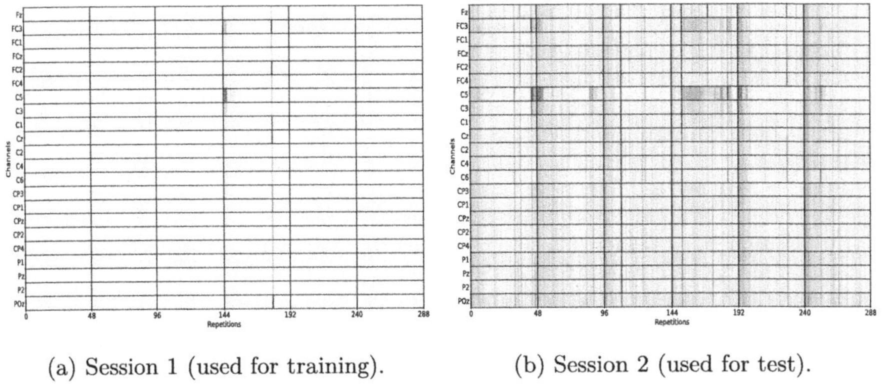

(a) Session 1 (used for training).   (b) Session 2 (used for test).

Fig. 3. Reconstruction errors in S4. Vertical axes represent EEG channels. Horizontal axes report repetition indices. Values range from 0 (light pink) to 10 or above (dark red). (Color figure online)

Furthermore, we explored the presence of artefacts of different shapes. We realized that S2 and S5's recordings were strongly affected by artefacts that were not reported by the domain expert. We found a general increase of the reconstruction error in their test sets and verified that such performance drop was associated with strong artefacts (see Fig. 4). Based on a well-established literature (see references in [5]), the average spectra of sessions 2 are recognized to show clear non-physiological behaviors. Particularly, S2's spectrum has large variability in the frequency band above 15 Hz, making us suspect the presence of muscular artefacts (see references in [5]). On the other hand, S5's spectrum is characterized by a predominant peak around 50 Hz, surprisingly revealing that the notch filter (supposed to be already applied) was not effective enough (or, mistakenly, not applied at all). Unexpectedly, in both cases, the domain expert did not record the presence of these artefacts. Nonetheless, many recent works used this dataset with no prior checks on the quality of the data, and used the above sessions to train and test, respectively, their DL architectures (see [11] for a short review on this piece of literature). This may have produced not fully reliable classification results and it deserves further investigation.

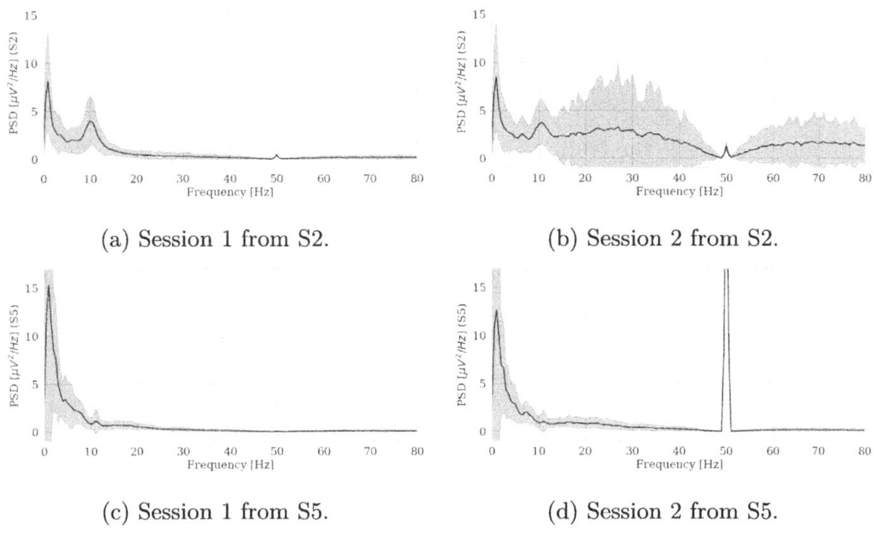

(a) Session 1 from S2.

(b) Session 2 from S2.

(c) Session 1 from S5.

(d) Session 2 from S5.

Fig. 4. Average power spectra for S2 and S5 (channel "C3") in both sessions. Average spectra are in black, standard deviation in grey. (Color figure online)

On the other hand, we highlight how our architecture can be effectively used as an anomaly detector, as it returns large reconstruction errors for S2 and S5 in correspondence to those artefacts. This can be also explained by plotting the statistical distributions of the EEG samples in the two sessions for S2 and S5 (Fig. 5), which appear different from each other. In line with [3], we quantified such difference by computing the Wasserstein distance between the samples belonging to session 1 and session 2 in the two subjects, separately.

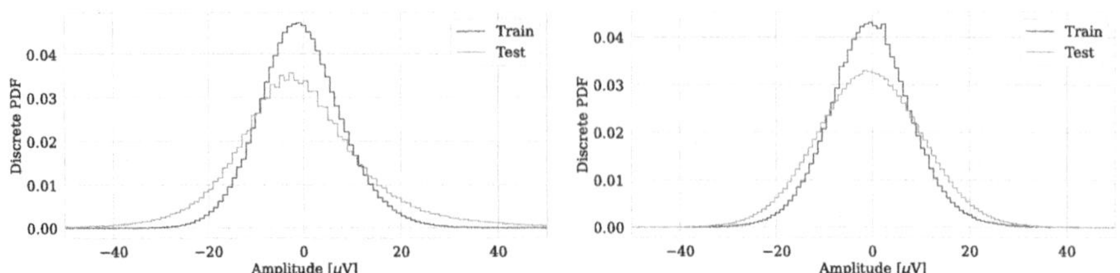

(a) Training and test sets distributions for S2. Wasserstein distance = 2280.84.

(b) Training and test sets distributions for S5. Wasserstein distance = 1847.06.

Fig. 5. Statistical distributions for session 1 (green lines) and session 2 (red lines) of (a) S2 and (b) S5. The plots are obtained using a histogram function with 200 bins. (Color figure online)

4   Conclusions

hvEEGNet, as any other VAE, has the ability to learn the probability distribution of the input dataset (the training set) and generate an output that has the same distribution. *hvEEGNet* has a particular architecture that enables high-fidelity reconstruction of multi-channel EEG recordings, as shown in our previous work [5]. EEG is known to be prone to instrumental intereferences and physiological artefacts. Also, it is generally challenging to obtain a large clean EEG dataset, thus making the learning process of DL models particularly ambitious. In this paper, we showed that *hvEEGNet* can well tolerate the consistent presence of artefacts, as labeled by a domain expert in the benchmark *dataset 2a*. At the same time, we found that there were cases where the reconstruction error was significantly large, and those EEG portions were associated to artefacts that were mis-labeled. These preliminary results led us to promote our *hvEEGNet* as a very good candidate for anomaly detection in multi-channel EEG data. Besides, we suggest other researchers to pay attention to the quality of the input dataset: specifically, many recent works used this dataset as a benchmark for their newly developed DL models, without prior check on the quality of the training set. This might have introduced learning biases or lower classification performances due to a consistent difference between the statistical distributions of the test set with respect to the training set. In the future, further investigations are needed to quantify the statistical differences between the training and the test set (the so-called *dataset shift problem*), the representativeness of the training set and its relationship with the learning process of the model, including its generalization ability.

Acknowledgements. This work was partially supported by the Italian Ministry of University and Research (MUR) under the grant "Dipartimenti di Eccellenza 2023–2027" of the Department of Informatics, Systems and Communication of the University of Milano-Bicocca, Italy. AZ acknowledges the financial support of PON 2014–2020 action IV.5 funded by MUR.

Availability of Software Code. The software is available at: https://github.com/jesus-333/Variational-Autoencoder-for-EEG-analysis/tree/hvEEGNet_paper

References

1. Al-amri, R., Murugesan, R.K., Man, M., Abdulateef, A.F., Al-Sharafi, M.A., Alkahtani, A.A.: A review of machine learning and deep learning techniques for anomaly detection in IoT data. Appl. Sci. **11**(12), 5320 (2021)
2. Al-Saegh, A., Dawwd, S.A., Abdul-Jabbar, J.M.: Deep learning for motor imagery EEG-based classification: a review. Biomed. Signal Process. Control **63**, 102172 (2021)
3. Barandas, M., et al.: Evaluation of uncertainty quantification methods in multi-label classification: a case study with automatic diagnosis of electrocardiogram. Inf. Fus. **101**, 101978 (2024)
4. Blankertz, B., Dornhege, G., Krauledat, M., Müller, K.R., Curio, G.: The non-invasive Berlin Brain-Computer Interface: fast acquisition of effective performance in untrained subjects. NeuroImage **37**, 539–50 (2007)
5. Cisotto, G., Zancanaro, A., Zoppis, I.F., Manzoni, S.L.: hvEEGNet: a novel deep learning model for high-fidelity EEG reconstruction. Front. Neuroinformatics **18**, 1459970 (2024)
6. Craik, A., He, Y., Contreras-Vidal, J.L.: Deep learning for electroencephalogram (EEG) classification tasks: a review. J. Neural Eng. **16**(3), 031001 (2019)
7. Gabardi, M., Saibene, A., Gasparini, F., Rizzo, D., Stella, F.A.: A multi-artifact EEG denoising by frequency-based deep learning. In: CEUR Workshop Proceedings, vol. 3576, pp. 28–41. CEUR-WS (2023)
8. Lawhern, V., Solon, A., Waytowich, N., Gordon, S., Hung, C., Lance, B.: EEGNet: a compact convolutional network for EEG-based Brain-Computer Interfaces. J. Neural Eng. **15** (2016)
9. Roy, Y., Banville, H., Albuquerque, I., Gramfort, A., Falk, T.H., Faubert, J.: Deep learning-based electroencephalography analysis: a systematic review. J. Neural Eng. **16**(5), 051001 (2019)
10. Vahdat, A., Kautz, J.: NVAE: a deep hierarchical variational autoencoder. In: Larochelle, H., Ranzato, M., Hadsell, R., Balcan, M., Lin, H. (eds.) Advances in Neural Information Processing Systems. vol. 33, pp. 19667–19679. Curran Associates, Inc. (2020)
11. Zancanaro, A., Cisotto, G., Zoppis, I., Manzoni, S.L.: vEEGNet: learning latent representations to reconstruct EEG raw data via variational autoencoders. In: International Conference on Information and Communication Technologies for Ageing Well and e-Health, pp. 114–129. Springer (2023)

Benchmark Study on Supervised Relevance-Redundancy Assessment for Feature Selection in Genomic Data

Simone Tomè[1,2(✉)] [iD], Silvia Cascianelli[1] [iD], Erika Salvi[2,3] [iD],
and Marco Masseroli[1,2] [iD]

[1] Department of Electronics, Information and Bioengineering (DEIB), Politecnico di
Milano, Milan, Italy
{simone.tome,silvia.xascianelli,marco.masseroli}@polimi.it
[2] Computational Multi-Omics of Neurological Disorders (MIND) Lab, Joint Research
Platform, Fondazione IRCCS Istituto Neurologico Carlo Besta, Milan, Italy
[3] Data Science Center, Fondazione IRCCS Istituto Neurologico Carlo Besta,
Milan, Italy
erika.salvi@istituto-besta.it

Abstract. Single variant Genome-Wide Association Studies (GWAS)
are the most common data-driven approach to discover genetic variants
associated to phenotypes. However, in complex diseases, single variants
often have no effect unless they coexist with other variants. Conversely,
machine learning (ML) can model potential interactions among variants, potentially addressing the missing heritability problem in these
diseases. Nevertheless, the curse of dimensionality must be considered,
given the tremendous number of variants in genomic datasets, requiring
feature selection techniques to reduce the number of features. This study
aims at a preliminary benchmark of the Relevance-Redundancy assessment (ReRa) feature selection method using a public genetic dataset
of a Parkinson's cohort. Obtained results demonstrated that ReRa can
achieve performances comparable to common filter-based feature selection techniques, with the benefit of building simpler models with fewer
features.

Keywords: Feature selection · Machine learning · Genotype ·
Supervised learning

1 Introduction

Parkinson's disease (PD) is a complex disease whose aetiology is influenced by
heterogeneous factors including multiple gene variants, rare mutations, environmental influences, lifestyle choices, comorbidities, and viral infections. Genome-
Wide Association Studies (GWAS) have been utilized to investigate the relationships between common genomic loci and observed traits, such as diabetes, cancer,
and Alzheimer's disease. However, the primary effects identified by GWAS do

L. Cerulo et al. (Eds.): CIBB 2024, LNBI 15276, pp. 256–265, 2025.
https://doi.org/10.1007/978-3-031-89704-7_20

not fully capture the complexity of the disease, as traditional statistical methods used in GWAS are inadequate for identifying complex, non-linear epistatic interactions among gene variants. Epistasis refers to the interactions between genes that contribute to a phenotype not observable when considering only individual variants. Therefore, it is essential to integrate these interactions, which classical GWAS methods overlook, into genotype-trait association studies. In the literature, machine learning (ML) models are emerging as a successful alternative able to capture complex interactions among features [5–7,10], improving the ability to associate the sample genotype (collection of measured variants) with the phenotype of interest. The ability of ML to capture non-linearities among features comes at the cost of interpretability, as the individual parameters learned during the ML model training phase lack the straightforward interpretation of the effect, unlike that provided by a Fisher's test odds ratio. The ability of ML models is also jeopardized by the curse of dimensionality, which happens when the number of features that characterize a sample is much greater than the number of evaluated individuals; unfortunately, this is a common scenario in high-throughput omics analysis. Thus, the curse of dimensionality imposes the necessity to adopt feature selection techniques to select a subspace of the initial measured features. Genotype data usually involves hundreds of thousands of initial features (gene variants); this constraints the suitable feature selection techniques, requiring effective and scalable algorithms. For instance, wrapper techniques, which are model-specific and search for an optimal subspace leveraging model performance as a cost function, are impractical due to their computational burden. Instead, filter techniques are generally less prone to overfit the observed data and more computationally efficient. Also, they are able to handle redundancies among features, something that could introduce harmful bias in the model. The aim of this work is to evaluate the application of the supervised Relevance-Redundancy (ReRa) [2] assessment to common filter-based feature selection methods in the context of genotype data, with the final objective of extracting predictive features and generating new hypotheses.

2 Data and Methods

2.1 Parkison's Progression Marker Initiative Dataset

The utilized dataset comes from the Parkinson's Progression Marker Initiative (PPMI) database. To be more specific, we focused on the classification between two cohorts: diagnosed PD and prodromal. Diagnosed PD are people with PD, and people with PD and pathogenic variants in GBA, LRRK2, SNCA genes. Instead, the prodromal cohort is composed of people with age older than 60, with REM sleep behavior disorder (RBD), hyposmia with dopamine transporter (DAT) deficit, and some cases have genetic risk variants (GBA, LRRK2, SNCA). Samples were genotyped using the *Illumina NeuroBooster*, an high-throughput custom-designed genotyping array aimed at screening for neurological disorder related variants across different populations. According to the PPMI documentation, the array contains a backbone of 1,914,934 variants (*Infinium Global Diversity Array*) complemented with custom content of 95,273 variants implicated in over 60 neurological conditions. Furthermore, it includes $> 10{,}000$ tagging variants to facilitate imputation and analyses of 199 neurodegenerative disease-related GWAS loci across populations. The dataset includes genotyping data for 896 samples, 338 prodromal and 558 diagnosed PD, and has undergone standard quality control steps [1] using PLINK v1.9. First, variants with more than 5% of missings calls have been excluded (655,227 variants). Then, the reported sex has been checked with respect to the obtained X chromosome variants and no samples reported any mismatch. Samples with a genotyping rate less than 95% have been excluded (18 samples). No samples with heterozygosity rate deviating from mean ± 3 standard deviations have been found and variants in linkage disequilibrium have been pruned using a window size of 50 kilo-bases, a step size of 5, and a r^2 threshold of 0.8. Then, only common single nucleotide polymorphisms with a minor allele frequency $\geq 1\%$ on autosome chromosomes, and with a genotype rate of at least 95% have been considered. At the end of the preliminary quality control steps, 878 samples (541 cases and 337 prodromal) remained, with 541,730 tracked variants codified as either 0 (homozygous wild type), 1 (heterozygous), or 2 (mutated homozygous); remaining missing genotypes have been imputed using k-Nearest Neighbors (using scikit-learn 1.4.2 default parameters).

2.2 Feature Selection

A machine learning model can be seen as a decider, whose accurate decisions in high dimensional manifolds require an amount of data that can teach the model an accurate approximation of the real underlying generating process. Curse of dimensionality happens when the feature space is much greater with respect to the sampled points. It is indeed unfeasible to correctly estimate performance metrics for any given model in a high dimensional space with few data points, as there are too many models that could return the same training error, even if their decision boundaries on the feature space are highly heterogeneous.

Typically, quality control steps still return a number of variants that is orders of magnitude greater than the number of samples, requiring further feature selection strategies. Feature selection methods can be categorized into three main taxonomies: filter, wrapped, and embedded techniques [9]. Filter strategies, either univariate or multivariate, are model-agnostic since are only based on data and done before any machine learning model training. Wrapped methods search for an optimal feature subspace leveraging the specific-model performance, in a closed feedback loop. Embedded methods are feature selection strategies integrated into the architecture of the model, which decides if a feature is relevant or not (e.g., the LASSO regularization).

In this study, we focused on filter methods, which are more suitable to the number of features as they guarantee a good trade-off between efficacy and computational complexity. In particular, we leveraged three supervised filters: χ^2-test, Mutual Information and F-test, all of which are implemented in the *scikit-learn* [8] feature selection module. All these methods do not account for redundancy of features, which can undermine model performances. To account for redundancy, we employed the supervised redundancy assessment of ReRa

[2], which can detect both global redundancy and class-specific redundancies using similarity metrics. ReRa has the advantage of being highly customizable, as it allows opting for correlation metrics suitable for the specific task, and it is designed to work in domains of high heterogeneity and unbalanced class distributions, where it is crucial to calculate local redundancy for each class. ReRa tests all the pairs of features initially selected using custom filtering: for each pair, it calculates first their pairwise correlation globally on the entire sample population. If the pair has a correlation below the pre-defined global threshold, it is appended to the temporary feature space (TFS). Otherwise, correlation pairwise metrics of the feature pair are calculated after splitting samples by class label: for each class, if a pairwise correlation is lower than the class local correlation threshold, the pair is considered locally dissimilar and added to the TFS; otherwise, only the feature with the highest mean deviation (calculated as the biggest difference between the class-specific mean values of the feature) is retained. The TFS is, therefore, constantly updated by removing any feature excluded by the sequential evaluation of all the feature pairs. Notably, global and class correlation distributions are also used in their entirety to estimate the best choice for the corresponding tuning of the thresholds. The pseudocode of ReRa is detailed in Algorithm 1.

2.3 Machine Learning

The task of interest is a binary classification able to distinguish between cases (Parkinson's samples) and prodromals. For this, several models have been evaluated, specifically Logistic Regression (with different regularization techniques), Random Forest (RF), and eXtreme Gradient Boosting (XGB) [3]. Feature selection methods have been applied over the whole pre-processed (undergone quality control reported in Sect. 2.1) dataset, and nested five-fold cross-validation has been used with exhaustive grid-search to tune the hyperparameters of the tested models. Specifically, an outer 5-fold cross-validation has been used for feature selection (using the training folds) and performance assessment, while the inner 5-fold has been used for hyperparameter tuning using an exhaustive grid search. This approach prevents information leakage as the feature selection and module tuning is agnostic to the information contained in the outer left-out fold. A schema of this approach is shown in Fig. 1.

3 Results

The redundancy assessment of ReRa has been tested downstream different supervised univariate filter feature selection techniques, namely the χ^2 test, the F-test, and the Mutual information. All these techniques are univariate in the sense that each feature is tested individually, w.r.t. multivariate techniques like ReliefF [11] that leverage possibly the whole feature space. The computational efficiency of filter techniques makes them appealing for datasets with hundreds of thousands of features. Different performance metrics have been used, specifically the ROC-AUC, the F1-score, the Matthews correlation coefficient [4][?], and the average

---

Algorithm 1: ReRa feature selection algorithm

Data: $X \in \mathbb{R}^{p \times n}$ (training dataset with p features), $y \in \mathbb{R}^n$ (labels), C (classes), q_g (global similarity quantile threshold), q_l (local similarity quantile threshold), n_c (indexes of samples of class c), f_i (i-th feature)

Result: Selected features stored in TFS

```
      /* Helper functions                                                    */
 1  Function filter_metric(f_i, label) is
        /* returns supervised relevance metric for feature f_i              */
 2  end
 3  Function arg_max(list, k) is
        /* returns k highest indexes of the list                           */
 4  end
 5  Function get_value(distribution, quantile) is
        /* returns value of distribution at quantile                       */
 6  end
 7  Function ReRa is
        /* ReRa feature selection algorithm                                 */
        /* Relevance step                                                   */
 8      for i = 1 to p do
 9      |   relevance[i] ← filter_metric(f_i, y);
10      end
        /* Take the top k relevant features                                 */
11      relevant_f ← arg_max(relevance, k);
        /* Calculate global and local similarity distributions              */
12      global_distribution ← similarity_distribution(X[relevant_f, :]);
13      for c ∈ C do
14      |   local_distribution[c] = similarity_distribution(X[relevant_f, n_c]);
15      end
16      TFS ← {};
        /* Redundancy assessment                                            */
        /* Test all pairs of relevant features                              */
17      for (i, j) ∈ relevant_f × relevant_f, i ≠ j do
18          local_counter ← FALSE;
            /* Test global similarity                                       */
19          if 1 − similarity(f_i, f_j) < get_value(global_distribution, q_g) then
20          |   TFS.append(f_i, f_j);
21          end
            /* Test local similarity                                        */
22          else
23              for c ∈ C do
24                  if 1 − similarity(f_i, f_j) > get_value(local_distribution[c], q_g) then
25                      TFS.append(f_i, f_j);
26                      local_counter ← TRUE;
27                  end
28              end
29              if local_counter == FALSE then
30                  for c ∈ C do
31                      mean_fi[c] ← mean(X[f_i, n_c]);
32                      mean_fj[c] ← mean(X[f_j, n_c]);
33                  end
34                  for (c_x, c_y) ∈ C × C, c_x ≠ c_y do
35                      I[c_x, c_y] = |mean_fi[c_x] − mean_fi[c_y]|;
36                      J[c_x, c_y] = |mean_fj[c_x] − mean_fj[c_y]|;
37                  end
38                  if arg_max_feature(I, J) == I then
39                  |   TFS.delete(f_j);
40                  end
41                  else
42                  |   TFS.delete(f_i);
43                  end
44              end
45          end
46      end
47  end
```

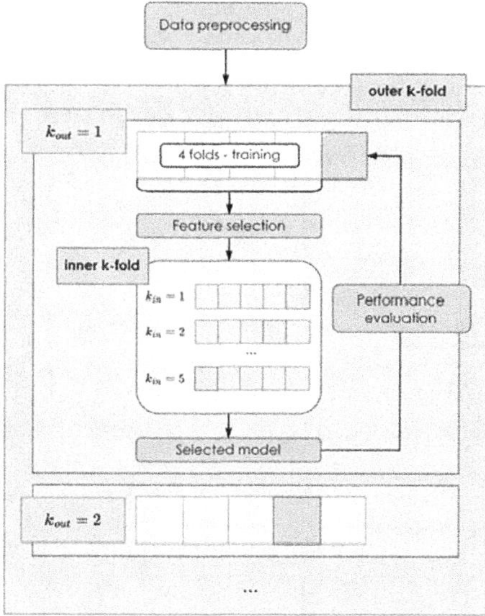

Fig. 1. Schema of the nested cross-validation approach. The outer 5-fold is used to evaluate performances of the obtained models and the outer 4 training folds are used to perform supervised feature selection. Model selection is performed with the inner 5-fold cross-validation.

weighted precision, which summarizes a precision-recall curve as the weighted mean of precisions achieved at each threshold of classification, weighting the precision with the increase in recall at each step.

For the redundancy assessment, different quantiles of similarity have been tested, as shown in Fig. 2. The similarity quantile is the threshold to define if two features are similar or not based on the global or local (intra-class) distribution. Therefore, higher quantiles imply a more stringent threshold as more features are considered similar. As we can observe from Fig. 2, performances at baseline - using all the 2500 features selected from the initial relevance step - are the highest for the vast majority of the cases across the different feature selection technique and metric in consideration. Still, it is interesting to observe that a significant loss in performances, especially for the Mutual information and F-test cases, is observed only after stringent quantiles (> 0.60), and this can be appreciated by observing the number of selected features across different similarity quantiles in Fig. 3. For instance, for the Mutual Information case, with the quantile parameter set to 0.6, we select only 330 features on average (across the 5 folds). That means that we obtain similar performances with 13.2% of the initial baseline feature set.

A more detailed picture of the obtained performances is shown in Table 1, where the obtained performances of three tested model algorithms (Random Forest, eXtreme Gradient Boosting, and Logistic Regression) are reported.

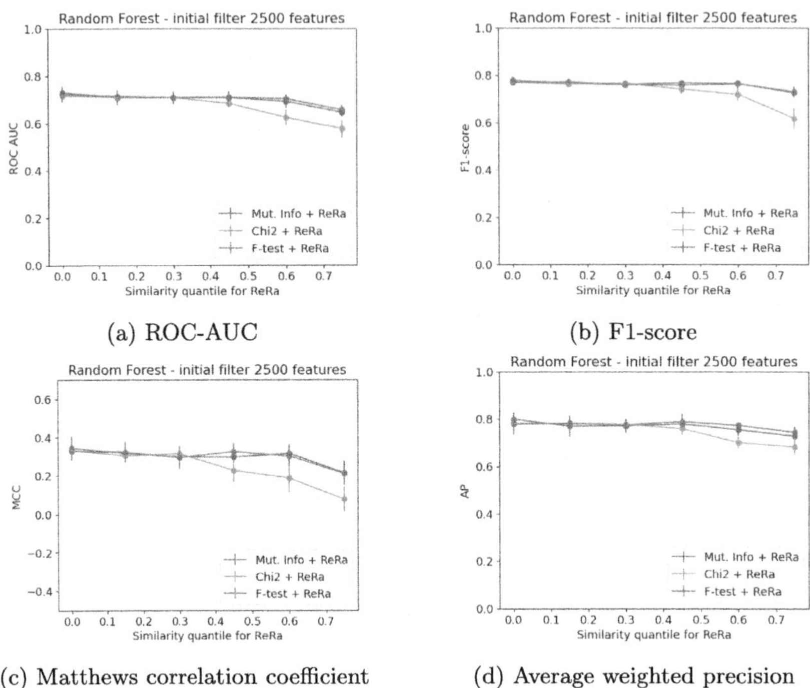

(a) ROC-AUC

(b) F1-score

(c) Matthews correlation coefficient

(d) Average weighted precision

Fig. 2. Performance metrics across different similarity quantiles for the Random Forest models. Each line is the average across the 5-fold evaluations of the outer fold cross-validation.

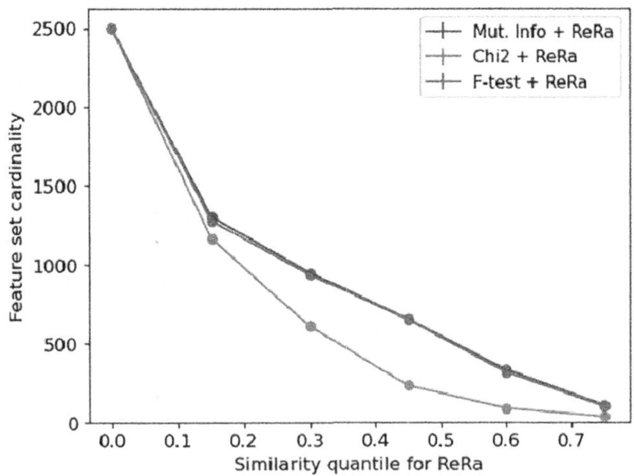

Fig. 3. Average number of selected features downstream ReRa across the five folds of the outer cross-validation.

Table 1. Obtained performances using different quantiles of similarity for the redundancy assessment. Reported metrics are the ROC-AUC, the F1-score, the Matthews correlation coefficient (MCC), and the average weighted precision (AP). Each score is the average across the 5 outer folds of the cross-validation $\pm$ 1 standard deviation.

Model	Filter	N_features	Quantile	ROC-AUC	F1-score	MCC	AP
RF	χ^2-**test**	2,500 $\pm$ 0	0.00	0.716 $\pm$ 0.025	0.770 $\pm$ 0.012	0.333 $\pm$ 0.043	0.782 $\pm$ 0.039
	F-test	2,500 $\pm$ 0	0.00	0.723 $\pm$ 0.035	0.769 $\pm$ 0.010	0.328 $\pm$ 0.042	0.778 $\pm$ 0.044
	MI	2,500 $\pm$ 0	0.00	0.731 $\pm$ 0.027	0.778 $\pm$ 0.017	0.343 $\pm$ 0.062	0.798 $\pm$ 0.029
	χ^2-**test**	1,165 $\pm$ 22	0.15	0.710 $\pm$ 0.021	0.763 $\pm$ 0.006	0.303 $\pm$ 0.036	0.781 $\pm$ 0.017
	F-test	1,278 $\pm$ 12	0.15	0.714 $\pm$ 0.027	0.771 $\pm$ 0.014	0.322 $\pm$ 0.052	0.780 $\pm$ 0.023
	MI	1,305 $\pm$ 10	0.15	0.709 $\pm$ 0.032	0.766 $\pm$ 0.006	0.317 $\pm$ 0.034	0.770 $\pm$ 0.044
	χ^2-**test**	229 $\pm$ 23	0.45	0.684 $\pm$ 0.016	0.741 $\pm$ 0.019	0.227 $\pm$ 0.062	0.758 $\pm$ 0.025
	F-test	648 $\pm$ 30	0.45	0.711 $\pm$ 0.024	0.769 $\pm$ 0.010	0.325 $\pm$ 0.044	0.788 $\pm$ 0.033
	MI	652 $\pm$ 26	0.45	0.709 $\pm$ 0.022	0.759 $\pm$ 0.007	0.297 $\pm$ 0.045	0.778 $\pm$ 0.027
	χ^2-**test**	85 $\pm$ 18	0.60	0.625 $\pm$ 0.032	0.719 $\pm$ 0.028	0.186 $\pm$ 0.072	0.700 $\pm$ 0.025
	F-test	314 $\pm$ 37	0.60	0.704 $\pm$ 0.013	0.764 $\pm$ 0.018	0.302 $\pm$ 0.059	0.772 $\pm$ 0.012
	MI	331 $\pm$ 27	0.60	0.692 $\pm$ 0.028	0.764 $\pm$ 0.010	0.316 $\pm$ 0.047	0.754 $\pm$ 0.024
XGB	χ^2-**test**	2,500 $\pm$ 0	0.00	0.689 $\pm$ 0.023	0.767 $\pm$ 0.019	0.334 $\pm$ 0.046	0.759 $\pm$ 0.019
	F-test	2,500 $\pm$ 0	0.00	0.679 $\pm$ 0.010	0.740 $\pm$ 0.019	0.247 $\pm$ 0.046	0.749 $\pm$ 0.020
	MI	2,500 $\pm$ 0	0.00	0.689 $\pm$ 0.018	0.750 $\pm$ 0.013	0.275 $\pm$ 0.042	0.761 $\pm$ 0.018
	χ^2-**test**	1,165 $\pm$ 22	0.15	0.688 $\pm$ 0.026	0.744 $\pm$ 0.029	0.268 $\pm$ 0.084	0.761 $\pm$ 0.026
	F-test	1,278 $\pm$ 12	0.15	0.698 $\pm$ 0.014	0.763 $\pm$ 0.014	0.322 $\pm$ 0.026	0.775 $\pm$ 0.016
	MI	1,305 $\pm$ 10	0.15	0.686 $\pm$ 0.017	0.755 $\pm$ 0.013	0.287 $\pm$ 0.033	0.749 $\pm$ 0.019
	χ^2-**test**	229 $\pm$ 23	0.45	0.667 $\pm$ 0.035	0.728 $\pm$ 0.029	0.247 $\pm$ 0.063	0.749 $\pm$ 0.038
	F-test	648 $\pm$ 30	0.45	0.674 $\pm$ 0.019	0.734 $\pm$ 0.017	0.237 $\pm$ 0.047	0.749 $\pm$ 0.030
	MI	652 $\pm$ 26	0.45	0.677 $\pm$ 0.026	0.740 $\pm$ 0.023	0.252 $\pm$ 0.072	0.755 $\pm$ 0.022
	χ^2-**test**	85 $\pm$ 18	0.60	0.627 $\pm$ 0.047	0.701 $\pm$ 0.035	0.177 $\pm$ 0.087	0.716 $\pm$ 0.034
	F-test	314 $\pm$ 37	0.60	0.667 $\pm$ 0.019	0.741 $\pm$ 0.015	0.265 $\pm$ 0.048	0.752 $\pm$ 0.010
	MI	331 $\pm$ 27	0.60	0.670 $\pm$ 0.028	0.742 $\pm$ 0.014	0.271 $\pm$ 0.041	0.745 $\pm$ 0.024
LR	χ^2-**test**	2,500 $\pm$ 0	0.00	0.684 $\pm$ 0.025	0.746 $\pm$ 0.015	0.273 $\pm$ 0.051	0.760 $\pm$ 0.029
	F-test	2,500 $\pm$ 0	0.00	0.700 $\pm$ 0.017	0.757 $\pm$ 0.008	0.310 $\pm$ 0.033	0.762 $\pm$ 0.027
	MI	2,500 $\pm$ 0	0.00	0.707 $\pm$ 0.021	0.755 $\pm$ 0.013	0.301 $\pm$ 0.048	0.770 $\pm$ 0.032
	χ^2-**test**	1,165 $\pm$ 22	0.15	0.703 $\pm$ 0.019	0.746 $\pm$ 0.022	0.259 $\pm$ 0.060	0.781 $\pm$ 0.014
	F-test	1,278 $\pm$ 12	0.15	0.690 $\pm$ 0.032	0.754 $\pm$ 0.022	0.298 $\pm$ 0.078	0.746 $\pm$ 0.026
	MI	1,305 $\pm$ 10	0.15	0.685 $\pm$ 0.030	0.758 $\pm$ 0.030	0.300 $\pm$ 0.103	0.749 $\pm$ 0.023
	χ^2-**test**	229 $\pm$ 23	0.45	0.671 $\pm$ 0.034	0.729 $\pm$ 0.024	0.249 $\pm$ 0.051	0.762 $\pm$ 0.025
	F-test	648 $\pm$ 30	0.45	0.677 $\pm$ 0.021	0.739 $\pm$ 0.009	0.264 $\pm$ 0.054	0.762 $\pm$ 0.016
	MI	652 $\pm$ 26	0.45	0.664 $\pm$ 0.027	0.720 $\pm$ 0.003	0.204 $\pm$ 0.028	0.754 $\pm$ 0.027
	χ^2-**test**	85 $\pm$ 18	0.60	0.635 $\pm$ 0.044	0.697 $\pm$ 0.034	0.164 $\pm$ 0.074	0.722 $\pm$ 0.033
	F-test	314 $\pm$ 37	0.60	0.664 $\pm$ 0.034	0.724 $\pm$ 0.023	0.238 $\pm$ 0.058	0.749 $\pm$ 0.028
	MI	331 $\pm$ 27	0.60	0.659 $\pm$ 0.020	0.735 $\pm$ 0.015	0.255 $\pm$ 0.046	0.737 $\pm$ 0.024

4   Conclusion

In conclusion, the redundancy assessment of ReRa does not show a consistent behavior that could lead to a firm conclusion in terms of performance metrics. For the tested models, all the metrics tend to be maintained, while they are subject to huge drops at high similarity quantiles. This behavior is somewhat expected, as information with fewer features should decrease if we have already reduced the signal-to-noise ratio with filter-based techniques. We expect that this behavior might change with different datasets; therefore, it is important to treat this feature selection algorithm as a hyperparameter that needs to be tuned in the context of the bias-variance trade-off. The tunable hyperparameter should include both the similarity quantiles (local and global) and the distance metric that measures the pairwise feature correlations. Notably, fewer features imply a simpler model that could be helpful in data scarcity domains, especially in high-dimensional omics datasets. As future work, more datasets (possibly including multi-omics scenarios) and multiclassification settings could be tested with a wider collection of machine learning models, using the similarity quantiles and distance metrics as tunable hyperparameters. Furthermore, the biological relevance of selected features could be tested, to determine whether the redundancy assessment improves the enrichment of prior knowledge on the selected features.

Acknowledgments. Data used in the preparation of this article were obtained on May, 16th 2024 from the Parkinson's Progression Markers Initiative (PPMI) database (www.ppmi-info.org/access-data-specimens/download-data), RRID:SCR 006431. For up-to-date information on the study, visit www.ppmi-info.org. PPMI - a public-private partnership - is funded by The Michael J. Fox Foundation for Parkinson's Research and funding partners. List of full names of all the PPMI funding partners can be found at https://www.ppmi-info.org/about-ppmi/who-we-are/study-sponsors.

Funding Information. This work has been partially supported by the Italian Ministry of Health (R.R.C.).

Disclosure of Interests. Authors have no conflict of interest to declare.

Availability of data and software code. PPMI dataset is available at: www.ppmi-info.org/access-data-specimens/download-data.

References

1. Anderson, C.A., Pettersson, F.H., Clarke, G.M., Cardon, L.R., Morris, A.P., Zondervan, K.T.: Data quality control in genetic case-control association studies. Nat. Protoc. **5**(9), 1564–1573 (2010). https://doi.org/10.1038/nprot.2010.116, http://dx.doi.org/10.1038/nprot.2010.116
2. Cascianelli, S., Galzerano, A., Masseroli, M.: Supervised Relevance-Redundancy assessments for feature selection in omics-based classification scenarios. J. Biomed. Inform. **144**(104457), 104457 (2023)

3. Chen, T., Guestrin, C.: XGBoost: a scalable tree boosting system. In: Proceedings of the 22nd ACM SIGKDD International Conference on Knowledge Discovery and Data Mining. KDD '16, vol. 11, pp. 785–794. ACM (2016). https://doi.org/10.1145/2939672.2939785, http://dx.doi.org/10.1145/2939672.2939785
4. Chicco, D., Jurman, G.: The Matthews correlation coefficient (MCC) should replace the roc AUC as the standard metric for assessing binary classification. BioData Min. **16**(1), 4 (2023). https://doi.org/10.1186/s13040-023-00322-4
5. Drouin, A., Letarte, G., Raymond, F., Marchand, M., Corbeil, J., Laviolette, F.: Interpretable genotype-to-phenotype classifiers with performance guarantees. Sci. Rep. **9**(1), 4071 (2019)
6. Enoma, D.O., Bishung, J., Abiodun, T., Ogunlana, O., Osamor, V.C.: Machine learning approaches to genome-wide association studies. J. King Saud Univ. Sci. **34**(4), 101847 (2022)
7. Nicholls, H.L., John, C.R., Watson, D.S., Munroe, P.B., Barnes, M.R., Cabrera, C.P.: Reaching the end-game for GWAS: machine learning approaches for the prioritization of complex disease loci. Front. Genet. **11**, 350 (2020)
8. Pedregosa, F., et al.: Scikit-learn: machine learning in Python. J. Mach. Learn. Res. **12**, 2825–2830 (2011)
9. Pudjihartono, N., Fadason, T., Kempa-Liehr, A.W., O'Sullivan, J.M.: A review of feature selection methods for machine learning-based disease risk prediction. Front. Bioinform. **2** (2022). https://doi.org/10.3389/fbinf.2022.927312
10. Romagnoni, A., Jégou, S., Van Steen, K., Wainrib, G., Hugot, J.P.: International Inflammatory Bowel Disease Genetics Consortium (IIBDGC): comparative performances of machine learning methods for classifying Crohn Disease patients using genome-wide genotyping data. Sci. Rep. **9**(1), 10351 (2019)
11. Urbanowicz, R.J., Meeker, M., La Cava, W., Olson, R.S., Moore, J.H.: Relief-based feature selection: introduction and review. J. Biomed. Inform. **85**, 189–203 (2018). https://doi.org/10.1016/j.jbi.2018.07.014

Computational Intelligence in Personalized Medicine

Group Discovery in a Clinical Database of Patients with Psychosis Who Have Undergone Metacognitive Training

Caroline König[1(✉)], Wafaa Guendouz[1], Pedro Copado[1], Cecilio Angulo[2], Àngela Nebot[1], Susana Ochoa[3,4], Maria Lamarca[3,4,5], Rabea Fischer[6], Fabrice Berna[7], Łucasz Gawęda[8], Vanessa Acuña[9], and Alfredo Vellido[1]

[1] Soft Computing Research Group (SOCO) at Intelligent Data Science and Artificial Intelligence (IDEAI-UPC) Research Centre, Universitat Politècnica de Catalunya (UPC Barcelona Tech), Jordi Girona 1−3, Barcelona 08034, Spain
`{ckonig,pcopado,angela,avellido}@cs.upc.edu`
[2] Knowledge Engineerig Research Group (GREC) at Intelligent Data Science and Artificial Intelligence (IDEAI-UPC) Research Centre, Universitat Politècnica de Catalunya (UPC Barcelona Tech), Jordi Girona 1−3, Barcelona 08034, Spain
`cecilio.angulo@upc.edu`
[3] Parc Sanitari Sant Joan de Déu. Group MERITT, Institut de Recerca Sant Joan de Déu, Sant Boi de Llobregat, 08830 Barcelona, Spain
[4] Consorcio de Investigación Biomédica en Red de Salud Mental (CIBERSAM), Instituto de Salud Carlos III, Madrid, Spain
[5] Clinical and Health Psychology Department, School of Psychology, Universitat Autònoma de Barcelona, 08193 Barcelona, Bellaterra, Spain
[6] Department of Psychiatry and Psychotherapy, University Medical Center Hamburg-Eppendorf, 20246 Hamburg, Germany
[7] University of Strasbourg, University Hospital of Strasbourg, Inserm, 67091 Strasbourg, France
[8] Experimental Psychopathology Lab, Institute of Psychology, Polish Academy of Sciences, Warsaw, Poland
[9] Departamento de Psiquiatría, Escuela de Medicina, Facultad de Medicina, Universidad de Valparaíso, Valparaíso, Chile

Abstract. In this study, a clinical dataset of patients with psychosis who have undergone Metacognitive Training (MCT) is analyzed using unsupervised machine learning (ML) methods to discover patient groups that are medically significant. The Positive and Negative Syndrome Scale (PANSS) is used in this study to assess the patient's health state before and after MCT treatment as well as to measure treatment effectiveness. Four well-separated patient groups are found based on an exploratory data analysis with K-means clustering and outlier detection using Local Outlier Factor (LOF) algorithm. The statistical analysis of the improvement in positive, negative, and general symptoms for these patient groups provides meaningful insights about the patient's profiles with regard to the effectiveness of the therapy. The results of the study confirm that the retrospective patient database under study comprises several differently-behaving groups of patients with a substantial variation in their data distribution, which needs to be addressed in the development of the predictive model for MCT effectiveness in future research.

Keywords: Personalized medicine · Metacognitive training · PANSS · Clustering · Data exploration

1   Introduction

Schizophrenia is a serious mental disorder that poses a substantial public health challenge in Europe affecting about 2% of the population [11]. The disease causes considerable disability and at an economic level high costs related to work absence and early retirement [15]. Schizophrenia shows especially symptoms such as delusions and hallucinations but also other symptoms influencing the affective domain. Treatment with antipsychotic medication is common, but often does not improve functional outcomes. Psychological treatment is now recommended as a supplementary therapy with additional effects. In particular, Metacognitive Training (MCT) [12,13], a psychologically oriented intervention has resulted efficient for the treatment of psychosis. MCT is based on Cognitive Behavioral Therapy (CBT), with a psychoeducational approach to reduce cognitive biases [14]. Furthermore, machine-learning (ML) and predictive medicine approaches are often used in clinical psychology [3,4], such as for diagnosis, prognosis [9], or personalization of treatments.

In this study and within the scope of the European research project entitled *'Towards a Personalized Medicine Approach to Psychological Treatment of Psychosis'* (PERMEPSY), retrospective data from approximately 700 patients who received MCT are analyzed to discover patterns regarding MCT effectiveness. The purpose of the PERMEPSY project is the construction of a data-centered support predictive model for personalized MCT delivery, using machine learning algorithms. For such a challenging goal, the first steps of the analytical pipeline involve evaluating data quality and understanding the data [7]. In this work, we present some preliminary results concerning the discovery of patient profiles related to their symptoms based on their evaluation using one of the most commonly used psychotic symptom questionnaires, namely the *Positive and Negative Syndrome Scale* (PANSS) [8]. The PANSS scale identifies the presence and severity of psychotic symptoms.

The methodological approach for data analysis is unsupervised clustering and outlier analysis, with the goal of discovering patients' groupings according to PANSS scores. Clustering is a common approach to discover patient profiles and disease subtypes in medical data [5]. Insights about the effectiveness of MCT are derived by statistically assessing the patients' evolution of in terms of positive, negative and general symptoms with the PANSS subscales. For this, the percentage variation between PANSS scores in the pre-evaluation and post-evaluation is taken into account, and detailed results are provided for the entire dataset and at the level of the patients' groups discovered through cluster analysis. The information derived from this study represents a first contribution towards the data-centric study for personalizing MCT delivery, providing the first insights about the typologies of patients who participated in the retrospective studies, and about the effectiveness of the treatment based on its assessment using the PANSS. These results are not only relevant for data understanding and exploration, but are meant to be the basis for the ML model developed in future work for the prediction of MCT effectiveness.

The outline of this article comprises a description of the data analyzed in Sect. 2, the explanation of the analytical methods in Sect. 3, and a presentation and discussion of the results of the experiments in Sect. 4.

2   Materials

The PERMEPSY dataset under study contains information about 698 patients diagnosed with different disorders in the psychotic spectrum having received MCT treatment. MCT comprises 10 modules that focus on improving problematic aspects such as monocausal attributions and cognitive biases such as jumping to conclusions, improving social cognition, false memories, stigma, and self-esteem, among others. The dataset also includes information about the patients' sociodemographic characteristics and the evaluation of their clinical characteristics with several psychological indicators, such as PANSS scores, Beck's Cognitive Insight Scale (BCIS), the Psychotic Symptom Rating Scale (PSYRATS), or the Rosenberg Self-Esteem Scale (RSES), among others.

The indicators of interest to this study are the PANSS scores, which comprise the PANSS positive scale (PANSS_P) for positive symptoms, the negative scale for negative symptoms (PANSS_N), and the general psychopathology scale (PANSS_G). The PANSS scale identifies the presence and severity of psychotic symptoms in terms of positive, negative, and general symptoms. Positive symptoms represent an excess or distortion of normal functions such as delusions, hallucinations, for example, while negative symptoms refer to a decrease or loss of basic functions such as emotional withdrawal, lack of normal thoughts or actions, and loss of spontaneity. General symptoms evaluate other types of less specific symptoms such as anxiety, feelings of guilt, or lack of attention. PANSS positive and negative scales comprise 7 items each, yielding scores in the range of 7 to 49. The general PANSS scale comprises 16 items yielding scores in the range of 16 to 112. Lower scores represent a lower level of symptoms. The evolution of the patient's psychotic symptoms is assessed by a pre-evaluation (before the start of treatment) and a post-evaluation (after having finished MCT).

Changes in the patient's symptoms are assessed deriving the ratio of change in the three PANSS scores for the respective scale according to Eqs. (1a)–1c. The ratio of change of the scores takes into account the absolute range of $N_P = N_N = 42$ scores for the positive and negative scale and the absolute range of

$N_G = 96$ scores for the general symptoms scale.

$$\Delta PANSS\_P = \frac{Pre(PANSS\_P) - Post(PANSS\_P)}{42} \tag{1a}$$

$$\Delta PANSS\_N = \frac{Pre(PANSS\_N) - Post(PANSS\_N)}{42} \tag{1b}$$

$$\Delta PANSS\_G = \frac{Pre(PANSS\_G) - Post(PANSS\_G)}{96} \tag{1c}$$

3   Methods

The data under study are analyzed with unsupervised ML methods, namely PANSS scores in pre-evaluation with K-Means clustering [6] for data segmentation and Local Outlier Factor (LOF) [2] for the detection of abnormal observations. Figure 1 describes the steps of data analysis done in this study.

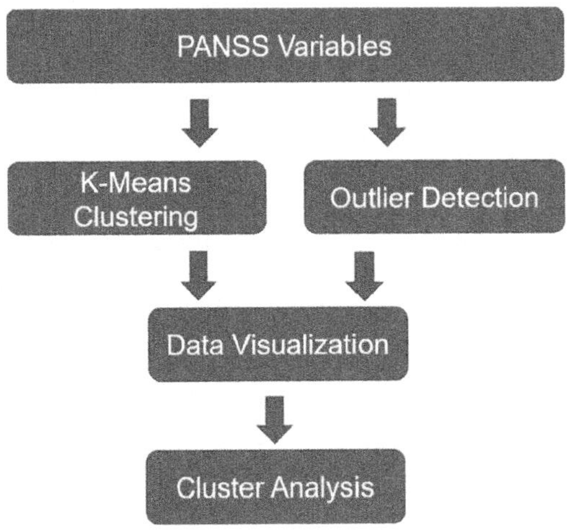

Fig. 1. Experimental setup of data analysis.

LOF is an unsupervised anomaly detection method that measures the local density of a given observation relative to its neighbors. The fundamental idea behind LOF is to single out observations that have a notable lower density than their neighbors. The basic steps of LOF are as follows:

1. **k-distance calculation**: For each point p in the dataset, find its k-nearest neighbors. The k-distance of p is the distance between p and its k-th nearest neighbor.
2. **Reachability distance**: The reachability distance of a point p from a point o is defined as:

$$\text{reachability\_dist}_k(p, o) = \max(k\text{-distance}(o), d(p, o))$$

where $d(p, o)$ is the distance between points p and o.

3. **Local Reachability Density (LRD)**: The local reachability density of point p is the inverse of the average reachability distance of p from its k-nearest neighbors. It is defined as:

$$\mathrm{lrd}_k(p) = \left(\frac{\sum_{o \in N_k(p)} \mathrm{reachability\_dist}_k(p, o)}{|N_k(p)|} \right)^{-1}$$

   where $N_k(p)$ denotes the set of k-nearest neighbors of p.
4. **LOF Score**: The LOF score of a point p is the average ratio of the local reachability density of p and those of its k-nearest neighbors. It is computed as:

$$\mathrm{LOF}_k(p) = \frac{\sum_{o \in N_k(p)} \frac{\mathrm{lrd}_k(o)}{\mathrm{lrd}_k(p)}}{|N_k(p)|}$$

A LOF score approximately equal to 1 indicates that the point has a local density similar to its neighbors, while a score significantly greater than 1 indicates that the point is an outlier.

K-means is a popular partitional algorithm for multivariate data analysis which aims to group observations into the same clusters based on their similarity. The algorithm is known for its simplicity and efficiency. It is an iterative algorithm that partitions a dataset into K clusters, each represented by a centroid, which equals to the approximated cluster centers. The goal is to minimize the within-cluster sum of squares (WCSS), also known as inertia [10], during the iterative partitioning process. The detailed step-by-step process is as follows:

1. Initialize K centroids randomly.
2. For each data point x_i, calculate the Euclidean distance to each centroid μ_k.
3. Assign each data point to the nearest centroid:

$$c_i = \arg \min_k \|x_i - \mu_k\|^2$$

   where c_i is the cluster assignment for data point x_i
4. Recalculate the centroids by taking the mean of all data points assigned to each cluster:

$$\mu_k = \frac{1}{|C_k|} \sum_{x_i \in C_k} x_i$$

   where C_k is the set of points assigned to cluster k and μ_k is the updated centroid.
5. Repeat the assignment and centroid update steps (step 2–4) until convergence, i.e. when the change of centroids is negligible.

K-Means aims to minimize the within-cluster variance, making it effective for spherical and evenly sized clusters. However, K-Means requires fixing the number of clusters in advance to accordingly select the cluster centers (centroids) at the initialization of the algorithm. The selection of the optimal number of clusters is based on inertia analysis for a wide range of numbers of clusters [16]. Inertia can be defined as the sum of squared distances of observations to their closest cluster center (centroid c) and quantifies how compact the clusters are by measuring the total variance within each cluster (Eq. (2)). The mean intracluster distance [1] is another measure for cohesion on individual clusters assessing the Euclidean distance between the observations of a cluster C (Eq. (3)).

$$\text{Inertia} = \sum_{i=1}^{n} \|x_i - c\|^2 \tag{2}$$

$$\text{Mean intracluster distance cluster C} = \frac{1}{|C| \cdot (|C| - 1)} \sum_{i=1}^{|C|} \sum_{j=i+1}^{|C|} d(x_i, x_j) \tag{3}$$

4 Results

This section describes the results of the analysis of PANSS scores with K-Means clustering using the dataset comprising PANSS_P, PANSS_N and PANSS_G scores. Figure 2 shows the evaluation of the Inertia parameter for different numbers of initial centroids and reveals that the within-cluster sum-of-square reaches the optimal trade-off between the number of clusters and inertia for $K = 4$.

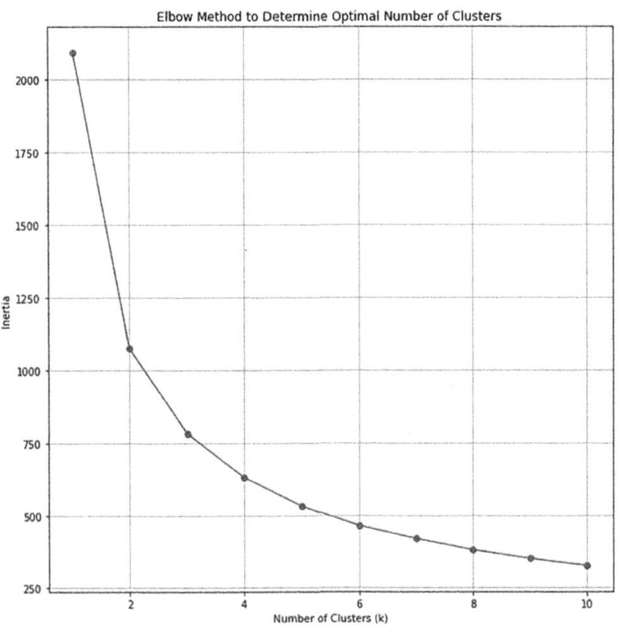

Fig. 2. Inertia plot for or K-Means Inertia according to the number of clusters.

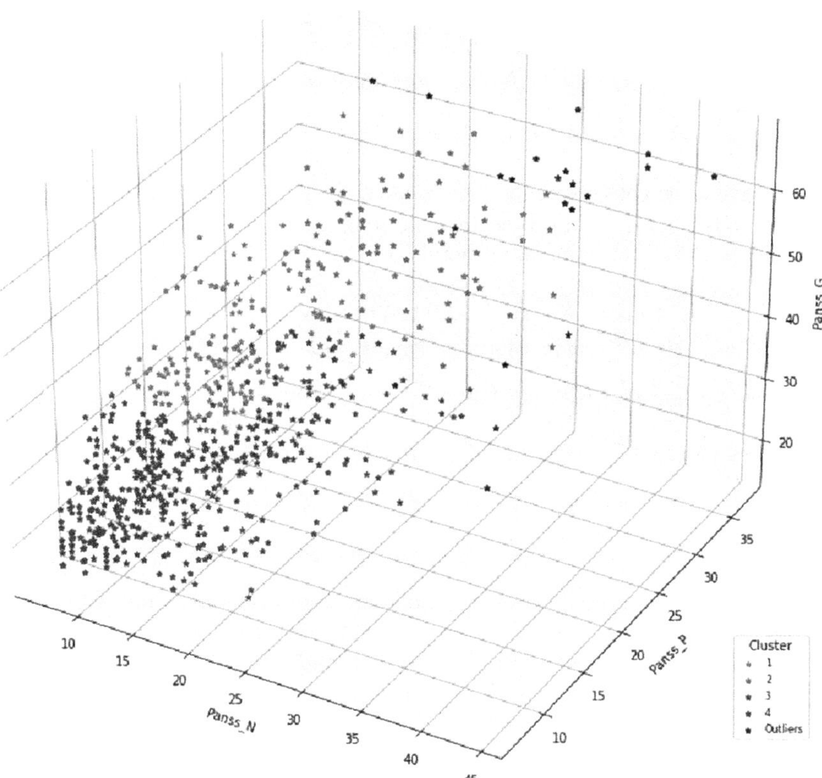

Fig. 3. 3-D visualization of clusters. PANSS scores clustering results for $K = 4$ and outlier analysis.

Consequently, the K-Means clustering algorithms were run for a number of four clusters, deriving the clustering results shown in Fig. 3.

Table 1 shows a description of the clusters by detailing the number of patients in the cluster (Num), the intra-cluster distance (Cohesion), and the characteristics of the centroids of the cluster in terms of their PANSS scores. The clustering of the PANSS scores reveals the existence of four differentiated groups as depicted in Fig. 3. According to the intracluster distance, cluster 4 is the most compact group, while cluster 2 shows the lowest cohesion between patients. The largest group includes 289 observations (cluster 4). It consists of patients with low PANSS scores in all the three subscales. On average, the positive, negative, and general PANSS score values are approximately 10.5, 10.8, and 24.6, respectively. Cluster 1 and cluster 3 describe large groups of patients with slightly higher PANSS scores in each scale. The main difference between patients in cluster 1 and in cluster 3 is between positive and negative scores in each case. For cluster 1, the negative PANSS score is low, while it has high values for positive and global scores. Cluster 3 comprises 147 patients with a low positive score and a high negative and general score. Finally, cluster 2 comprises a smaller group of patients with high scores on all three PANSS subscales. On average, the positive, negative, and general PANSS scores were rounded to 28.8, 25.0, and 54.3, respectively.

Table 1. Mean PANSS scores. For each category (cluster or outlier), the number of observations (*Num*), the mean intracluster distance (cohesion), and the mean value for PANSS positive, negative, and general scores are shown.

Cluster	Num	Cohesion	Positive	Negative	General
1	177	0.6	20.0	12.7	30.2
2	64	1.2	28.8	25.0	54.3
3	147	0.7	15.5	21.7	34.8
4	289	0.4	10.5	10.8	24.6
Outliers	21	-	28.8	33.0	59.3

Overall for the whole dataset, patients considered in this study show a moderate level of symptoms according to their PANSS scores. Outlier analysis with Local Outlier Factor (LOF) [2] nevertheless highlights the existence of 21 atypical patients that, under scrutiny, reveal unusually high PANSS scores, especially for the negative and general PANSS subscales.

Concerning the evolution of the patient's psychotic symptoms, the variation of the respective PANSS scores was compared from pre-evaluation to post-evaluation, i.e., before starting MCT and after finishing it, assessing the percentage change in the respective scores. Table 2 provides an overview of this variation for the three PANSS subscales. Interestingly, these statistics show an enhancement concerning the positive symptoms assessed with the positive PANSS scores. Specifically, the positive scores have changed on average by 6. 1%, and the interquartile range (IQR) ranges from 0.0% to 11.9%. The negative PANSS scores show a mainly positive variation with an average improvement of 2.7%, comprising 50% of observations in an IQR of -1.7% to 7.1%. Therefore, for negative PANSS scores the variation has been less pronounced. For general PANSS scores, the variation represents an improvement comprising 50% of observations in a positive IQR (0. 0% to 9. 4%) and a mean of 4. 2%.

Figure 4 describes the variation of the PANSS scores using a boxplot representation that focuses on the different patient groups identified with the cluster analysis. Table 3 shows the respective mean values of variation for each cluster.

Cluster 4 represents the group of patients with the lowest PANSS scores in pre-evaluation and low percentage variation on average (ΔPANSS_P= 1.5%, ΔPANSS_N=-0.1%, ΔPANSS_G= 1.8%). These low variations might be attributed to the already low initial PANSS scores for this group in pre-evaluation. Focusing on the IQR (Q3-Q1) we observe mostly cases of improve-

Table 2. Variation of PANSS scores. For each PANSS score, the statistical summary of variation between pre-evaluation and post-evaluation is shown. The ratio of change is expressed in percentage values.

Score	Mean	Std	Min	25%	50%	75%	Max
Δ**PANSS_P**	6.1	9.7	-21.4	0.0	4.7	11.9	54.7
Δ**PANSS_N**	2.7	8.9	-40.4	-1.7	2.4	7.1	45.2
Δ**PANSS_G**	4.2	7.4	-17.7	0.0	4.2	9.4	29.1

Table 3. Mean variation of PANSS scores by cluster. The ratio of change is expressed in percentage values.

Cluster	1	2	3	4
Δ**PANSS_P**	9.0	15.9	6.3	1.5
Δ**PANSS_N**	-1.9	7.9	6.6	-0.1
Δ**PANSS_G**	6.8	9.3	3.0	1.8

ment for positive and general symptoms, but also cases of worsening in symptoms (Q1 ΔPANSS_P=-1.4%, Q1 ΔPANSS_N=-2.4%, Q1 ΔPANSS_G=-1.0%).

Cluster 1, which has intermediate levels of PANSS scores in the pre-evaluation, shows a high variation in the PANSS positive and general scores representing mainly improvements (Q1 ΔPANSS_P= 2.4%, Q1 ΔPANSS_G=1.0%), while negative PANSS scores have a significantly lower percentage variation. Although there is on average an improvement of 1.8% in negative symptoms, there are also to some lower extent cases of worsening of symptoms (Q1 ΔPANSS_N=-2.4% and Q3 ΔPANSS_N=7.1%).

Cluster 3 describes patients with higher negative and general PANSS scores and lower positive PANSS scores in pre-evaluation. For this group, positive and negative symptoms experience a high improvement of approximately 6.0%, while general symptoms only improve on average 3.0%. According to IQR, most patients have an improvement in PANSS scores.

Cluster 2 comprises patients with the highest PANSS scores in the pre-evaluation for the three subscales and shows the highest percentage variation, reaching on average an improvement of 15.9%, 7.9%, and 9.3% for positive, negative, and general symptoms. According to the IQR, most patients experience an improvement in PANSS scores.

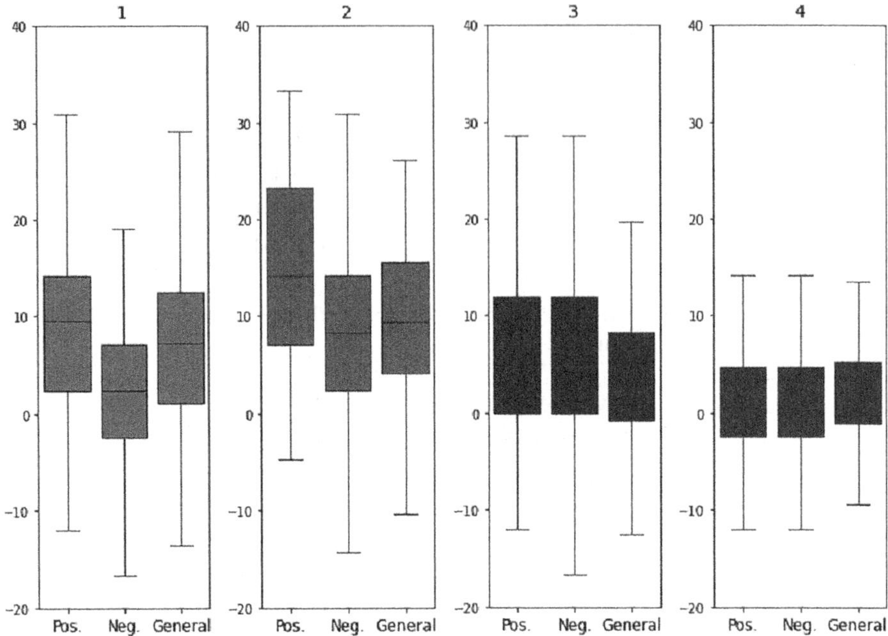

Fig. 4. Boxplot visualization of PANSS scores variation by cluster. For each cluster, the percentage variation in all three PANSS scores is shown.

5   Conclusion

The analysis of PANSS scores from the PERMEPSY dataset with unsupervised clustering methods has revealed insights about the patients' profiles concerning their clinical symptoms. Currently, clinicians know that not all patients respond well to treatment, but they do not know which patients benefit from treatment and who do not. So far there is tentative evidence that those with severe delusional symptoms may be less responsive and need tailored interventions. In consequence, the results of the cluster analysis of this study are useful for clinicians to gain insights about different patient profiles and possibly different disease subtypes. The cluster analysis has shown that patients considered in the study correspond to four well-separated groups with different severity regarding positive, negative, and general symptoms. The change in PANSS scores after psychological treatment shows that patients participating in this study improved on average by 6.1%, 2.7%, and 4.2% in PANSS positive, negative, and general scores. Nevertheless, the statistical analysis also shows a greater deviation of the data with regard to the respective mean values. The patient groups revealed by cluster analysis show differences with regard to the variations of symptoms. This information helps understand the nuances in the variation of symptoms, which seems to depend on the initial state of symptoms and probably other not yet explored characteristics of patients. Concerning data quality assessment and data understanding, results confirm that the retrospective patient database under study comprises several differently-behaving groups of patients with a substantial variation in their data distribution, which needs to be addressed in the development of the predictive model for MCT effectiveness. In future work, we plan to expand the cluster analysis to account for sociodemographic variables and other relevant psychological indicators, such as those from the PSYRATS and BCIS scales.

Acknowledgments. The PERMEPSY team would like to thank and acknowledge the contribution to the retrospective database by researchers from the Spanish Metacognition Group and the following researchers: Ryan Balzan, Javier-David Lopez-Morinigo, Masayoshi Kobayashi, Ritta Kuokkanen, Linda Swanson, Mustafa Yildiz, Zita Fekete, Miguel Simón-Expósito, Christina Andreou, Jérôme Favrod, Lara Guedes de Pinho, Rita Roncone, Suzanne So, Hiroki Tanoue, Brooke Viertel, Erna Erawati, Ryotaro Ishikawa.

Funding Information. This work is part of the European ERAPERMED 2022-292 research project entitled 'Towards a Personalized Medicine Approach to Psychological Treatment of Psychosis' (PERMEPSY) funded by European Union - NextGenerationEU supported as grant AC22/0053 by the Instituto de Salud Carlos III (ISCIII) in Spain.

The PERMEPSY project was supported under the frame of ERA PerMed by: Instituto de Salud Carlos III (ISCIII), Spain; German Federal Ministry of Education and Research (BMBF), Germany; Agence Nationale de la Recherche (ANR), France; National Centre for Research and Development (NCBR), Poland; Agencia Nacional de Investigación y Desarrollo (ANID), Chile.

Conflict of interests. The authors declare not to have any potential conflicts of interests.

References

1. Binu Jose, A., Das, P.: A multi-objective approach for inter-cluster and intra-cluster distance analysis for numeric data. In: Kumar, R., Ahn, C.W., Sharma, T.K., Verma, O.P., Agarwal, A. (eds.) Soft Computing: Theories and Applications, pp. 319–332. Springer Nature Singapore, Singapore (2022)
2. Breunig, M.M., Kriegel, H.P., Ng, R.T., Sander, J.: LOF: identifying density-based local outliers. In: Proceedings of the 2000 ACM SIGMOD International Conference on Management of Data, pp. 93–104 (2000)
3. Dwyer, D.B., Falkai, P., Koutsouleris, N.: Machine learning approaches for clinical psychology and psychiatry. Annu. Rev. Clin. Psychol. **14**(1), 91–118 (2018)
4. Fusar-Poli, P., Schultze-Lutter, F.: Predicting the onset of psychosis in patients at clinical high risk: practical guide to probabilistic prognostic reasoning. BMJ Ment. Health **19**(1), 10–15 (2016)
5. Grant, R.W., et al.: Use of latent class analysis and k-means clustering to identify complex patient profiles. JAMA Netw. Open **3**(12), e2029068–e2029068 (2020)
6. Hartigan, J.A., Wong, M.A.: Algorithm as 136: a k-means clustering algorithm. J. R. Stat. Soc. Ser. C (applied statistics) **28**(1), 100–108 (1979)

7. Jain, A., et al.: Overview and importance of data quality for machine learning tasks. In: Proceedings of the 26th ACM SIGKDD International Conference on Knowledge Discovery & Data Mining, pp. 3561–3562 (2020)

8. Kay, S.R., Fiszbein, A., Opler, L.A.: The positive and negative syndrome scale (PANSS) for schizophrenia. Schizophrenia Bull. **13**(2), 261–276 (1987). https://doi.org/10.1093/schbul/13.2.261

9. Lee, T.Y., et al.: Prediction of psychosis: model development and internal validation of a personalized risk calculator. Psychol. Med. **52**(13), 2632–2640 (2022). https://doi.org/10.1017/S0033291720004675

10. MacQueen, J.: Some methods for classification and analysis of multivariate observations. In: Proceedings of the Fifth Berkeley Symposium on Mathematical Statistics and Probability, Volume 1: Statistics, pp. 281–297. University of California Press (1967)

11. Mancuso, A., et al.: Economic burden of schizophrenia: the European situation. A scientific literature review. Eur. J. Public Health **24** (2014). https://doi.org/10.1093/eurpub/cku166.129

12. Moritz, S., Menon, M., Balzan, R., Woodward, T.S.: Metacognitive training for psychosis (MCT): past, present, and future. Eur. Arch. Psychiatry Clin. Neurosci. **273**(4), 811–817 (2023)

13. Moritz, S., Woodward, T.S., Balzan, R.: Is metacognitive training for psychosis effective? Expert Rev. Neurother. **16**, 105–107 (2016). https://doi.org/10.1586/14737175.2016.1135737

14. Ochoa, S., et al.: Randomized control trial to assess the efficacy of metacognitive training compared with a psycho-educational group in people with a recent-onset psychosis. Psychol. Med. **47**(9), 1573–1584 (2017)

15. Pares-Badell, O., Barbaglia, G., Jerinic, P., Gustavsson, A., Salvador-Carulla, L., Alonso, J.: Cost of disorders of the brain in Spain. PLoS ONE **9**(8), e105471 (2014). https://doi.org/10.1371/journal.pone.0105471

16. Rykov, A., De Amorim, R.C., Makarenkov, V., Mirkin, B.: Inertia-based indices to determine the number of clusters in k-means: an experimental evaluation. IEEE Access (2024)

Hierarchical Clustering with an Ensemble of Principle Component Trees for Interpretable Patient Stratification

Bastian Pfeifer[1]([✉]) [iD], Marcus D. Bloice[1] [iD], Christel Sirocchi[2] [iD], and Markus Loecher[3] [iD]

[1] Institute for Medical Informatics, Statistics and Documentation,
Medical University of Graz, Graz, Austria
`bastian.pfeifer@medunigraz.at`
[2] Department of Pure and Applied Sciences, University of Urbino, Urbino, Italy
[3] Berlin School of Economy and Law, Berlin, Germany

Abstract. Patient stratification plays a crucial role in personalized medicine by identifying distinct subgroups of patients based on their molecular and/or clinical characteristics. However, many unsupervised machine learning-based stratification techniques fail to identify the essential biomarker traits associated with each patient group. In this paper, we present a novel approach for interpretable patient stratification using hierarchical ensemble clustering. Our method leverages feature sampling in conjunction with principal component analysis (PCA) to capture the most significant patterns and contributing biomarkers. We demonstrate the effectiveness of our approach using machine learning benchmark datasets and real-world data from The Cancer Genome Atlas (TCGA), showcasing the improved interpretability of the detected patient clusters.

Keywords: hierarchical clustering · principal component analysis · feature importance · disease subtyping

1 Motivation

Clustering for patient stratification or disease subtyping is a valuable strategy that employs computational methods to categorize patients based on shared characteristics, such as clinical features, genetic profiles, or patterns of biomarker expression [7]. By harnessing sophisticated clustering algorithms, researchers can detect distinct subgroups within a disease, leading to a more refined comprehension of its underlying molecular fingerprints [10,13]. This subtyping approach holds significant promise for personalized medicine, as it facilitates tailored treatments and targeted interventions, ultimately improving patient outcomes.

While numerous clustering methods have been developed and applied for patient stratification, less attention has been devoted to making such unsupervised models interpretable, for instance, by identifying the features most relevant for assigning patients to each cluster. Algorithms to verify feature importance

L. Cerulo et al. (Eds.): CIBB 2024, LNBI 15276, pp. 281–291, 2025.
https://doi.org/10.1007/978-3-031-89704-7_22

contribute to interpretable machine learning by providing insight into how models make decisions [4]. By identifying which features have the greatest impact on model predictions, these algorithms help clarify the relationships between input data and outcomes. This transparency allows practitioners to validate the model's behavior and detect biases. Moreover, feature importance analysis enhances trust and accountability in machine learning, making it easier to understand and justify the model's predictions, especially in critical applications like healthcare [11]. However, feature importance has predominantly been explored in supervised machine learning, while it has received less attention in unsupervised settings such as clustering. This proposed approach contributes to addressing this gap.

In disease subtyping, such investigations could reveal that certain risk groups exhibit a higher relevance of particular genes or biomarkers, underscoring their importance in understanding the risk associated with those subgroups. Identifying and prioritizing these distinctive genetic factors within clusters is crucial for targeted interventions and personalized risk assessments. The methodologies presented in this study aim to rectify this gap in previous research by elucidating the features driving patient stratification for enhanced clustering interpretability.

The remainder of the paper is organized as follows: Sect. 2 introduces our new approach to interpretable disease subtyping. Section 2.1 presents the algorithmic details of our ensemble clustering technique, Sect. 2.2 describes the capacity of our approach to handle multi-omics data. A possible extension for longitudinal data is discussed in Sect. 2.3, while Sect. 2.4 outlines our novel strategy to compute cluster-specific feature importance values. In Sect. 3 we describe our evaluation strategy and the results are discussed in Sect. 4. We conclude with an outlook for future work in Sect. 5.

2 New Approach

The proposed methodology is called TAPIO (**T**ree **A**ffinities with **P**rincipal Components) and consists of two novel developments. First, instead of performing clustering on the entire dataset, including the full set of features, we suggest that a certain amount of features are sampled without replacement and to compute the first principal component based on this data subset. Hierarchical clustering is then executed on this embedded single feature to construct a dendrogram, which ultimately is split at multiple levels to derive multiple cluster partitions. These partitions are utilized to generate an affinity matrix. This process is iterated multiple times, resulting in an ensemble of affinity matrices, which are aggregated into a single matrix on which clustering is performed. Second, based on the above described scheme, we evaluate cluster-specific feature relevance by computing the contribution of each feature to the principal components. For each ensemble member, the feature contribution to the principal component is scaled by the pair-wise affinity of the elements of the considered cluster. Subsequently, these contributions are aggregated across ensemble members.

2.1 Hierarchical Ensemble Learning

Let's denote the input tabular data as X, where X is an $n \times p$ matrix, with n samples (patients) and p features. Each row of X represents a patient, and each column represents a feature. Our developed algorithm consists of five steps:

1. Feature Sampling: Randomly select a subset of k features from the dataset X to construct a new $n \times k$ matrix termed X'.
2. Principal Component Analysis (PCA): Reduce the dimensionality of the matrix X' via PCA. This entails performing eigendecomposition of the covariance matrix $\Sigma' = \frac{1}{n} X'^T X'$ to obtain its eigenvectors v_i and corresponding eigenvalues λ_i (with i ranging from 1 to k), selecting the q eigenvectors corresponding to the q largest eigenvalues to form the matrix V_q, and projecting the matrix X' onto the subspace spanned by the first q principal components: $Z = X'V_q$. Here, Z is the $n \times q$ matrix of the first q principal components of X'. In this context, $q = 1$ as only the principal component is considered.
3. Hierarchical Clustering: Apply hierarchical clustering on the reduced data Z. Let D denote the resulting dendrogram obtained from hierarchical clustering.
4. Dendrogram Cutting: Cut the dendrogram D at the first d levels, resulting in a set of d partitions $\mathcal{C}^e$, with e ranging from 1 to d, where each partition $\mathcal{C}^e = \{C_1, C_2, .., C_p\}$ assigns each of the n data samples to a cluster C. Then, $C^e(i) = C_f$ indicates that, at level e, patient i is assigned to the cluster $C_f \in \mathcal{C}^e$.
5. Affinity Matrix Construction: Construct the $n \times n$ affinity matrix A based on the resulting partitions, such that each entry A_{ij} represents the strength of the affinity between patient i and patient j across the partitions. This is computed as the proportion of samples that fall in the same cluster:

$$A_{ij} = \frac{1}{d} \sum_{e=1}^{d} \begin{cases} 1 & \text{if } C^e(i) = C^e(j) \\ 0 & \text{otherwise} \end{cases} \tag{1}$$

The above described procedure is executed M times for the construction of an ensemble consisting of M affinity matrices $\mathcal{A} = \{A^{(1)}, A^{(2)}, \ldots, A^{(M)}\}$. We aggregate these affinity matrices into a single matrix A using element-wise summation:

$$A_{agg} = \sum_{m=1}^{M} \mathcal{A}^{(m)}, \tag{2}$$

The aggregated affinity matrix A_{agg} is clustered to obtain the final partitioning of the data.

2.2 Integrating Multi-omics Data

The availability of multiple omics datasets for the same group of patients offers the potential to improve patient stratification, enabling the identification of more precise molecular subtypes. Here, we utilize Multiple Factor Analysis (MFA) when multi-omics data is available. MFA is a direct and straightforward extension of PCA to accommodate multiple data types. The key algorithmic extension lies in a pre-processing step of matrix X'. Let's assume X' consists of a list of termed input matrices $\mathcal{X}' = \{X_1', X_2', \ldots, X_L'\}$, where L denotes the total number of available omic data tables. Balance among these data types is achieved by normalizing each of the data types in $\mathcal{X}'$, before stacking them and passing them on to PCA. The normalization is accomplished by dividing the principal component decomposition of each termed matrix by its first eigenvalue:

$$\mathcal{X}' = [X_1'/\lambda_1^{(1)}, X_2'/\lambda_1^{(2)}, \ldots, X_L'/\lambda_1^{(L)}], \tag{3}$$

where $\lambda_1^{(i)}$ is the first eigenvalue of the principal component decomposition of X_i', and $i \in \{1, 2, \ldots, L\}$. Following this normalization and column-wise concatenation step, we apply PCA to $\mathcal{X}'$ and proceed as described in Sect. 2.1.

2.3 Longitudinal Data

In longitudinal studies, variables are measured repeatedly over time, resulting in observations that are both clustered and correlated. This type of data structure, known as repeated measures data sets, is prevalent not only in medical research [15] but also in other fields, such as marketing. For instance, marketing research often involves weekly household purchase data and responses to panel survey questionnaires [1]. Clustering longitudinal data presents unique challenges due to the necessity of grouping based on the similarity of individual trajectories while accommodating sparse and irregular times of observation [19]. These complexities underscore the importance of developing effective methods for clustering longitudinal data, which can provide valuable insights across various domains.

In [3], the *KmL3D* software package was introduced, employing a specialized version of k-means to cluster joint trajectories. The distance metric between individuals is defined as the Minkowski distance between their respective trajectory matrices. [19] introduces a hierarchical agglomerative clustering approach that uses a dissimilarity metric to quantify the merging cost between two groups of curves, represented by B-splines for repeatedly measured data.

None of the outlined methods can be applied to our setting in a straightforward manner. Instead, we propose to add one more sampling step to the algorithm described in Sect. 2.1 to handle longitudinal data:

At each iteration we sample not only k features but also one row (time of measure) from each patient. For sufficiently large numbers of M iterations we would actually "cover" all the multiple time trajectories from each user. In this way, the clustering algorithm would take the entire variability of each user into account and not just the first moment. It is evident that averaging replicated data for each subject, or even the first principal component, results in a loss of information as intra-subject variation is effectively smoothed out. The most similar approach to our method that we found in the literature is the "subject-level bootstrapping strategy" [5,6], where bootstrap resampling is conducted at the subject level, and all observations from the chosen subjects are treated as in-bag samples.

Implementing this proposed strategy would be beyond the scope of this paper.

2.4   Feature Importance

Feature importance is derived from the feature contribution to the principal components used in each ensemble member. In PCA, the loading matrix $\mathbf{V}$ contains the loading vectors as its columns. In the case of p original features and q principal components, then $\mathbf{V}$ is a $p \times q$ matrix where each column represents the loading vector for a principal component. Within the constructed ensemble, $\mathbf{V}^{(m)}$ is the loading matrix obtained for the mth member of the ensemble. In this context, $q = 1$ as only the principal component is considered, so $\mathbf{V}$ is a $p \times 1$ vector. Then, for a specific cluster cl the feature importance $\mathcal{I}_{cl}$ is the $p \times 1$ vector, with the ith element of the vector corresponding to the cluster-specific importance of the ith feature of the dataset, computed as follows:

$$\mathcal{I}_{cl} = \sum_{m=1}^{M} \sum_{i,j \in cl} A_{i,j}^{(m)} \cdot \mathbf{V}^{(m)}. \tag{4}$$

The values in $\mathcal{I}_{cl}$ are normalized by the maximum value across all cluster-specific importance values $\mathbf{max}_{i,j}\{\mathcal{I}_{cl_1}, ..., \mathcal{I}_{cl_K}\}$, where K refers to the number of clusters.

3   Evaluation Strategy

Our evaluation approach was two-fold. First, we have validated the proposed methods based on six machine learning datasets (see Table 1), in terms of accuracy and cluster interpretability. Since we have the ground-truth in these cases we could asses the performance of TAPIO utilizing the Adjusted Rand Index (ARI). We performed the clustering with Ward's linkage [9,18] 100 times for each set-up to study the robustness while varying the number of trees within the ensemble. The number of features randomly sampled for building a dendrogram was set to $\lfloor \sqrt{p} \rfloor$. The number of levels to cut the dendrograms was set to 30. We compared the results to a baseline, where Ward's linkage clustering is applied to the full dataset. Furthermore, we have computed feature importance values for each variable within each cluster according to Eq. 4.

Second, we retrieved real-world gene expression data from The Cancer Genome Atlas (TCGA) and compared TAPIO to three state-of-the-art disease subtyping methods, SNF [17], NEMO [14], and HCFUSED [12]. We conducted a comparison with the competing approaches based on gene expression data. The data table consisted of 208 patients and gene expression values from 20,531 genes. We randomly subsampled 100 patients from the data pool and assessed the performance using the c-index [16] and the log-rank test statistic. This evaluation procedure was repeated 30 times. Additionally, we executed the aforementioned

Table 1. Benchmark data sets

Datasets	#Samples	#Features	#Classes	#Samples per Class
Glass	214	9	4	70,76,17,51
Wine	178	13	3	59,71,48
Iris	150	4	3	50,50,50
Ionosphere	351	34	2	225,126
WDBC	569	30	2	212,357
Zoo	219	16	5	41,20,13,8,10

evaluation procedure using multi-omics data. Along with gene expression levels, we incorporated DNA methylation data, consisting of 5,000 markers, and microRNA data, which included 1,046 biological features.

Finally, we performed patient stratification on the full dataset and analyzed feature importance for the detected risk clusters.

4   Results and Discussion

The evaluation based on the machine learning benchmark datasets underscores the potential of feature sampling in conjunction with Principal Component Analysis (see Fig. 1). In five out of six cases TAPIO outperforms the baseline approach. We could show that increasing the number of trees within the ensemble has an overall stabilizing effect on the clustering. For each detected cluster we display the cluster-specific feature importance values, allowing researchers to leverage the key features for further in-depth analysis. As can be seen from Fig. 1, the feature importance values can differ substantially among the clusters.

The application of our approach on real-world kidney cancer data revealed competitive results when compared to the state-of-the-art in detecting disease subtypes. Evaluation limited to gene expression revealed log-rank p-values below the significance level for all studied methods (see Fig. 2). When integrating multi-omics data, however, TAPIO outperforms the competing methods (see Fig. 3), and almost all sub-sampling iterations result in significant log-rank p-values. An increase in performance can also be observed for HCFUSED. Interestingly, NEMO performs worse when multi-omics data is incorporated, compared to using gene expression only (Fig. 2).

We also assessed TAPIO on liver cancer, and the results presented in Fig. 4 demonstrate that TAPIO performs competitively with HCFUSED, while outperforming NEMO and SNF. In general, both TAPIO and HCFUSED appear to perform substantially better on kidney cancer compared to liver cancer. However, TAPIO provides techniques for calculating cluster-specific feature importance, whereas HCFUSED focuses solely on reporting the contribution of a specific molecular modality to the clustering solution. Future work could combine these two approaches to further improve cluster interpretability.

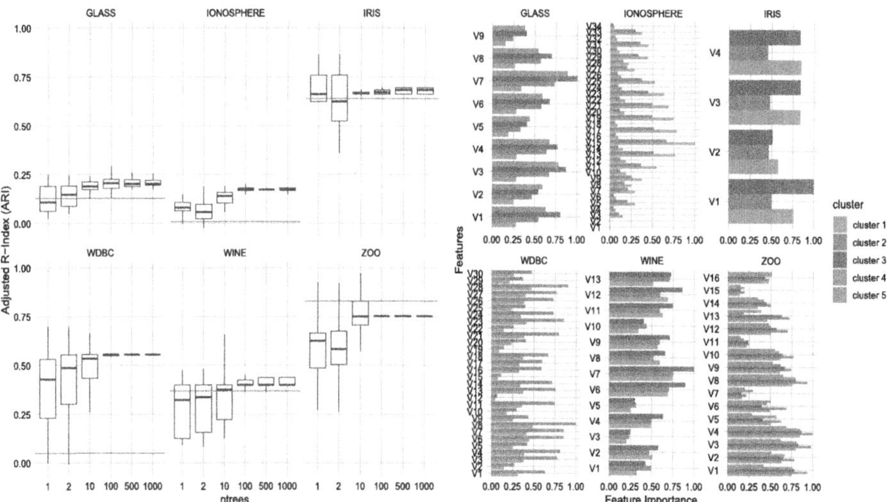

Fig. 1. Varying the number of trees. *Left panel*: The ARI performance of our method is displayed for six machine learning benchmark datasets while subsequently increasing the number of trees. The horizontal red lines refer to the performance of the baseline approach. *Right panel*: The cluster-specific feature importance values for each of the six analyzed machine learning benchmark datasets are displayed. (Color figure online)

Finally, we performed patient stratification on the full kidney cancer dataset using TAPIO. We detected four clusters of size [cl1 = 64, cl2 = 55, cl3 = 73, cl4 = 16] patients with a significant log-rank p-value of $p = 0.0053$. The corresponding survival curves are displayed in Fig. 5. Clusters 3 and 4 are high risk clusters with substantially lower survival rates than patient clusters 1 and 2.

Furthermore, we calculated the 15 relevant genes based on the mean across all cluster-specific feature importance values. The gene "RBL2" emerged as the most relevant, consistently ranking as the top priority across all patient clusters. "RBL2" is structurally similar to the gene "RB1", which is a known tumor suppressor gene. However, it is not yet fully investigated whether it functions in a similar way, thereby providing an opportunity for further investigation [2]. The results shown in Fig. 5 contain some interesting results. For high-risk cluster 4, the 15 most relevant genes exhibit significantly elevated importance values relative to the other clusters, suggesting a distinct and crucial function within this particular cluster. In addition, we noted slight variations in the distribution of importance levels among the selected genes within clusters 1 to 3. However, to derive final and medically relevant conclusions further in-depth analyses of the detected genes is required.

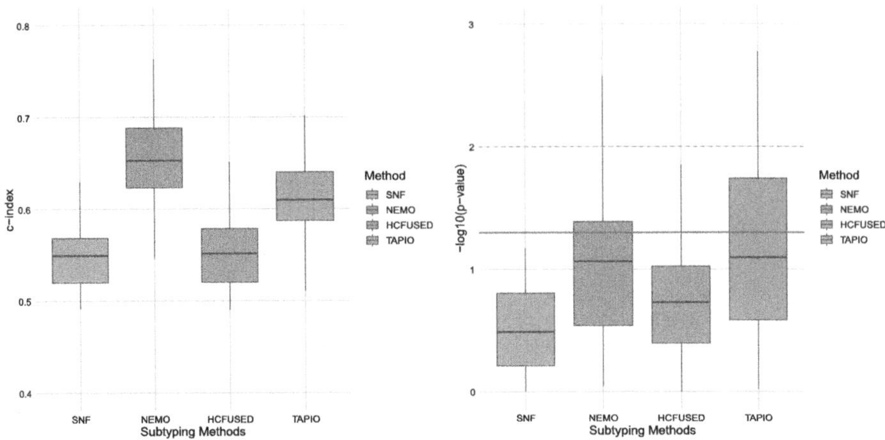

Fig. 2. Application on TCGA kidney cancer gene expression data. Our clustering approach TAPIO in comparison with three state-of-the-art disease subtyping methods, NEMO, HCFUSED, and SNF. *Left panel*: Results based on the c-index are displayed. *Right panel*: The log-rank p-values on a logarithmic scale. The red line refers to the 0.05 significance level; the dashed red line refers to the 0.01 significance level. Values above these lines are significant. (Color figure online)

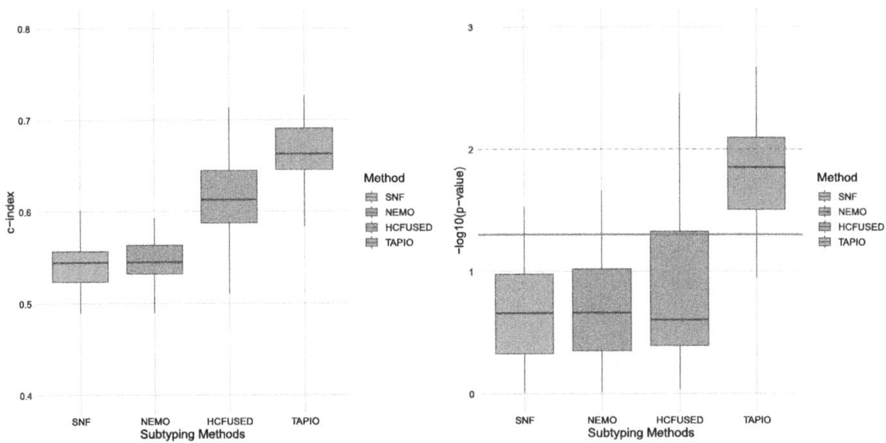

Fig. 3. Application on TCGA kidney cancer multi-omics data. Our clustering approach TAPIO in comparison with three state-of-the-art disease subtyping methods, NEMO, HCFUSED, and SNF. *Left panel*: Results based on the c-index are displayed. *Right panel*: The log-rank p-values on a logarithmic scale. The red line refers to the 0.05 significance level; the dashed red line refers to the 0.01 significance level. Values above these lines are significant. (Color figure online)

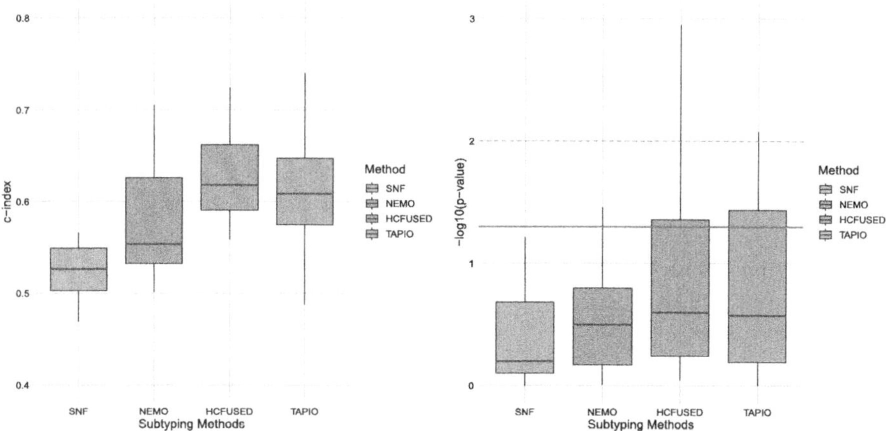

Fig. 4. Application on TCGA liver cancer multi-omics data. Our clustering approach TAPIO in comparison with three state-of-the-art disease subtyping methods, NEMO, HCFUSED, and SNF. *Left panel*: Results based on the c-index are displayed. *Right panel*: The log-rank p-values on a logarithmic scale. The red line refers to the 0.05 significance level; the dashed red line refers to the 0.01 significance level. Values above these lines are significant. (Color figure online)

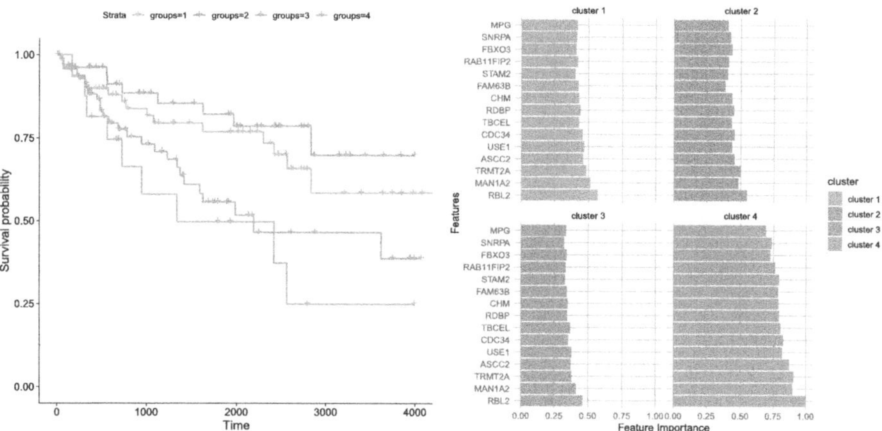

Fig. 5. Application on TCGA kidney cancer gene expression data. *Left panel*: Survival curves of the four detected patient groups based on the full kidney cancer dataset. The log-rank p-value is $p = 0.0053$. *Right panel*: Feature importance values of each detected patient group from a selected set of the 15 most relevant genes. (Color figure online)

5   Summary and Future Work

We have presented a novel clustering algorithm for enhancing the interpretability in patient stratification. We could show superior performance of our method over a baseline approach based on six machine learning benchmark datasets. The application on real-world cancer data revealed competitive results in patient stratification when compared to the state-of-the-art. Overall, the interpretability of the obtained clusters could be substantially improved.

In future work, the use of UMAP (Uniform Manifold Approximation and Projection) [8] could be explored as an alternative to PCA for dimensionality reduction. While PCA effectively captures linear relationships and global variance, UMAP has the advantage of preserving both local and global structures in non-linear data. Investigating UMAP's potential may enhance the understanding of the detected disease subtypes and improve downstream tasks that rely on the reduced feature space.

Moreover, we plan to incorporate the fusion algorithm implemented in HCFUSED. As a consequence, TAPIO will not only provide insights into the importance of the analyzed features, but also will report on the contribution of each omic type to the final clustering. Furthermore, we intend to analyze the dynamic nature of these clusters over time, considering longitudinal data to capture temporal changes in patient subtypes and disease progression. This longitudinal perspective will offer valuable insights into the evolution of disease subtypes and the effectiveness of interventions tailored to specific cluster profiles, thus advancing our understanding of personalized medicine's potential impact on long-term patient outcomes.

Acknowledgments. This work has been funded by the European Union's Horizon Europe research and innovation programme as part of the PoCCardio project, grant no. 101095432.

Conflict of Interests. The authors declare no conflicts of interests.

Availability of Data and Software Code. Our software code and the analyzed datasets are available at https://github.com/pievos101/TAPIO.

References

1. Bradlow, E.T.: Exploring repeated measures data sets for key features using principal components analysis. Int. J. Res. Mark. **19**(2), 167–179 (2002)
2. Flores, M., Goodrich, D.W.: Retinoblastoma protein paralogs and tumor suppression. Front. Genet. **13**, 818719 (2022)
3. Genolini, C., et al.: Kml3d: a non-parametric algorithm for clustering joint trajectories. Comput. Methods Programs Biomed. **109**(1), 104–111 (2013)
4. Guyon, I., Elisseeff, A.: An introduction to variable and feature selection. J. Mach. Learn. Res. **3**(3), 1157–1182 (2003)
5. Hu, J., Szymczak, S.: A review on longitudinal data analysis with random forest. Brief. Bioinform. **24**(2), bbad002 (2023)
6. Karpievitch, Y.V., Hill, E.G., Leclerc, A.P., Dabney, A.R., Almeida, J.S.: An introspective comparison of random forest-based classifiers for the analysis of cluster-correlated data by way of rf++. PLoS ONE **4**(9), e7087 (2009)

7. Leng, D., et al.: A benchmark study of deep learning-based multi-omics data fusion methods for cancer. Genome Biol. **23**(1), 1–32 (2022)
8. McInnes, L., Healy, J., Melville, J.: Umap: Uniform manifold approximation and projection for dimension reduction. arXiv preprint arXiv:1802.03426 (2018)
9. Murtagh, F., Legendre, P.: Ward's hierarchical agglomerative clustering method: Which algorithms implement Ward's criterion? J. Classif. **31**(3), 274–295 (2014)
10. Pfeifer, B., Bloice, M.D., Schimek, M.G.: Parea: multi-view ensemble clustering for cancer subtype discovery. J. Biomed. Inform. **143**, 104406 (2023)
11. Pfeifer, B., Holzinger, A., Schimek, M.G.: Robust random forest-based all-relevant feature ranks for trustworthy AI. Stud. Health Technol. Inform. **294**, 137–138 (2022)
12. Pfeifer, B., Schimek, M.G.: A hierarchical clustering and data fusion approach for disease subtype discovery. J. Biomed. Inform. **113**, 103636 (2021)
13. Rappoport, N., Shamir, R.: Multi-omic and multi-view clustering algorithms: review and cancer benchmark. Nucleic Acids Res. **46**(20), 10546–10562 (2018)
14. Rappoport, N., Shamir, R.: NEMO: cancer subtyping by integration of partial multi-omic data. Bioinformatics **35**(18), 3348–3356 (2019)
15. Schüssler-Florenza Rose, S.M., et al.: A longitudinal big data approach for precision health. Nat. Med. **25**(5), 792–804 (2019)
16. Uno, H., Cai, T., Pencina, M.J., D'Agostino, R.B., Wei, L.J.: On the c-statistics for evaluating overall adequacy of risk prediction procedures with censored survival data. Stat. Med. **30**(10), 1105–1117 (2011)
17. Wang, B., Mezlini, A.M., Demir, F., Fiume, M., Tu, Z., Brudno, M., Haibe-Kains, B., Goldenberg, A.: Similarity network fusion for aggregating data types on a genomic scale. Nat. Methods **11**(3), 333–337 (2014)
18. Ward, J.H., Jr.: Hierarchical to optimize an objective function. J. Am. Stat. Assoc. **58**(301), 236–244 (1963)
19. Zhou, J., Zhang, Y., Tu, W.: clustermld: an efficient hierarchical clustering method for multivariate longitudinal data. J. Comput. Graph. Stat. **32**(3), 1131–1144 (2023)

Computational Structural Bioinformatics

ESMCrystal : Enhancing Protein Crystallization Prediction Through Protein Embeddings

Jayanth Kumar[1(✉)] [iD], Kavya Jayakumar[2] [iD], and Jayaprakash Sundararaj[1] [iD]

[1] Computer Science, Indian Institute of Technology Bombay, Mumbai, India
`jayanth8506@iitbombay.org`, `jayaprakash12@cse.iitb.ac.in`
[2] Tata Institute of Fundamental Research, Hyderabad, India
`jkavya@tifrh.res.in`

Abstract. Protein crystallization is a critical yet challenging step in determining protein structures, crucial for advancing our understanding of biological mechanisms. This study introduces ESMCrystal, a novel approach leveraging protein embeddings derived from the advanced Meta ESMFold2 architecture to predict protein crystallization. By integrating transfer learning techniques, ESMCrystal models demonstrate enhanced predictive performance across various datasets, highlighting the potential of deep learning in structural biology. This research not only improves the predictability of protein crystallization but also sets the stage for broader applications of machine learning in understanding complex biological systems. The standalone source code and models, along with the inference server are available at https://huggingface.co/jaykmr/ESMCrystal_t6_8M_v1 and https://huggingface.co/jaykmr/ESMCrystal_t12_35M_v2.

Keywords: Protein crystallization · Structural biology · Protein Embeddings · ESMCrystal Models · ESMFold2 · Machine learning · Protein Structure Analysis

1 Introduction

The quest to determine protein structures has long been pivotal in the field of biochemistry and molecular biology, fundamentally inspiring advancements in medicine, genetics, and various biotechnologies. Traditionally, the determination of these structures is primarily facilitated through X-ray crystallography. However, this method poses significant challenges, notably its high failure rate and the substantial costs associated with producing diffraction-quality crystals. While the success rates of obtaining such crystals range only between 2–10% [16], the costs attributed to unsuccessful attempts account for more than 70% [15] of total expenses. This inefficiency underscores an urgent need for innovative approaches to predict protein crystallization success more accurately and efficiently.

© The Author(s), under exclusive license to Springer Nature Switzerland AG 2025
L. Cerulo et al. (Eds.): CIBB 2024, LNBI 15276, pp. 295–306, 2025.
https://doi.org/10.1007/978-3-031-89704-7_23

Historically, several machine learning and statistical methods have been developed to predict protein crystallization propensity from sequence data [4,18]. These models including CrystalP2 [9], PPCpred [1], PredPPCrys [19], XtalPred-RF [6], TargetCrys [5], Crysalis [18], Crysf [17], fDETECT [12], DeepCrystal [3], and BCrystal [2], which span from early statistical analyses to more complex machine learning frameworks, have primarily hinged on feature extraction techniques to identify critical biological and physiochemical features from protein sequences. However, these methods often require intricate feature selection and substantial computational resources, which can be a bottleneck for scalability and practical application in both academic and industrial settings.

Our paper presents ESMCrystal, a model that utilizes deep learning and protein embeddings [14] to predict protein crystallization, aiming to significantly reduce the computational and financial costs associated with traditional methods. Our study assesses these models across various standard datasets including DeepCrystal Test, Balanced Test, SP Test, and TR Test, employing comprehensive evaluation metrics such as confusion matrices, accuracy, precision, recall, F1-score, PR-AUC and ROC-AUC. The results obtained not only demonstrate high accuracy levels but also highlight the robustness of protein embeddings in enhancing the predictability of successful protein crystallization under diverse experimental conditions.

2 Data and Methods

2.1 Datasets

We perform our experiments on publicly available datasets, specifically from BCrystal [2] and DeepCrystal [3] dataset. Furthermore, the training dataset was completely based on the DeepCrystal dataset, resulting in 26,821 training samples, of which 4,420 are crystallizable and 22,401 are non-crystallizable. Out of the 26,821 samples, 5% (1342 samples) were used for the validation dataset, picked randomly before training. Additionally, there are 1,898 test samples, with 898 being crystallizable and 1,000 non-crystallizable.

We treat the crystallization prediction problem as a binary classification problem, distinguishing diffraction-quality crystals from the rest. The positive class or label 1 denotes crystallizable protein sequence, while the negative class or label 0 denotes non-crystallizable protein sequence.

Additionally, we use two independent test sets for further validation and comparison. These datasets, SP final (SwissProt) and TR final (Tremble) from [17], contain sequences with <25% sequence similarity with the training set. We also, use a fairly balanced test set from [3], consisting of 891 crystallizable and 896 non-crystallizable proteins for evaluation.

In the SP final dataset, there are 148 proteins belonging to the positive class, while the remaining 89 sequences are non-crystallizable. In the TR final dataset, there are 374 crystallizable proteins and 638 proteins belonging to the negative class.

Table 1. Dataset Summary. Details of Datasets Used in the Study.

Dataset	Total Sequences	Crystallizable	Non-crystallizable
DeepCrystal Train	26281	4420	22401
DeepCrystal Test	1898	898	1000
Balanced Test	1787	891	896
SP Test	237	148	89
TR Test	1012	374	638

These datasets, as per the Table 1, contain a diverse range of protein sequences annotated based on their crystallization outcome. This diversity is crucial for testing the robustness and generalizability of our prediction models across different experimental scenarios and protein types.

The imbalance in the training data, with a higher proportion of non-crystallizable proteins, compared to the more balanced test datasets introduces challenges in interpreting the model's performance. Models trained on imbalanced data are prone to bias towards the majority class, potentially affecting their ability to generalize effectively on balanced test datasets. This may result in inflated precision and recall for the majority class during training but could lead to a drop in performance for the minority class when evaluated on more balanced data. To mitigate this effect, measures such as evaluating performance on a range of metrics, and employing balanced test datasets were implemented. While these strategies help address the issue, the inherent imbalance in the training data remains a factor that could influence the robustness of the results and highlights the need for careful interpretation of the model's predictive capabilities.

2.2   Label Semantics

For this study, proteins are classified into two categories:

- Label 0: Non-crystallizable (Negative)
- Label 1: Crystallizable (Positive)

This binary classification forms the foundational basis for supervising our training and testing processes within this work.

2.3   Training Procedure

The models were trained using a transfer learning approach as shown in the Fig. 1, starting with pre-trained embeddings from the ESMFold2 architecture [10], which were then adapted to the specific task of protein crystallization prediction. This method leverages the general protein structure understanding of ESMFold2, fine-tuning it further to predict crystallization outcomes from protein sequences.

2.4   Models

We trained two state-of-the-art machine learning models based on the Meta ESMFold architecture, fine-tuning it specifically for protein crystallization prediction:

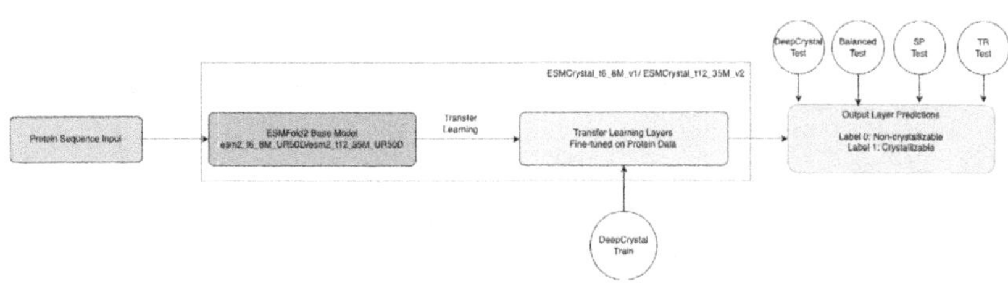

Fig. 1. Transfer Learning Architecture for Protein Crystallization Prediction. This diagram illustrates the transfer learning process utilizing the ESMFold2 base model to predict protein crystallization. Starting with an input protein sequence, the architecture employs the foundational layers of Meta's ESMFold2, which are fine-tuned with specific protein crystallization datasets. The predictive model outputs two labels: 'Non-crystallizable' (Label 0) and 'Crystallizable' (Label 1), demonstrating an application of advanced machine learning techniques in structural biology.

1. ESMCrystal_t6_8M_v1: This smaller model is fine-tuned on the esm2_t6_8M_UR50D data, with 6 hidden layers and 8 million parameters. The model size is approximately 31.4MB. It is likely faster to train and requires less computational resources but may capture less complex patterns compared to the larger model.
2. ESMCrystal_t12_35M_v2: This larger and more complex model is fine-tuned on the esm2_t12_35M_UR50D, featuring 12 hidden layers and 35 million parameters, with a total size of approximately 136MB. Since it has more hidden layers, it is more capable of capturing complex patterns in the protein sequence.

The dimensionality of the hidden states in smaller model is set to 320 while in the larger model is set to 480, which determines the size of the vector representations learned by the model. Within each transformer block, the dimensionality of the intermediate layer in the feedforward network is set to 1280 in the smaller model and 1920 in the larger model, which processes the output of the attention mechanism. These configurations are used as default from the ESMFold2 architecture.

Both models have 20 attention heads, which means they are equally capable of parallelizing the process of attending to different parts of the input sequence. This feature is particularly useful in tasks like protein sequence analysis where different segments of the sequence might have various functional implications.

We disabled dropout for attention probabilities and hidden layers in both the models as model robustness is not much of an issue due to less variation in protein sequence data. We use "rotary" position embeddings, ideal for maintaining relative positional information, crucial in protein sequences. We also, enable token dropout, which helps improve generalization by randomly dropping tokens during training.

2.5 Evaluation Metrics

Confusion matrices were generated for a thorough assessment of true positives (TP), false positives (FP), false negatives (FN), and true negatives (TN) across all test datasets. To assess the performance of our models, several metrics were generated and compared based on confusion metrics, such as precision (PRE), recall (REC), F-score (F), accuracy (ACC), Matthews Correlation Coefficient (MCC), negative predictive value (NPV), receiver operating characteristic - area under curve (ROC), and Precision-Recall Area Under Curve (PR-AUC). A detailed definition of these sets and the importance of each of these evaluation metrics are provided in [3, 8, 13].

Based on the testing across the specified datasets, each model's performance was documented, focusing on their predictive accuracy.

ROC-AUC (as in Figs. 2 and 3) and PR-AUC (as in Figs. 4 and 5) curves were plotted for both models across all tests, providing visual insights into model performance and the trade-offs between sensitivity and specificity that each model exhibits.

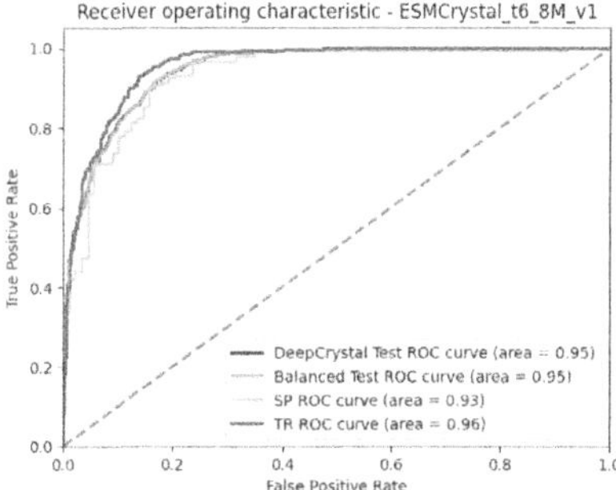

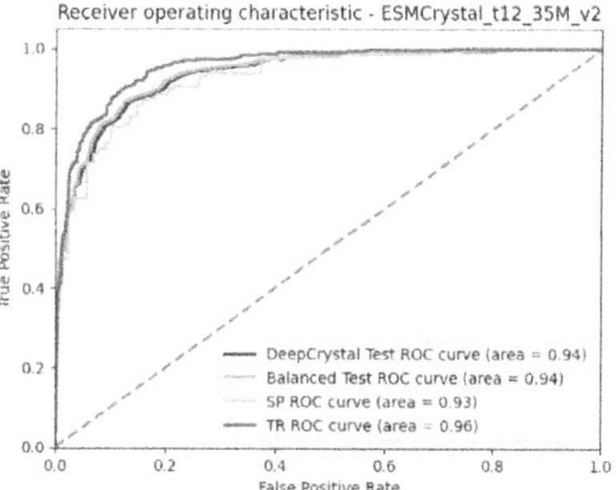

Fig. 2. ROC curve for ESMCrystal_t6_8M_v1. This diagram plots the Receiver Operating Characteristic curve for ESMCrystal_t6_8M_v1 on the different test datasets.

Fig. 3. ROC curve for ESMCrystal_t12_35M_v2. This diagram plots the Receiver Operating Characteristic curve for ESMCrystal_t12_35M_v2 on the different test datasets.

3   Results

The experimental results obtained from employing the ESMCrystal_t6_8M_v1 and ESMCrystal_t12_35M_v2 models on various test datasets have provided significant insights into the efficacy of using advanced machine learning techniques for predicting protein crystallization. Below, we detail the performance outcomes and findings for each of these models across our testing environments in the Tables 2 and 3.

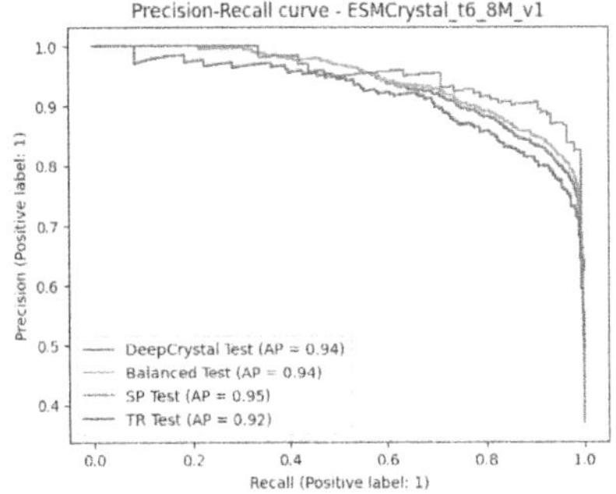

Fig. 4. PR curve for ESMCrystal_t6_8M_v1. This diagram plots the Precision-Recall curve for ESMCrystal_t6_8M_v1 on the different test datasets.

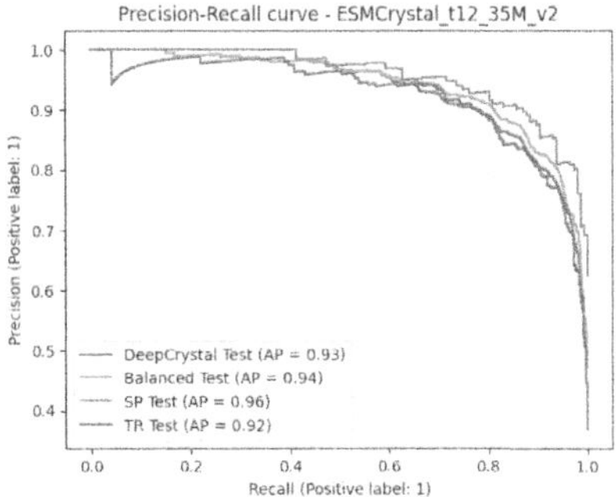

Fig. 5. PR curve for ESMCrystal_t12_35M_v2. This diagram plots the Precision-Recall curve for ESMCrystal_t12_35M_v2 on the different test datasets.

Table 2. Performance of ESMCrystal_t6_8M_v1. Performance of the smaller model on different test datasets

Dataset	TP	FP	FN	TN	Precision	Recall	F1	Acc	ROC	MCC	NPV
DeepCrystal	532	362	34	966	0.5951	0.9399	0.7288	0.7909	0.9467	0.6119	0.9660
Balanced	531	360	31	865	0.5960	0.9448	0.7309	0.7812	0.9396	0.6045	0.9654
SP	80	68	4	85	0.5405	0.9524	0.6897	0.6962	0.9328	0.5017	0.9551
TR	207	167	16	622	0.5535	0.9283	0.6935	0.8192	0.9562	0.6153	0.9749

The ESMCrystal_t12_35M_v2 model illustrated a noticeable improvement over the 6-layer model, particularly in handling the dataset complexities more effectively, which is attributed to its more hidden layers and more parameters, i.e. deeper learning structure.

Table 3. Performance of ESMCrystal_t12_35M_v2. Performance of the larger model on different test datasets

Dataset	TP	FP	FN	TN	Precision	Recall	F1	Acc	ROC	MCC	NPV
DeepCrystal	579	319	30	970	0.6448	0.9507	0.7684	0.8161	0.9403	0.6575	0.9700
Balanced	573	318	30	866	0.6431	0.9502	0.7671	0.8053	0.9396	0.6446	0.9665
SP Test	97	51	5	84	0.6554	0.9510	0.7760	0.7637	0.9293	0.5861	0.9438
TR Test	225	149	14	624	0.6016	0.9414	0.7341	0.8389	0.9562	0.6588	0.9781

The comparison between ESMCrystal_t6_8M_v1 and ESMCrystal_t12 _35M_v2 models clearly shows the advantage of a deeper neural network architecture, as evidenced by the consistently higher performance metrics across all test datasets by the latter.

Both models have shown high predictive reliability, emphasized by high ROC-AUC scores across datasets. The ESMCrystal_t12_35M_v2 model generally reported better precision and recall rates, especially evident in the TR Test dataset.

Despite the varied nature of the test datasets, the robustness of ESMFold-based models under different experimental conditions was commendable. It suggests that transfer learning from a pre-trained model on a broad dataset allows effective generalization across divergent protein sequences.

3.1   Comparative Analysis

The performance of ESMCrystal models was compared against several state-of-the-art sequence-based protein crystallization predictors, including BCrystal, DeepCrystal, Crysf, Crysalis I and II, fDETECT, TargetCrys, XtalPred-RF, PPCPred and CrystalP2. The comparison with Crysf was conducted only on the SP and TR datasets as Crysf required Uniprot ids as input, which were available only for these two datasets.

These results in the Tables 4, 5 and 6 underline the significant potential of using deep learning techniques, specifically those harnessing robust pre-trained models like Meta's ESMFold, in advancing protein crystallization predictions.

In contrast to traditional feature-based methods that rely on intricate feature selection and significant computational resources, ESMCrystal leverages protein embeddings from the ESMFold2 architecture to eliminate the need for manual feature engineering. This approach not only reduces computational overhead but also enhances scalability, making it particularly suitable for large-scale and diverse datasets.

In conclusion, while ESMCrystal demonstrates strong predictive capabilities and highlights the potential of leveraging protein sequence embeddings, it does not outperform all existing methods, notably BCrystal. This difference in performance can be attributed to the fundamental approaches of the two models: BCrystal employs a comprehensive set of features, including sequence-derived,

Table 4. Comparision of ESMCrystal models on balanced dataset. ESMCrystal performs comparably with other protein crystallization predictors on the balanced test data

Models	Accuracy	MCC	AUC	F-score	Recall	Precision	NPV
PPCpred	0.672	0.359	0.754	0.616	0.528	0.740	0.635
fDETECT	0.646	0.355	0.778	0.504	0.360	0.840	0.593
Crysalis I	0.777	0.556	0.865	0.767	0.738	0.799	0.758
Crysalis II	0.804	0.610	0.888	0.796	0.767	0.828	0.784
XtalPred-RF	0.650	0.301	0.710	0.654	0.663	0.645	0.655
TargetCrys	0.627	0.255	0.637	0.637	0.656	0.619	0.593
CrystalP2	0.585	0.177	0.608	0.627	0.700	0.568	0.613
DeepCrystal	0.828	0.658	0.903	0.822	0.795	0.851	0.809
BCrystal	0.954	0.908	0.981	0.954	0.970	0.939	0.969
ESMCrystal_t6_8M_v1	0.7812	0.6045	0.9396	0.7309	0.9448	0.5960	0.9654
ESMCrystal_t12_35M_v2	0.8053	0.6446	0.9396	0.7671	0.9502	0.6431	0.9665

Table 5. Comparision of ESMCrystal models on SP dataset. ESMCrystal performs comparably with other protein crystallization predictors on the SP test data

Models	Accuracy	MCC	AUC	F-score	Recall	Precision	NPV
Crysf	0.700	0.426	0.811	0.727	0.641	0.840	0.572
PPCpred	0.666	0.403	0.784	0.675	0.554	0.863	0.535
fDETECT	0.616	0.381	0.837	0.580	0.425	0.913	0.494
Crysalis I	0.725	0.448	0.835	0.763	0.709	0.826	0.609
Crysalis II	0.751	0.505	0.851	0.783	0.722	0.856	0.633
XtalPred-RF	0.451	0.149	0.449	0.548	0.553	0.564	0.288
TargetCrys	0.611	0.223	0.641	0.659	0.601	0.729	0.486
CrystalP2	0.658	0.257	0.696	0.734	0.756	0.713	0.550
DeepCrystal	0.759	0.530	0.874	0.788	0.716	0.876	0.637
BCrystal	0.894	0.774	0.951	0.919	0.966	0.877	0.932
ESMCrystal_t6_8M_v1	0.6962	0.5017	0.9328	0.6897	0.9524	0.5405	0.9551
ESMCrystal_t12_35M_v2	0.7637	0.5861	0.9293	0.7760	0.9510	0.6554	0.9438

structural, and physicochemical properties, to enhance its prediction accuracy. In contrast, ESMCrystal relies solely on protein sequence embeddings generated by the ESMFold2 architecture, which eliminates the need for manual feature engineering and streamlines the prediction process. This distinction underscores a trade-off between model complexity and performance, with ESMCrystal offering a more scalable and efficient approach while BCrystal benefits from additional features that may provide it with an edge in certain scenarios. This compari-

Table 6. Comparision of ESMCrystal models on TR dataset. ESMCrystal performs comparably with other protein crystallization predictors on the TR test data

Models	Accuracy	MCC	AUC	F-score	Recall	Precision	NPV
Crysf	0.841	0.663	0.887	0.747	0.631	0.918	0.817
PPCpred	0.748	0.448	0.819	0.640	0.606	0.677	0.782
fDETECT	0.750	0.447	0.847	0.548	0.411	0.823	0.733
Crysalis I	0.787	0.546	0.870	0.715	0.724	0.707	0.836
Crysalis II	0.816	0.603	0.892	0.748	0.740	0.756	0.849
XtalPred-RF	0.451	0.040	0.525	0.452	0.537	0.390	0.651
TargetCrys	0.634	0.325	0.693	0.614	0.788	0.503	0.733
CrystalP2	0.581	0.241	0.673	0.577	0.775	0.460	0.780
DeepCrystal	0.841	0.657	0.910	0.781	0.762	0.800	0.864
BCrystal	0.963	0.922	0.988	0.951	0.970	0.933	0.982
ESMCrystal_t6_8M_v1	0.8192	0.6153	0.9562	0.6935	0.9283	0.5535	0.9749
ESMCrystal_t12_35M_v2	0.8389	0.6588	0.9562	0.7341	0.9414	0.6016	0.9781

son highlights areas for future improvements, such as integrating complementary features to further enhance the predictive power of ESMCrystal.

Furthermore, the study advocates for additional explorations into refining these models, with particular attention on enhancing their ability to manage datasets marked by high variability and complexity.

4   Conclusion

This study highlights the potent capabilities of advanced machine learning, specifically utilizing Meta's ESMFold [10,11] architecture, to predict protein crystallization from sequence data. The ESMCrystal models, especially the ESM-Crystal_t12_35M_v2, have demonstrated exceptional accuracy in forecasting crystallization potential, showcasing the adaptability and robustness of deep learning in tackling complex biological problems. These results validate the effectiveness of these sophisticated models in structural biology and suggest broader applications for these techniques.

The superior performance of the ESMCrystal_t12_35M_v2 model, owing to its deeper architecture and extensive training, underscores the potential for further advancements in this technology. This research opens pathways for refining these models using even Alphafold [7], enhancing their accuracy and reliability. Moreover, the successful application of transfer learning methodologies within this study advocates for their expanded use across the life sciences, reducing the need for extensive manual intervention and facilitating more efficient, high-throughput predictions.

The ability to predict crystallization potential directly from protein sequences can significantly accelerate drug discovery by identifying crystallizable therapeutic targets, enabling faster structure-based drug design. Moreover, this method has potential utility in structural biology for high-throughput analysis of proteins, reducing the reliance on labor-intensive and costly experimental techniques.

In conclusion, the findings from this study set a solid foundation for the use of deep learning in predicting protein crystallization, emphasizing the transformative impact of AI and machine learning in structural biology. As we continue to innovate and refine these computational strategies, integrating them with broader biological data, we pave the way for significant advancements in understanding protein structures and their complex functions. This ongoing evolution in computational biology promises to accelerate scientific discoveries, pushing the boundaries of our knowledge and capabilities.

5   Availability of Data and Software Code

Our software code is available at the following URL:

- ESMCrystal_t6_8M_v1 : https://huggingface.co/jaykmr/ESMCrystal_t6_8M_v1
- ESMCrystal_t12_35M_v2 : https://huggingface.co/jaykmr/ESMCrystal_t12_35M_v2

The dataset and the inference API is also, available for prediction on the above URLs.

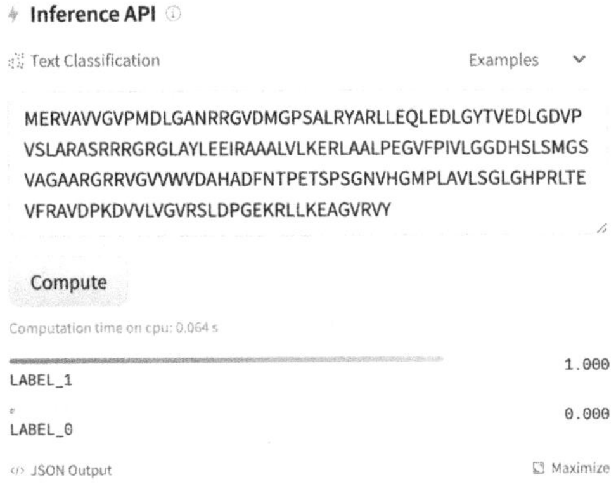

Fig. 6. Inference API Input the protein sequence and predict crystallization.

Acknowledgments. The authors extend gratitude to Meta ESMFold2 for making their protein embeddings openly available, to Huggingface for hosting the code and providing inference capabilities, and to Paperspace Compute Cloud for supplying the GPU servers essential for computational tasks.

References

1. Charoenkwan, P., Shoombuatong, W., Lee, H.C., Chaijaruwanich, J., Huang, H.L., Ho, S.Y.: Scmcrys: predicting protein crystallization using an ensemble scoring card method with estimating propensity scores of p-collocated amino acid pairs. PLoS ONE **8**(9), e72368 (2013)
2. Elbasir, A., et al.: Bcrystal: an interpretable sequence-based protein crystallization predictor. Bioinformatics **36**(5), 1429–1438 (2020)
3. Elbasir, A., Moovarkumudalvan, B., Kunji, K., Kolatkar, P.R., Mall, R., Bensmail, H.: Deepcrystal: a deep learning framework for sequence-based protein crystallization prediction. Bioinformatics **35**(13), 2216–2225 (2019)
4. Gao, J., et al.: Survey of predictors of propensity for protein production and crystallization with application to predict resolution of crystal structures. Curr. Protein Pept. Sci. **19**(2), 200–210 (2018)
5. Hu, J., Han, K., Li, Y., Yang, J.Y., Shen, H.B., Yu, D.J.: Targetcrys: protein crystallization prediction by fusing multi-view features with two-layered SVM. Amino Acids **48**, 2533–2547 (2016)
6. Jahandideh, S., Jaroszewski, L., Godzik, A.: Improving the chances of successful protein structure determination with a random forest classifier. Acta Crystallogr. D Biol. Crystallogr. **70**(3), 627–635 (2014)
7. Jumper, J., et al.: Highly accurate protein structure prediction with alphafold. Nature **596**(7873), 583–589 (2021)
8. Khurana, S., Rawi, R., Kunji, K., Chuang, G.Y., Bensmail, H., Mall, R.: Deepsol: a deep learning framework for sequence-based protein solubility prediction. Bioinformatics **34**(15), 2605–2613 (2018)
9. Kurgan, L., Mizianty, M.J., et al.: Sequence-based protein crystallization propensity prediction for structural genomics: review and comparative analysis. Nat. Sci. **1**(02), 93 (2009)
10. Lin, Z., et al.: Language models of protein sequences at the scale of evolution enable accurate structure prediction. bioRxiv (2022)
11. Meier, J., Rao, R., Verkuil, R., Liu, J., Sercu, T., Rives, A.: Language models enable zero-shot prediction of the effects of mutations on protein function. bioRxiv (2021). https://doi.org/10.1101/2021.07.09.450648, https://www.biorxiv.org/content/10.1101/2021.07.09.450648v1
12. Meng, F., Wang, C., Kurgan, L.: fdetect webserver: fast predictor of propensity for protein production, purification, and crystallization. BMC Bioinform. **18**, 1–11 (2017)
13. Rawi, R., Mall, R., Kunji, K., Shen, C.H., Kwong, P.D., Chuang, G.Y.: Parsnip: sequence-based protein solubility prediction using gradient boosting machine. Bioinformatics **34**(7), 1092–1098 (2018)
14. Rives, A., et al.: Biological structure and function emerge from scaling unsupervised learning to 250 million protein sequences. In: PNAS (2019)
15. Service, R.: Structural genomics, round 2 (2005)
16. Terwilliger, T.C., Stuart, D., Yokoyama, S.: Lessons from structural genomics. Annu. Rev. Biophys. **38**, 371–383 (2009)

17. Wang, H., Feng, L., Webb, G.I., Kurgan, L., Song, J., Lin, D.: Critical evaluation of bioinformatics tools for the prediction of protein crystallization propensity. Brief. Bioinform. **19**(5), 838–852 (2018)
18. Wang, H., Feng, L., Zhang, Z., Webb, G.I., Lin, D., Song, J.: Crysalis: an integrated server for computational analysis and design of protein crystallization. Sci. Rep. **6**(1), 21383 (2016)
19. Wang, H., Wang, M., Tan, H., Li, Y., Zhang, Z., Song, J.: Predppcrys: accurate prediction of sequence cloning, protein production, purification and crystallization propensity from protein sequences using multi-step heterogeneous feature fusion and selection. PLoS ONE **9**(8), e105902 (2014)

TARNAS, a TrAnslator for RNA Secondary Structure Formats

Michela Quadrini$^{(\boxtimes)}$ ⓘD, Piero Hierro Canchari, Piermichele Rosati,
and Luca Tesei ⓘD

School of Sciences and Technology, University of Camerino, Camerino, MC, Italy
`{michela.quadrini,luca.tesei}`@unicam.it

Abstract. RNAs are single-stranded molecules that fold into themselves, determining a complex shape to perform their biological functions. Considering the chemical bonds established, such shapes can be abstracted into secondary structures, which are tractable from a computational point of view and encode valuable biological information. The analysis of such structures, including comparison and classification, plays a fundamental role in different biological studies. Unfortunately, the available tools take secondary structures as input using different formats, making the translation among different them a necessary step in every analysis.

In this work, we propose TARNAS, a software that permits the translation of secondary structure formats, including BPSEQ, CT, Dot-Bracket, RNAML, FASTA (only primary structure) and Arc-annotated Sequence. TARNAS also allows the abstraction of RNA secondary structures into three views, namely Core, Core Plus and Shape. Finally, TAR-NAS permits to delete or retain comments, blank lines and headers of the files. TARNAS is developed as a standalone desktop application and as a web app. The tool, developed in Java, is available as a standalone application at https://github.com/bdslab/TARNAS or as a web application at https://bdslab.unicam.it/tarnas/. The standalone version allows the processing of large sets of RNA secondary structures in a batch fashion, whereas the web version translates one molecule at a time.

Keywords: RNA Secondary Structure Formats · Translation of RNA formats · Batch Translation of RNA datasets

1 Introduction

RNA is a single-strand molecule made of four different types of nucleotides, i.e., Adenine (A), Guanine (G), Cytosine (C) and Uracil (U). Such a single-stand, named *primary structure*, folds back on itself by determining complex shapes. RNA molecules can be described in terms of secondary structures, i.e., the sequence of nucleotides equipped with the hydrogen bonds established during the folding process. Such hydrogen bonds can be Watson-Crick-Franklin base pairs (G-C and A-U), or non-canonical base pairs, including wobble base

© The Author(s) 2025
L. Cerulo et al. (Eds.): CIBB 2024, LNBI 15276, pp. 307–316, 2025.
https://doi.org/10.1007/978-3-031-89704-7_24

pairs (*G-U*). Secondary structures are an abstraction that represents an intermediate level between the primary sequence and the three-dimensional shape. Figure 1-a and 1-b show an example of RNA secondary structure. An RNA secondary structure can be *pseudoknot-free* or *pseudoknotted*. An RNA secondary structure is said to be *pseudoknot-free* if its arc-diagram does not present crossings among arcs (Fig. 1-a) and it is called *pseudoknotted* otherwise (Fig. 1-b). Such an abstraction encodes valuable biological information and is tractable from a computational point of view. Therefore, RNA secondary structure analysis, including comparison and classification, plays a fundamental role in facing different problems, such as the prediction of RNA functions and the study of regulating gene expression. In the literature, some approaches, based on comparison or classification techniques, exist for the RNA secondary structure analysis. Comparison methods apply alignment algorithms to quantify the differences between two molecules. An approach based on edit distance is implemented in the tool RNAdistance [33] and PSMAlign [5], while another using tree alignment is developed in RNAforester [10,11]. Other tools that exploit such approaches are MiGaL [1], TreeMatching [20], Gardenia [3], RNAStrAT [8], and ASPRAlign [29,30]. PskOrder [37] and RAG-2D [6], take advantage of concepts and approaches from topology and graph theory, respectively. Only some of the listed approaches accept as input molecules with pseudoknots. In [31] a framework for evaluating comparison approaches with or without pseudoknots is defined. More in general, the aforementioned tools do not take RNA secondary structures as input in the same formats due to the lack of an input standard to represent them. As an example, PSMAlign takes molecules in dot-bracket formats, while RAG2D requires the structures in CT format. Other tools, such as ASPRAlign, take molecules in different formats as input. Finally, public databases of RNA secondary structures may provide the molecules in different formats. Therefore, translation of RNA secondary structures is often required when dealing with different sources of data and different tools to process them.

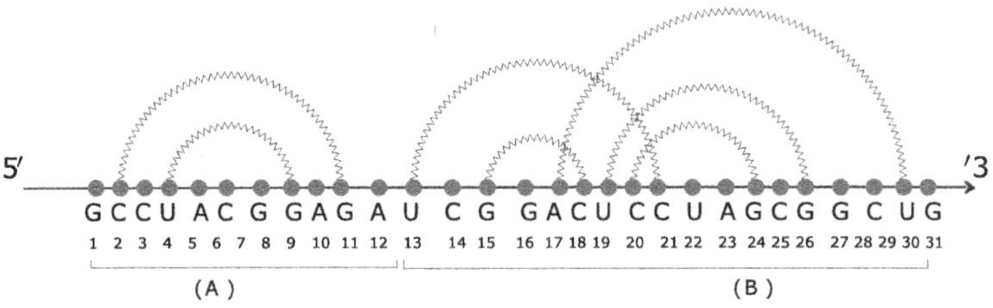

Fig. 1. The arc diagram of an RNA secondary structure. The red balls abstract the nucleotides and the zigzag arcs identify the base pairs. The motif in part **a** is pseudoknot-free, while the one in part **b** is pseudoknotted. (Color figure online)

In this work, we propose TARNAS, a software that allows one to transform secondary structure formats such as BPSEQ, CT, Dot-Bracket, RNAML, FASTA (only primary structure) and arc-annotated sequence. Such formats are usually codified as simple text that consists of two parts: the header and the body. The header contains information related to the molecules, such as the name of the organism or the id code, while the body contains precise information about the chemical bonds. This division is not present in RNAML. TARNAS permits to retain or eliminate the header while translating. It is developed in Java and takes advantage of the ANTLR4 tool [22] to generate a parser from a grammar describing the syntactic structure of the various formats. For RNAML, it uses the W3C Document Object Model API Reference [35].

The paper is organised as follows. Section 2.1 describes the formats managed by TARNAS, Sect. 2.2 illustrates the abstractions that can be computed by TARNAS and Sect. 2.3 presents the implementation details of TARNAS. Finally, Sect. 3 contains conclusions and future works.

2   Data and Methods

Several formats have been defined to represent RNA secondary structures. Section 2.1 presents them in detail, Sect. 2.2 presents RNA abstractions that can be computed by TARNAS and Sect. 2.3 discusses the methodologies and approaches employed in the development of TARNAS.

2.1   Data Formats

BPSEQ [9] is a text format in which every line of the body corresponds to a nucleotide in the RNA primary sequence. Each line lists the position of the base (with the leftmost position being 1), the name of the base (A, C, G, U, or other alphabetical characters according to the IUPAC codes [14]) and the position number of the base with which it is paired. A 0 indicates that the base is unpaired.

CT [19] is a text format where the first line of the body contains the sequence length and the other lines are one per nucleotide. The i-th line starts with i, then the letter denoting the i-th nucleotide, then the $5'$-connecting base index ($i - 1$), then the $3'$-connecting base index ($i + 1$), then the paired base index (or 0 if unpaired), and finally the base index in the original sequence.

Figure 2-a and Fig. 2-b show the RNA secondary structure of the structure in Fig. 1, formalized in terms of BPSEQ and CT formats, respectively.

In the (extended) dot-bracket format [12], the sequence is given first, from the $5'$ end to the $3'$ end. The structure is then given as a sequence of dots (denoting unpaired bases), parentheses such as "(,)", "[,]", "{,}", or uppercase (opening) and lowercase (closing) letters. The use of square and curly brackets or of uppercase/lowercase letters denotes the presence of pseudoknots. The position of an open bracket (or opening letter) indicates that the corresponding nucleotide in the primary sequence has a bond with the nucleotide at the position of the matching closed bracket (or closing letter).

A similar format is arc-annotated sequence [29,30]. The body of this format consists of the sequence of nucleotides, given from $5'$ to $3'$ end, followed by a list

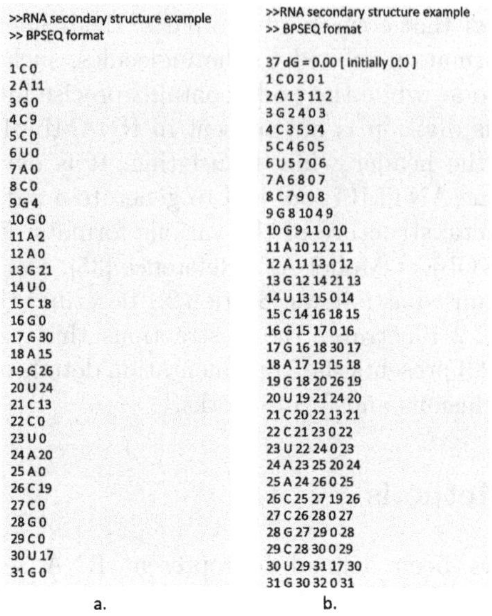

```
>>RNA secondary structure example
>> BPSEQ format

1 C 0
2 A 11
3 G 0
4 C 9
5 C 0
6 U 0
7 A 0
8 C 0
9 G 4
10 G 0
11 A 2
12 A 0
13 G 21
14 U 0
15 C 18
16 G 0
17 G 30
18 A 15
19 G 26
20 U 24
21 C 13
22 C 0
23 U 0
24 A 20
25 A 0
26 C 19
27 C 0
28 G 0
29 C 0
30 U 17
31 G 0
```

a.

```
>>RNA secondary structure example
>> BPSEQ format

37 dG = 0.00 [ initially 0.0 ]
1 C 0 2 0 1
2 A 1 3 11 2
3 G 2 4 0 3
4 C 3 5 9 4
5 C 4 6 0 5
6 U 5 7 0 6
7 A 6 8 0 7
8 C 7 9 0 8
9 G 8 10 4 9
10 G 9 11 0 10
11 A 10 12 2 11
12 A 11 13 0 12
13 G 12 14 21 13
14 U 13 15 0 14
15 C 14 16 18 15
16 G 15 17 0 16
17 G 16 18 30 17
18 A 17 19 15 18
19 G 18 20 26 19
20 U 19 21 24 20
21 C 20 22 13 21
22 C 21 23 0 22
23 U 22 24 0 23
24 A 23 25 20 24
25 A 24 26 0 25
26 C 25 27 19 26
27 C 26 28 0 27
28 G 27 29 0 28
29 C 28 30 0 29
30 U 29 31 17 30
31 G 30 32 0 31
```

b.

Fig. 2. BPSEQ, in a, CT, in b, formats of the RNA secondary structure depicted in Fig. 1.

of base pairs $(i_1, j_1); (i_2, j_2); \ldots; (i_n, j_n)$, where each index i_k, j_k corresponds to the base index of the molecule.

The FASTA format codifies only the sequence of nucleotides (the so-called primary structure) in IUPAC codes [23]. Figure 3-a, 3-b and 3-c show the RNA secondary structure of Fig. 1, formalized in terms of dot-bracket, arc-annotated sequence and FASTA, respectively.

RNAML [18,34] is an XML format specifically designed to express data related to RNA sequences and structures. Following XML's markup-oriented approach, all data in an RNAML document is included as parsable CDATA, with markup used to categorize the data within a hierarchy. Moreover, many RNAML elements are optional. Figures 3-a show the RNAML format of the RNA secondary structure of Fig. 1.

2.2   RNA Secondary Structure Abstractions

Core, **Core Plus** and **Shape** are three *abstractions* of RNA secondary structures. Shapes, introduced by Huang and Reidys [13], are derived from the arc diagram of a secondary structure. This process involves disregarding the nucleotide sequence, removing unpaired nucleotides, and collapsing the remaining parallel arcs into a single arc. Figure 4-a illustrates the shape of the RNA secondary structure depicted in Fig. 1. The abstractions Core and Core Plus follow a similar approach by preserving more information. Core is obtained by first removing unpaired nucleotides and then collapsing all the parallel arcs into one. Instead,

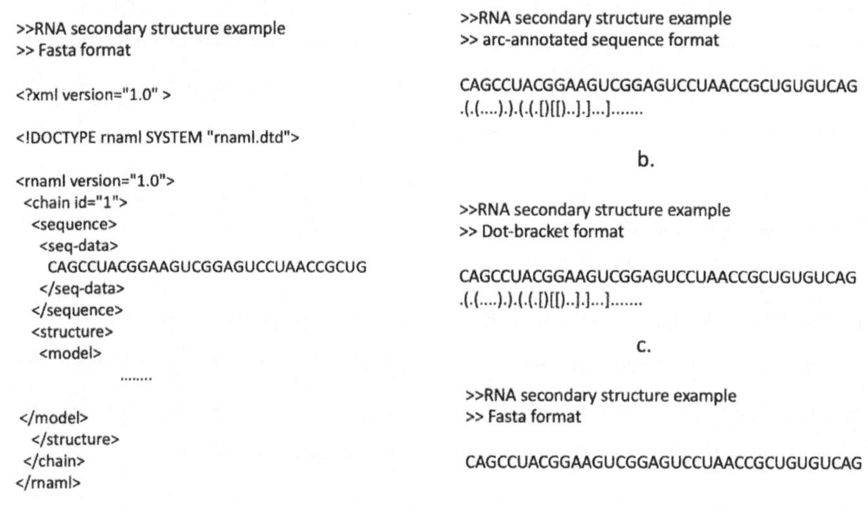

Fig. 3. RNAML, in a, arc-annotated-sequence, in b, dot-backet, in c, and FASTA, in d, formats of the RNA secondary structure depicted in Fig. 1.

Core plus is obtained by *first* collapsing all the parallel arcs and *then* removing unpaired nucleotides. Figures 4-b and Figs. 4-c show the Core and Core Plus of the structure in Fig. 1.

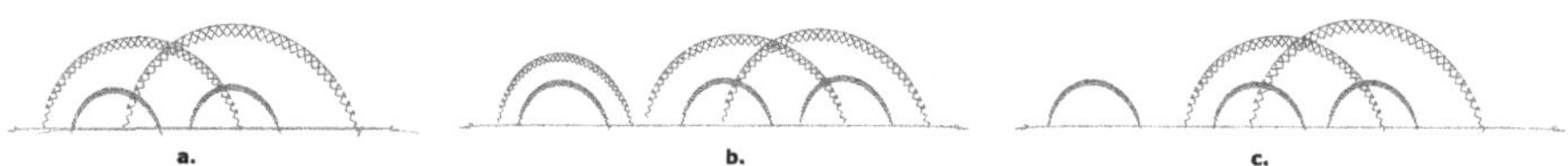

Fig. 4. The Shape, in a, the Core, in b, and Core Plus, in c., of the RNA secondary structure depicted in Fig. 1.

2.3   Methods

ANother Tool for Language Recognition (ANTLR) [21] is a parser generator used for reading, processing, executing, or translating structured text or binary files [22]. It is commonly employed in the construction of languages, tools, and frameworks. Given a grammar, ANTLR can generate a parser capable of constructing parse trees, and it also produces a listener interface (or visitor), simplifying the process of responding to the recognition of phrases of interest. In TARNAS, we used ANTLR version 4 to create a parser of the supported formats with target language Java.

TARNAS has been implemented in a two-layer architecture with separate back-end and front-end. The back-end is unique while two different front-ends are available: a web application and a desktop standalone application. The architecture of the back-end relies on the design pattern MVC, which consists of a model, a view, and a controller. The model provides methods for accessing useful application data. The view shows the data contained in the model and guarantees interaction with the user, while the controller is responsible for connecting the model and the view.

In both versions, after loading (or editing, in the web app) the molecules as file(s), the tool checks the correctness of the content and identifies the format. The user can decide to translate into another format, according to Table 1, retaining or not the entire header (containing the comments of the molecules) of the text file. The user can also decide to cancel the empty lines or just the header lines that contain a word given as input. Alternatively, the user can ask to abstract the RNA molecule(s) into Core, Core Plus or Shape.

Table 1. Conversion table between two formats. The "no seq." version of Dot-Bracket and AAS formats consists of the format in which the primary sequence is absent and only a structure is given. Formats with only the structure can be translated only into formats without the primary sequence and formats with only the primary sequence (FASTA) can not be translated to formats with structure.

	Dot-bracket no seq	Dot-bracket	RNAML	CT	BPSEQ	AAS	AAS no seq	FASTA
Dot-bracket no seq	−	×	×	×	×	×	✓	×
Dot-bracket	✓	−	✓	✓	✓	✓	✓	✓
RNAML	✓	✓	−	✓	✓	✓	✓	✓
CT	✓	✓	✓	−	✓	✓	✓	✓
BPSEQ	✓	✓	✓	✓	−	✓	✓	✓
AAS	✓	✓	✓	✓	✓	−	✓	✓
AAS no seq		×	×	×	×	×	−	×
FASTA	×	×	×	×	×	×	×	−

On the home page of TARNAS, both in the Web App and in the Standalone versions, such functionalities are grouped into "RNA 2D STRUCTURE UPLOAD", "TRANSATION", "EDITING", and "ABSTRACTION" sections, respectively. The two home pages are shown in Fig. 5 and Fig. 6.

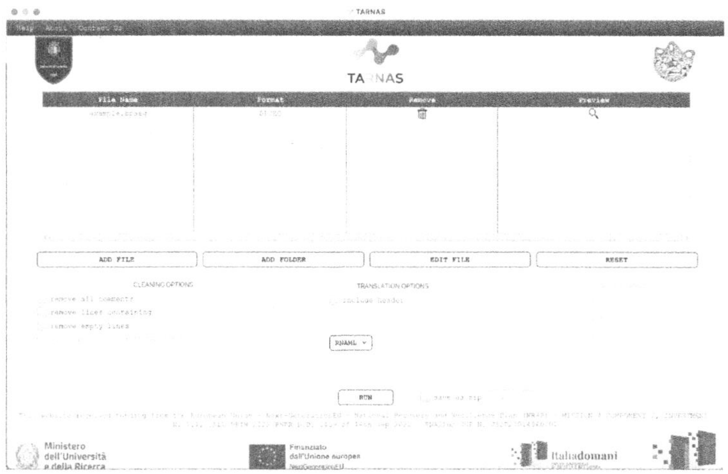

Fig. 5. Home page of TARNAS Web App.

Fig. 6. Home page of the TARNAS Standalone version.

3   Conclusion

In this work, we have introduced TARNAS, a Java tool for translating RNA secondary structures into different formats, including BPSEQ, CT, RNAML, Dot-Bracket, FASTA (only primary structure) and arc-annotated sequence. Moreover, TARNAS allows us to abstract RNA secondary structures into three views, namely Core, Core Plus and Shape. Finally, TARNAS permits to delete or retain comments, blank lines and headers of the files. TARNAS is developed as a standalone desktop application and as a web app. The desktop application takes as input several files allowing one to process a big amount of molecules in one step.

As future works, we intend to integrate methods, such as 3DNA/DSSR [17], RNAView [36], MC-Annotate [36] and FR3D [32], for determining RNA secondary structures starting from their 3D structures. We will integrate a functionality for visualizing RNA secondary structures by considering external tools like PseudoViewer [4] and R-chie [15]. Moreover, we will also add other approaches to classify RNA secondary structures based on intersection graph, structural patterns [24,26,28] or take into account other abstractions like the one introduced in [27]. For RNA classification, another interesting improvement is to add methods to encode RNA structures into images considering only sequence by applying chaos game representation [16], or considering sequence and structural information by extending approach proposed for other other domains like [25]. Finally, another direction, for spreading the use of TARNAS, is to make it available as a Docker image in BioContainers [2] and in pegi3S Bioinformatics Docker Images project [7].

Acknowledgements. This work was supported by the European Union - Next-GenerationEU - National Recovery and Resilience Plan (NRRP) - MISSION 4 COMPONENT 2, INVESTMENT N. 1.1, CALL PRIN 2022 PNRR D.D. 1409 of 14th Sep 2022 - RNA2FUN CUP N. J53D23014960001. MQ is a member of the "GNCS - INdAM".

Disclosure of Interests. The authors declare no conflict of interest.

Availability of Data and Software Code. Our software code is available at the following URL: https://github.com/bdslab/TARNAS. The corresponding web application is available at https://bdslab.unicam.it/tarnas/.

References

1. Allali, J., Sagot, M.F.: A multiple layer model to compare RNA secondary structures. Softw. Pract. Exp. **38**(8), 775–792 (2008)
2. BioContainers: Biocontainers. https://biocontainers.pro/. Accessed 20 Oct 2024
3. Blin, G., Denise, A., Dulucq, S., Herrbach, C., Touzet, H.: Alignments of RNA structures. IEEE/ACM Trans. Comput. Biol. Bioinf. **7**(2), 309–322 (2010)
4. Byun, Y., Han, K.: PseudoViewer3: generating planar drawings of large-scale RNA structures with pseudoknots. Bioinformatics **25**(11), 1435–1437 (2009)
5. Chiu, J., Chen, Y.: Pairwise RNA secondary structure alignment with conserved stem pattern. Bioinformatics **31**(24), 3914–3921 (2015)
6. Gan, H.H., et al.: RAG: RNA-as-graphs database-concepts, analysis, and features. Nutr. Health **5**(1–2), 1285–1291 (1987)
7. Group, P.E.: Bioinformatics docker images project. https://pegi3s.github.io/dockerfiles/. Accessed 20 Oct 2024
8. Guignon, V., Chauve, C., Hamel, S.: RNA StrAT: RNA structure analysis toolkit. In: 16th Annual International Conference on Intelligent Systems for Molecular Biology (ISMB 2008), p. D31. Citeseer (2008)
9. Gutell, R., Cannone, D., Schnare, J., Shu, Z., Hertz, B., Parsch, K.P.: BPSEQ file format (2000)

10. Hochsmann, M., Toller, T., Giegerich, R., Kurtz, S.: Local similarity in RNA secondary structures. In: Computational Systems Bioinformatics. CSB2003. Proceedings of the 2003 IEEE Bioinformatics Conference, pp. 159–168. IEEE (2003)

11. Hochsmann, M., Voss, B., Giegerich, R.: Pure multiple RNA secondary structure alignments: a progressive profile approach. IEEE/ACM Trans. Comput. Biol. Bioinf. **1**(1), 53–62 (2004)

12. Hofacker, I.L., Fontana, W., Stadler, P.F., Bonhoeffer, S., Tacker, M., Schuster, P.: Fast folding and comparison of RNA secondary structures. Monatshefte für Chemie / Chem. Mon. **125**, 167–188 (1994). https://doi.org/10.1007/BF00818163

13. Huang, F.W., Reidys, C.M.: Shapes of Topological RNA Structures. arxiv:1403.2908 (2014)

14. International Union of Pure and Applied Chemistry: IUPAC nucleotide code. https://www.bioinformatics.org/sms/iupac.html. Accessed 20 Oct 2024

15. Lai, D., Proctor, J.R., Zhu, J., Meyer, I.M.: R-CHIE: a web server and R package for visualizing RNA secondary structures. Nucleic Acids Res. **40**(12), e95–e95 (2012)

16. Löchel, H.F., Heider, D.: Chaos game representation and its applications in bioinformatics. Comput. Struct. Biotechnol. J. **19**, 6263–6271 (2021)

17. Lu, X.J., Olson, W.K.: 3DNA: a software package for the analysis, rebuilding and visualization of three-dimensional nucleic acid structures. Nucleic Acids Res. **31**(17), 5108–5121 (2003)

18. Masuya, H., Griffiths-Jones, S., Bateman, A., Quang, T.T., Gaudin, N., Lowe, T.M.: RNAML: a standard syntax for exchanging RNA information. Bioinformatics **22**(5), 628–636 (2006). https://doi.org/10.1093/bioinformatics/btk030

19. Mathews, D.H., Sabina, J., Zuker, M., Turner, D.H.: Expanded sequence dependence of thermodynamic parameters improves prediction of RNA secondary structure. J. Mol. Biol. **288**(5), 911–940 (1999). https://doi.org/10.1006/jmbi.1999.2700

20. Ouangraoua, A., Ferraro, P., Tichit, L., Dulucq, S.: Local similarity between quotiented ordered trees. J. Discret. Algorithms **5**(1), 23–35 (2007)

21. Parr, T.: ANTLR website. https://www.antlr.org/. Accessed 20 Oct 2024

22. Parr, T.: The definitive ANTLR 4 reference. The Pragmatic Bookshelf (2013)

23. Pearson, W.R., Lipman, D.J.: Improved tools for biological sequence comparison. Proc. Natl. Acad. Sci. **85**(8), 2444–2448 (1988). https://doi.org/10.1073/pnas.85.8.2444

24. Quadrini, M., Culmone, R., Merelli, E.: Topological classification of RNA structures via intersection graph. In: International Conference on Theory and Practice of Natural Computing, pp. 203–215. Springer, Cham (2017)

25. Quadrini, M., Daberdaku, S., Blanda, A., Capuccio, A., Bellanova, L., Gerard, G.: Stress detection from wearable sensor data using Gramian angular fields and CNN. In: International Conference on Discovery Science, pp. 173–183. Springer, Cham (2022)

26. Quadrini, M., Merelli, E.: Loop-loop interaction metrics on RNA secondary structures with pseudoknots. In: Proceedings of the 13th International Joint Conference on Biomedical Engineering Systems and Technologies - BIOINFORMATICS, pp. 29–37 (2018)

27. Quadrini, M., Merelli, E., Piergallini, R.: Label core for understanding RNA structure. In: International Meeting on Computational Intelligence Methods for Bioinformatics and Biostatistics, pp. 171–179. Springer, Cham (2019)

28. Quadrini, M., Merelli, E., Piergallini, R.: Loop grammars to identify RNA structural patterns. In: Proceedings of the 12th International Joint Conference on Biomedical Engineering Systems and Technologies - BIOINFORMATICS, pp. 302–309. SciTePress (2019)

29. Quadrini, M., Tesei, L., Merelli, E.: An algebraic language for RNA pseudoknots comparison. BMC Bioinform. **20**, 1–18 (2019)
30. Quadrini, M., Tesei, L., Merelli, E.: ASPRAlign: a tool for the alignment of RNA secondary structures with arbitrary pseudoknots. Bioinformatics **36**(11), 3578–3579 (2020)
31. Quadrini, M., Tesei, L., Merelli, E.: Automatic generation of pseudoknotted RNAs taxonomy. BMC Bioinform. **23**(Suppl 6), 575 (2022)
32. Sarver, M., Zirbel, C.L., Stombaugh, J., Mokdad, A., Leontis, N.B.: FR3D: finding local and composite recurrent structural motifs in RNA 3D structures. J. Math. Biol. **56**, 215–252 (2008)
33. Shapiro, B.A., Zhang, K.: Comparing multiple RNA secondary structures using tree comparisons. Bioinformatics **6**(4), 309–318 (1990)
34. Waugh, A., et al.: RNAML: a standard syntax for exchanging RNA information. RNA **8**(6), 707–717 (2002)
35. World Wide Web Consortium: W3C Document Object Model (DOM) API Reference: org.w3c.dom (2004). https://www.w3.org/TR/DOM-Level-3-Core/. Accessed 20 Oct 2024
36. Yang, H., et al.: Tools for the automatic identification and classification of RNA base pairs. Nucleic Acids Res. **31**(13), 3450–3460 (2003)
37. Zok, T., Badura, J., Swat, S., Figurski, K., Popenda, M., Antczak, M.: New models and algorithms for RNA pseudoknot order assignment. Int. J. Appl. Math. Comput. Sci. **30**(2), 315–324 (2020)

Short Papers

Novel Approaches for Spatially Resolving Gene Responses and Injection Site Localization in Transcriptomic Data

Antonio Ammendola$^{(\boxtimes)}$, Felicita Masone, Luigi Cerulo, and Francesco Napolitano

Department of Sciences and Technologies, University of Sannio, 82100 Benevento, Italy
a.ammendola@studenti.unisannio.it

Keywords: Spatial transcriptomics · Distance-Based Analysis · Treatment-Responsive Genes · Gene Expression Analysis

1 Background

Spatial transcriptomics (ST) enables the analysis of gene expression while preserving the spatial context within tissue samples. By mapping where specific genes are active within a tissue, ST provides a unique opportunity to explore how cells function in their native environments. Traditional methods of RNA analysis often require tissue dissociation, losing critical spatial information. Additionally, these methods can be challenging to integrate into standard histological workflows used in research and clinical diagnostics. The ability to maintain spatial context while generating detailed transcriptomic data makes ST an essential tool across numerous scientific and medical disciplines [1]. Its potential extends beyond individual disease areas, offering valuable insights into how diseases progress and affect tissues at a molecular level. ST allows researchers to uncover cellular interactions, identify biomarkers, and study the molecular basis of complex tissue structures. By mapping gene expression across tissues, ST enables the study of disease mechanisms such as inflammation, fibrosis, and tissue regeneration. It supports the development of personalized therapeutic strategies by providing precise molecular and spatial profiles of affected tissues. As ST technology evolves, its integration into routine research and diagnostic workflows is becoming increasingly feasible, potentially revolutionizing diagnostics by enabling more accurate disease classification and personalized treatment decisions [2].

In this context, we introduce RGDIST (Responsive Genes based on Distance from Injection SiTe), a novel computational method designed to identify treatment-responsive genes in ST data by analyzing gene expression patterns relative to an injection site. RGDIST operates without the need for reference controls, making it particularly useful in experimental settings where untreated controls are unavailable or impractical. By leveraging the spatial structure of ST data, RGDIST identifies genes whose expression correlates with proximity to the injection site, providing localized insights into treatment effects. This is achieved by leveraging the spatial structure of ST data to detect changes

L. Cerulo et al. (Eds.): CIBB 2024, LNBI 15276, pp. 319–322, 2025.
https://doi.org/10.1007/978-3-031-89704-7_25

in gene expression as a function of distance from the injection site, with the assumption that the treatment effect diminishes as the distance from the injection site increases.

2 Methods

The RGDIST workflow begins by normalizing the data using the ComBat method (R package sva v3.35.2) [3] to account for potential confounding factors, such as anatomical batch effects. It then calculates the distances of tissue spots from the injection site, grouping them into concentric rings based on their proximity. For each gene, Spearman's correlation is computed between its expression levels and these distance rings, enabling the identification of genes whose expression is significantly influenced by proximity to the injection site.

RGDIST identifies treatment-responsive genes by modeling spatial gradients in gene expression. Genes that show significant changes in expression relative to their distance from the injection site are considered treatment-responsive. This distance-dependent approach offers a more localized and precise assessment of treatment effects compared to bulk tissue analyses, providing deeper insights into molecular responses in complex biological systems.

In addition to identifying responsive genes, RGDIST can be used for detecting the injection site itself, based solely on spatial transcriptomic data. This is done through an application of the method where the distance rings are correlated with the log-fold change (logFC) in gene expression across all tissue spots. By calculating the median correlation for each spot with respect to the top 100 genes showing the highest absolute correlations, RGDIST can autonomously infer the most likely injection site. This eliminates the need for prior knowledge of the injection location and streamlines the analytical workflow, making RGDIST a versatile tool for diverse experimental designs.

To validate the effectiveness of the RGDIST method, we compared the gene rankings derived from our approach with those obtained from a baseline expression model and the well-established SpatialDE method, which is commonly used for identifying spatially variable genes. The results from RGDIST were then evaluated against a ground-truth set of genes that had been previously identified by Buzzi et al. [3] as being up- or down-regulated in response to heme treatment in comparison to a control sample. This provided a valuable benchmark for assessing the performance and accuracy of our method in detecting treatment-responsive genes.

3 Results and Conclusions

RGDIST represents a significant advancement in spatial transcriptomics, offering a powerful tool for studying treatment-responsive genes without the need for control groups. Its ability to autonomously detect the injection site based solely on spatial gene expression patterns makes it an invaluable resource for drug delivery studies, therapeutic targeting experiments, and disease modeling in spatially resolved tissues. The ST data used for our analysis are publicly available from the Gene Expression Omnibus website (accession number GSE182127).

3.1 Genes Detected by RGDIST

RGDIST initially identified many of the genes previously reported by Buzzi et al. [4], producing results consistent with the existing literature. Additionally, RGDIST led to the discovery of several genes that had not been identified in previous studies. For example, genes such as Cdkn1a (involved in cell cycle regulation), Il33 (involved in the inflammatory response), and Gjb1 (involved in intracellular communication) were found to exhibit significant, distance-dependent expression changes. These findings highlight previously unrecognized layers of biological response to treatment, underscoring RGDIST's ability to expand our understanding of treatment effects at the molecular level.

3.2 Injection Site Detection

A key innovation of RGDIST is its capacity to identify the injection site directly from spatial gene expression data. This is achieved by focusing on regions that exhibit the most pronounced transcriptional changes in relation to distance from the injection site. Correlation analysis revealed that the highest values were localized around the area identified as the injection site by previous studies (e.g., Buzzi et al.), validating the accuracy of our approach. The spatial distribution of the most strongly correlated genes aligns with the known injection site location, confirming RGDIST's ability to autonomously detect the treatment site. This feature eliminates the need for prior knowledge of treatment location or manual identification, making it especially useful in experimental settings where the injection site may not be readily accessible or well defined.

3.3 Applications in Disease and Treatment Studies

By combining spatial transcriptomic data with distance-based analysis, RGDIST offers a novel method for identifying treatment-responsive genes and localizing experimental interventions. This approach not only provides detailed insights into the localized molecular effects of treatments but also removes the need for control samples, offering a more flexible and scalable solution for a variety of experimental settings. In addition to its utility in studying treatment responses, RGDIST can also be extended to analyze disease progression within tissues. For instance, by applying the same distance-based analysis, researchers could examine how gene expression patterns evolve in response to diseases such as tumors or injuries. This capability enables the study of disease dynamics across tissue regions, improving diagnostic strategies and supporting the development of more targeted therapies.

3.4 Overall Impact

The RGDIST method represents a significant advancement in spatial transcriptomic analysis by providing a detailed, distance-based understanding of gene expression dynamics without relying on conventional control samples. Its ability to identify treatment-responsive genes based on spatial gradients offers a more precise view of how interventions influence gene activity within tissues. Additionally, its capacity to autonomously detect the injection site directly from spatial transcriptomic data makes it a versatile and valuable tool in the study of gene expression in both health and disease. This approach holds great promise for advancing spatial biology and molecular research, with broad applications ranging from therapeutic targeting to the study of disease progression and response to treatment.

References

1. Robles-Remacho, A., Sanchez-Martin, R.M., Diaz-Mochon, J.J.: Spatial transcriptomics: emerging technologies in tissue gene expression profiling. Anal. Chem. **95**(42), 15450–15460 (2023). https://doi.org/10.1021/acs.analchem.3c02029
2. Ståhl, P.L., Salmén, F., Vickovic, S., et al.: Visualization and analysis of gene expression in tissue sections by spatial transcriptomics. Science **353**(6294), 78–82 (2016). https://doi.org/10.1126/science.aaf2403
3. Zhang, Y., Parmigiani, G., Johnson, W.E.: ComBat-seq: batch effect adjustment for RNA-seq count data. NARGAB **2**(3), lqaa078 (2020). https://academic.oup.com/nargab/article/doi/10.1093/nargab/lqaa078/5909519
4. Buzzi, R.M., Akeret, K., Schwendinger, N., et al.: Spatial transcriptome analysis defines heme as a hemopexin-targetable inflam-matoxin in the brain. Free Radic. Biol. Med. (2022). PMID: 34793930

Deep Learning Approaches for Forensics DNA Profiling: a Replication Study

Mirella Sangiovanni[1,2(✉)], Willem-Jan van den Heuvel[1,2], and George Manias[1,2]

[1] Tilburg University, Tilburg, The Netherlands
m.sangiovanni@tilburguniversity.edu
[2] Jheronimus Academy of Data Science (JADS), Den Bosch, The Netherlands

Keywords: Artificial Intelligence (AI) · Deep Learning (DL) · DNA Profiling · Criminal Investigation

1 Introduction

Recent approaches in Artificial Intelligence (AI) offer advanced computational algorithms, with high levels of accuracy and consistency, on complex and crucial investigations [8]. The integration of such AI-based methods with traditional ones in the field of forensic analysis, holds the potential to leverage automated interpretation of large and unstructured datasets—e.g., as extracted from Genetic Labs—to generate predictive models [1]. DNA profiling provides evidence that may be used to convict criminals, utilizing various genetic markers classified into two main categories: bi-allelic markers, such as single-nucleotide polymorphisms (SNPs), and multi-allelic polymorphisms, including variable number of tandem repeats (VNTRs) and short tandem repeats (STRs) [5]. Our contribution is to conduct a replication study of AI-driven DNA profiling process, including the detection and serial slicing of alleles in strings. This study was initiated as a first attempt to exploit AI for interpreting and comparing the ML-generated results with the currently available baseline results gained from classical detection methods [7]. Forensics analysis focuses on biological evidence, such as samples (e.g. blood, tissues), using specific genetic markers to create a DNA dataset [3,5]. The data set, represented as a value extracted from the electropherogram, is used to recognize alleles (alternative forms of a gene) and the individualization of repeating sequences (short tandem repeats) in this sample of human DNA [6]. In our experiments, the dataset used was generated synthetically. This study emphasizes the growing need to expedite the DNA profiling process from forensic samples and highlights the crucial role of software–along with its intrinsic replicability limitations–in instrumenting such advancements [2]. Furthermore, accelerating the process of interpretation of forensics DNA analysis can be crucial for faster and more precise identification, especially in cases where time and accuracy are critical, and improving forensic analysis and decision-making.

L. Cerulo et al. (Eds.): CIBB 2024, LNBI 15276, pp. 323–326, 2025.
https://doi.org/10.1007/978-3-031-89704-7_26

2   Data and Methods

This study comes from a collaboration with the Netherlands Forensics Institute (NFI)[1], institute for forensics analysis and crime-scene investigation. The institute is also engaged in research to refine investigative techniques and the current target for exploitation of AI, is aimed at improving the extent and possibilities of analysis available to investigators on the job. This first study—conducted jointly with the Delft University of Technology in 2022— aimed to establish a baseline AI approach in DNA profiling[2].

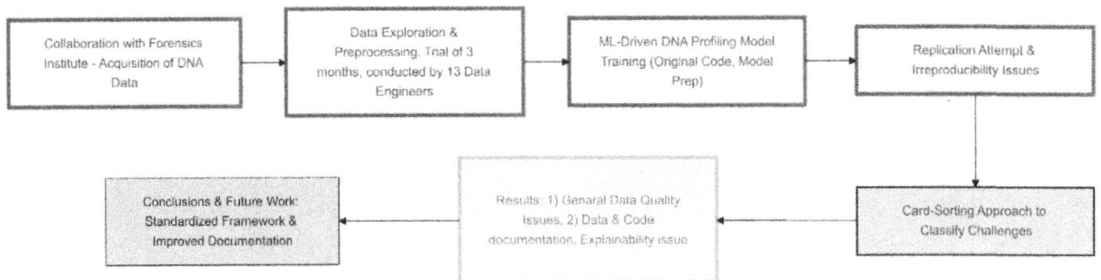

Fig. 1. High-Level Flowchart of replication study for DNA profiling.

The main goal of our research work was geared towards replication and analysis of said baseline approach, with the intent of rebooting the research line and possibly advancing it with experimentation of more refined and promising artificial intelligence approaches that have become available over the last few years. As demonstrated in the literature, artificial neural networks can be used to identify peaks on an electropherogram and classify them [9]. Figure 1 shows an overview of our research. This project was tasked to 13 Data Engineering young professionals, each with at least 2+ years of experience on the field. The methodology employed in this study was designed to rigorously evaluate the performance and reliability of machine learning models in the context of forensic DNA analysis. The process followed by said practitioners began with a thorough exploration of the existing data, which involved preprocessing steps to clean and prepare the datasets for analysis. These steps were crucial in ensuring that the data used was of high quality, eliminating inconsistencies and inaccuracies that could skew the results and allowing us to elicit limitations and challenges with the previous study. Once the data was cleaned, it was split into training and testing sets, with a larger portion allocated to training the model and a smaller subset reserved for evaluating its performance. Despite having a considerable training/testing score in its original study report, the replication effort concluded that the same score could not be attained, in fact, the training and model preparation code itself was not replicable. As a result, the only exercise possible was enacting a card-sorting approach [4][3]among the replication study practitioners to identify

---

[1] https://www.forensischinstituut.nl/.

[2] https://repository.tudelft.nl/record/uuid:f0a4d8ec-0d95-441c-ba35-65b087447929.

[3] https://miro.com/app/board/uXjVKkANp4E=/.

and classify the challenges and issues reported in the scope of the replication. The results of this exercise are reported in the next section.

3   Results

At the end of the replication trial (around three months), interesting aspects have been highlighted. We categorize them as reported in the Tables 1 and 2, which address *general data quality* issues reported during the replication as well as *data and code documentation or explainability* aspects, respectively. More in

Table 1. General Data Quality Issues in the Replication Study of DNA profiling.

Category	General Data Quality Observations
Data Inconsistencies	- Inconsistencies in data files, such as unclear or mismatched formats, made accurate data processing difficult. - Inconsistent structure and labelling across files led to confusion during analysis.
Incomplete Documentation	- Lack of clear documentation on the origin and structure of data hindered quality assessment and reliability checks. - Difficulty understanding how data was generated or organized impacts data quality assurance.
Data Shaping Issues	- Poor initial data quality complicated the process of shaping data into a usable form for analysis. - Challenges in transforming raw data into training and testing datasets negatively affected the overall analysis quality.

Table 2. Summary of Identified Issues in Data and Code Explainability in the Replication Study.

Category	Identified Issues
Data Input Clarity	-Unclear formats and origins of key files like 'TraceDataSet11.txt' made the data challenging to interpret. -Understanding the relationship between files in different folders was difficult due to insufficient documentation.
Code Documentation	-Lack of comments explaining critical functions and parameters (e.g., 'train = True/False', 'txt_read_sample'). -The purpose operations, such as assigning replicas in data and handling specific data segments, was not well-documented.
Data Transformation and Usage	-The transformation of tracedata files into training and testing data lacked transparency, making problematic the validation. -Specific decisions, such as the use of leftoffset and cutoff values, were not justified in the code.
Visualizations and Processing	-Visualization processes were inefficient, with some taking excessive time (e.g., 10 h), which is impractical for standard computational setups.

details, in the context of the replication study, the issue of general data quality was reflected in several key observations.

Moreover, several key issues were identified regarding Data and Code explainability. The summary of the challenges encountered are listed in the tab1.

4   Conclusion

While the original work reports F1-scores and an accuracy score of around 95%, demonstrating strong model performance, it lacked critical visualizations that would provide a more thorough understanding of the training process. Specifically, the work does not include visualizations of training and validation performance across epochs, which is essential to evaluate how neural networks learned over time. These missing visualizations would have helped to assess the model's progression, identify potential issues like overfitting, and ensure the stability and reliability of the final performance metrics while aiding in replicability. Conversely, our replication effort concluded that these deficiencies do not allow for a full replication, issuing the need for further confirmatory and exploratory work on the matter. We call out for more methodical and reproducible solutions, while we aim to develop a standardized framework and platform to help establish this.

References

1. Alotaibi, H., Alsolami, F., Abozinadah, E., Mehmood, R.: Tawseem: a deep-learning-based tool for estimating the number of unknown contributors in DNA profiling. Electronics **11**(4), 548 (2022)
2. Barash, M., McNevin, D., Fedorenko, V., Giverts, P.: Machine learning applications in forensic DNA profiling: a critical review. Forensic Sci. Int. Genet. 102994 (2023)
3. Bukyya, J.L., et al.: DNA profiling in forensic science: a review. Glob. Med. Genet. **8**(04), 135–143 (2021)
4. Fincher, S., Tenenberg, J.: Making sense of card sorting data. Expert. Syst. **22**(3), 89–93 (2005)
5. Haddrill, P.R.: Developments in forensic DNA analysis. Emerg. Top. Life Sci. **5**(3), 381–393 (2021)
6. McDonald, C., Taylor, D., Linacre, A.: PCR in forensic science: a critical review. Genes **15**(4), 438 (2024)
7. Sessa, F., et al.: Artificial intelligence and forensic genetics: current applications and future perspectives. Appl. Sci. **14**(5), 2113 (2024)
8. Singh, P.: Systematic review of data-centric approaches in artificial intelligence and machine learning. Data Sci. Manag. **6**(3), 144–157 (2023)
9. Taylor, D., Harrison, A., Powers, D.: An artificial neural network system to identify alleles in reference electropherograms. Forensic Sci. Int. Genet. **30**, 114–126 (2017)

Author Index

L. Cerulo et al. (Eds.): CIBB 2024, LNBI 15276, pp. 327–328, 2025.
https://doi.org/10.1007/978-3-031-89704-7

Made in the USA
Monee, IL
07 July 2026